Carbocation Chemistry
Applications in Organic Synthesis

NEW DIRECTIONS in ORGANIC and BIOLOGICAL CHEMISTRY

Series Editor: Philip Page

PUBLISHED TITLES

Activated Metals in Organic Synthesis
Pedro Cintas

The Anomeric Effect
Eusebio Juaristi and Gabriel Cuevas

C-Glycoside Synthesis
Maarten Postema

Capillary Electrophoresis: Theory and Practice
Patrick Camilleri

Carbocation Chemistry: Applications in Organic Synthesis
Jie Jack Li

Chemical Approaches to the Synthesis of Peptides and Proteins
Paul Lloyd-Williams, Fernando Albericio, and Ernest Giralt

Chiral Sulfur Reagents
M. Mikolajczyk, J. Drabowicz, and P. Kielbasiński

Chirality and the Biological Activity of Drugs
Roger J. Crossley

Concerted Organic and Bio-organic Mechanisms
Andrew Williams

Cyclization Reactions
C. Thebtaranonth and Y. Thebtaranonth

Dianion Chemistry in Organic Synthesis
Charles M. Thompson

Mannich Bases: Chemistry and Uses
Maurilio Tramontini and Luigi Angiolini

Modern NMR Techniques for Synthetic Chemistry
Julie Fisher

Organozinc Reagents in Organic Synthesis
Ender Erdik

Carbocation Chemistry
Applications in Organic Synthesis

edited by
Jie Jack Li
University of San Francisco
California, USA

CRC Press
Taylor & Francis Group
Boca Raton London New York

CRC Press is an imprint of the
Taylor & Francis Group, an **informa** business

CRC Press
Taylor & Francis Group
6000 Broken Sound Parkway NW, Suite 300
Boca Raton, FL 33487-2742

First issued in paperback 2019

© 2017 by Taylor & Francis Group, LLC
CRC Press is an imprint of Taylor & Francis Group, an Informa business

No claim to original U.S. Government works

ISBN-13: 978-0-4987-2908-6 (hbk)
ISBN-13: 978-0-367-87346-2 (pbk)

Visit the Taylor & Francis Web site at
http://www.taylorandfrancis.com

and the CRC Press Web site at
http://www.crcpress.com

Contents

Series Preface

NEW DIRECTIONS IN ORGANIC AND BIOLOGICAL CHEMISTRY

Organic and biological chemistry forms a major division of chemistry research and continues to develop rapidly, with new avenues of research opening up in parallel with exciting progress in existing activities. This book series encompasses cutting-edge research across the entire field, including important developments in both new and fundamental aspects of the discipline. It covers all aspects of organic and biological chemistry. The series is designed to be very wide in scope and provides a vehicle for the publication of edited review volumes, monographs, and *"How to"* manuals of experimental practice.

Examples of topic areas would include

Synthetic chemistry
Natural products chemistry
Materials chemistry
Supramolecular chemistry
Electrochemistry
Organometallic chemistry
Green chemistry, including catalysis and biocatalysis
Polymer chemistry
Protein–protein, protein–DNA, protein–lipid, and protein–carbohydrate interactions
Enzyme/protein mechanism, including cofactor–protein interactions and mechanisms of action
Pathways in cofactor (including metal) homeostasis
Synthetic biology—the engineering of natural systems toward novel functions
Development of physical methods and tools, for example, new spectroscopic methods, imaging techniques, and tools

The volumes highlight the strengths and weaknesses of each topic and emphasize the latest developments emerging from current and recent research. The series is of primary interest to academic and industrial chemists involved broadly in organic, materials, biological, and medicinal chemistry. It is also intended to provide research students with clear and accessible books covering different aspects of the field.

Philip C. Bulman Page, DSc, FRSC, CChem
Series Editor
School of Chemistry, University of East Anglia, Norwich, UK

Preface

Carbocation chemistry is one of the most important fields of organic chemistry. Yet, counter intuitively, the field is not represented by the number of books published. Indeed, the last book on carbocation chemistry was published a decade ago and the materials were mostly geared toward specialists in the field.

This book, in contrast, is written by and for synthetic chemists, who are more interested in how to apply carbocation chemistry in synthesis.

To that end, Chapter 1 summarizes the advancement of our understanding of the mechanisms of carbocation chemistry during the last decade. Chapters 2 and 3 cover the S_N1 and S_N2 reactions, respectively. The implication of carbocation intermediates for the S_N1 reactions is universal, the connection between carbocation intermediates and S_N2 reactions is somewhat less rigorous. Chapter 4 covers electrophilic addition to alkenes, and Chapter 5 covers electrophilic aromatic substitution, both of which are critical to our learning of the fundamentals of organic chemistry. Last but not least, Chapter 6 entails fragmentation and rearrangement reactions, which have historically shed much light on the mechanisms of carbocation chemistry and are now important to synthesis as well. I want to thank all the authors who painstakingly contributed to this manuscript. I am also indebted to John Hendrix for proofreading my chapters.

As a consequence, this book is not just for chemists whose expertise is carbocation chemistry. The experts could learn a thing or two from the advancement of carbocation chemistry in the last decade. More than that, any practitioner in organic synthesis will find this manuscript helpful in gauging the literature of the last decade. Therefore, this book is useful to senior undergraduate students, graduate students, professors teaching organic chemistry, and researchers in pharmaceutical and chemical companies.

Jie Jack Li
San Francisco

Editor

Jie Jack Li is an associate professor of chemistry at the University of San Francisco. Previously, he was a discovery chemist at Pfizer and Bristol-Myers Squibb, respectively. He has authored or edited over 20 books including *C−H Bond Activation in Organic Synthesis* published by Taylor & Francis/ CRC in 2015.

Contributors

Adam M. Azman
Department of Chemistry
Butler University
Indianapolis, Indiana

Yu Feng
Abide Therapeutics Inc.
San Diego, California

Safiyyah Forbes
Department of Chemistry
University of San Francisco
San Francisco, California

Micheal Fultz
Department of Chemistry
West Virginia State University
Institute, West Virginia

Jie Jack Li
Department of Chemistry
University of San Francisco
San Francisco, California

John W. Lippert III
Department of Chemistry
Cyclics Corporation
University Place
Rensselaer, New York

Sharon Molnar
Department of Chemistry
West Virginia State University
Institute, West Virginia

Hannah Payne
Department of Chemistry
West Virginia State University
Institute, West Virginia

1 Introduction

Jie Jack Li

CONTENTS

1.1 NOMENCLATURE, STRUCTURE, AND STABILITY

In sophomore organic chemistry, we learned that many organic compounds bear positive charges. For instance, bromonium ion was invoked as the putative intermediate when explaining the mechanism of electrophilic addition of bromine to alkenes. In 1985, Brown at Alberta isolated and determined the crystal structure of the bromonium ion **2** during the course of electrophilic bromination of adamantylideneadamantane (**1**).[1]

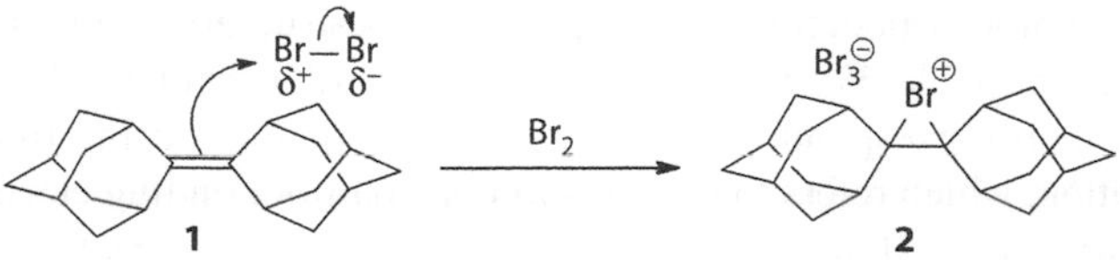

Other positive species in organic chemistry include mercurinium ion (**3**), hydronium ion (**4**), ammonium ion (**5**), nitronium ion (**6**), sulfonium ion (**7**), and phosphonium ion (**8**), just to name a few.

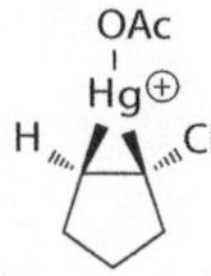

Mercurinium ion (**3**) 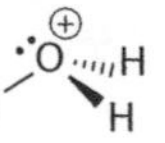   Hydronium ion (**4**) 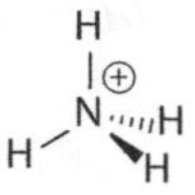   Ammonium ion (**5**)

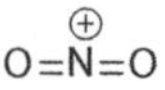

Nitronium ion (**6**) 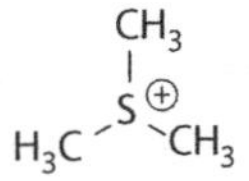   Sulfonium ion (**7**) 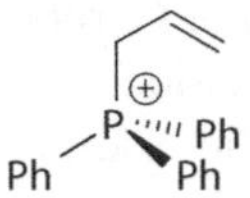   Phosphonium ion (**8**)

But none of the aforementioned -nium ions are as prevalent as carbonium ions. After all, organic chemistry is the study of carbon-containing compounds. As time passes, the term carbonium ions have been largely replaced by the more popular term, carbocations. This was probably inevitable since *carbocation* is one word and *carbonium ion* are two!

As a matter of fact, initially proposed by Olah[2] and later accepted by the IUPAC (International Union of Pure and Applied Chemistry), the term *carbocation* encompasses two species, one is the *carbenium* ion, which represents the "classical" trivalent ions with $CH_3^\oplus$ as the parent, and the other is the *carbonium* ion, which represents the "nonclassical" penta- (or higher) coordinate ions with $CH_5^\oplus$ as the parent.

This book titled *Carbocation Chemistry* will cover both carbonium ions and carbenium ions. And in this chapter, carbocation and carbenium ion are used interchangeably because nearly all of the carbocations mentioned here are carbenium ions.

Carbocations depicted below as **9**, with a cation on the carbon atom, is one of the important reaction intermediates in organic chemistry. The other intermediates include carbanions (**10**), radicals (**11**), carbenes (**12**), etc. For a trivalent carbocation, it is sp^2 hybridized with a geometry of trigonal planar and all three bond angles are 120°.

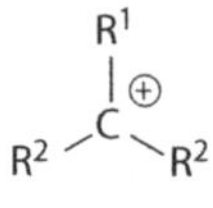

Carbocation (**9**) 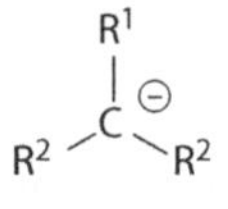   Carbanion (**10**) 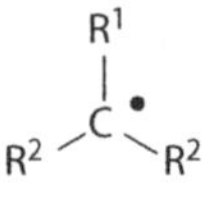   Free radical (**11**) 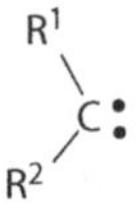   Carbene (**12**)

The stability of a carbocation is determined by its structure: the more alkyl substituents, the more stable. For tertiary (3°), secondary (2°), primary (1°), and methyl carbocations, their corresponding stability may be explained using the concept of hyperconjugation, which refers to the donation of a pair of bonding electrons into an unfilled or partially filled orbital. As depicted by ethyl cation (**13**), the two bonding

electrons of the C–H σ-bond adjacent to the carbocation may donate part of its electron cloud to the carbocation's empty p orbital. As a consequence, cation **13** is more
stabilized than it would if the α C–H bond did not exist. In all, ethyl cation **13** has
three α C–H bonds, thus three hyperconjugations (a short hand to describe three pos-
sibilities of having hyperconjugation).

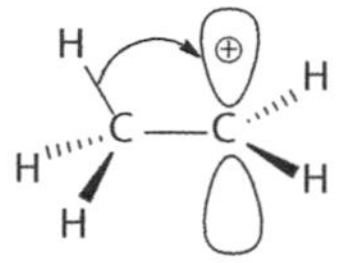

13

As clearly depicted below, *t*-butyl cation (**14**) as a representative of tertiary (3°)
carbocations possesses nine α C–H bonds, therefore it has nine hyperconjugations.
Meanwhile, isopropyl cation (**15**) as a representative of secondary (2°) carbocations
possesses six α C–H bonds, therefore it has six hyperconjugations. In the same vein,
primary (1°) cation **16** has three hyperconjugations and the methyl cation (**17**) has
none. We can thus conclude that tertiary (3°) carbocations are the most stable in
this series, followed by secondary (2°), primary (1°), and methyl carbocations. The
stability of cations decreases in this sequence from the left to the right. This trend is,
not surprisingly, consistent with experimental data generated from their dissociation
energies in the gas phase.[3]

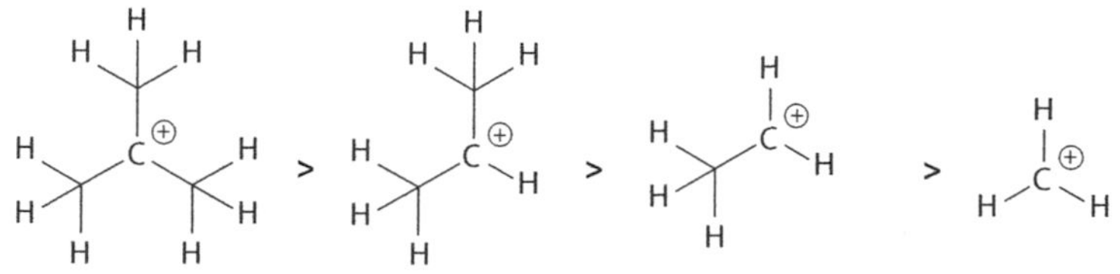

3° Carbocation (**14**)   2° (**15**)   1° (**16**)   Methyl cation (**17**)

In addition to evoking the concept of hyperconjugation, the stability of carbocations may be explained using the concept of inductive effects. The carbon atom is
more electronegative than the hydrogen atom with electronegativity values of 2.55
and 2.20, respectively. Therefore, the methyl group is considered *electron donating*.
The *t*-butyl cation (**14**) has three electron-donating methyl groups; isopropyl cation (**15**) has two electron-donating methyl groups; ethyl cation **16** has one electron-
donating methyl group; and methyl cation (**17**) has none. The more electron-donating
groups a carbocation has, the more stable it is. As a consequence, the same trend of
stability can be arrived at: (**14**) > (**15**) > (**16**) > (**17**).

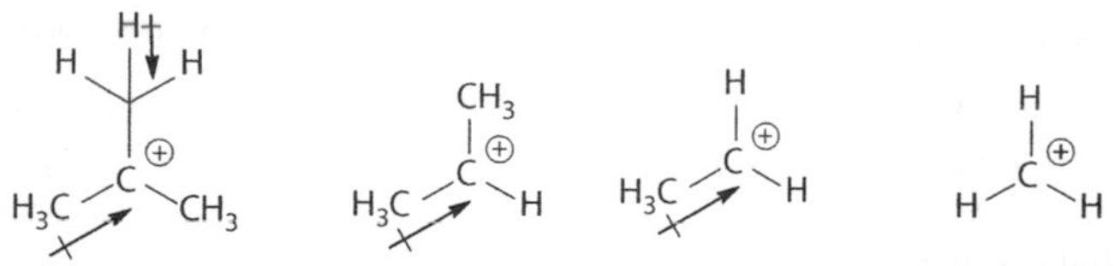

3° Carbocation (**14**)   2° (**15**)   1° (**16**)   Methyl cation (**17**)

However, hyperconjugation/inductive effect is *not* the only force playing here in determining the stability of carbocations. The other factor is resonance as in resonance structures. For instance, the dissociation energies in kJ/mol for generating the following cations in gas phase are listed below.[4]

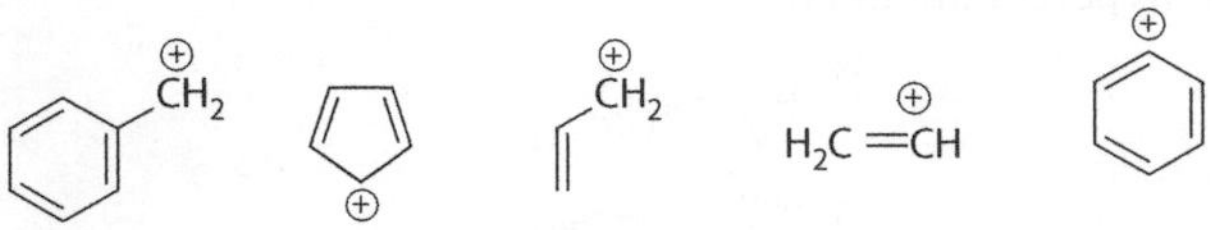

18	19	20	21	22
996	1030	1070	1200	1230 kJ/mol

Since a lower dissociation energy correlates to a more stable cation, benzyl cation (**18**) is more stable than cyclopentyl cation (**19**). This coincides with the number of resonance structures that we can draw: five for benzyl cation (**18**) and three for cyclopentyl cation (**19**). The allylic cation (**20**) is the least stable of the trio because it only has two resonance structures. There is no additional resonance structure for both ethylene cation (**21**) and benzene cation (**22**).

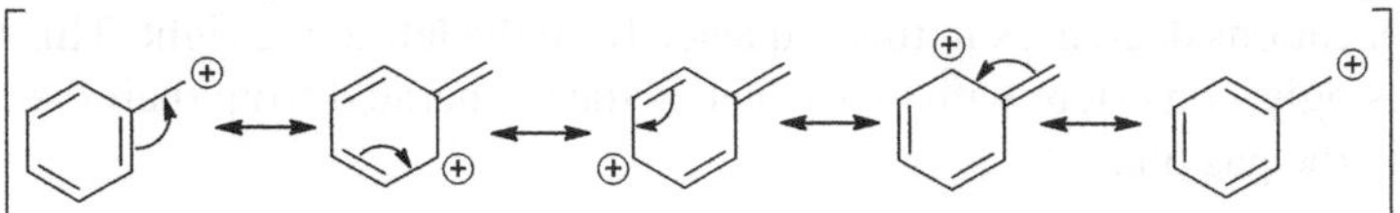

18-2   **18-2**   **18-3**   **18-4**   **18-5**

19-1   **19-2**   **19-3**

20-1   **20-2**

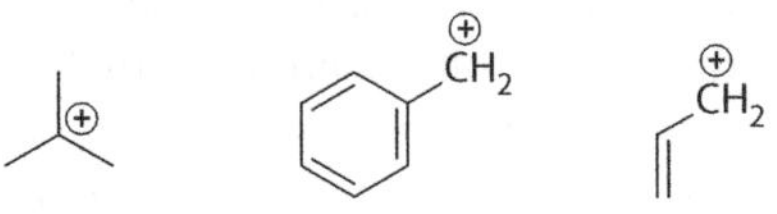

14	18	20
970	996	1070 kJ/mol

According to the dissociation energies in gas phase shown below,[3,4] the stability of the *t*-butyl cation (**14**) is greater than those of benzyl cation (**18**) and allylic cation (**20**). This is probably the interplay among all three factors including resonance, hyperconjugation, and inductive effects, especially the electron-donating effect of the three methyl groups.

1.2 GENERATION OF CARBOCATIONS

One of the popular methods of generating carbocations is acidic dehydration of secondary or tertiary alcohols. When colorless triphenylmethanol (21) was treated with concentrated sulfuric acid, a deep yellow solution of triphenylmethyl cation (trityl, 22) was obtained.[5] As a matter of fact, that was how one of the first carbocations, trityl cation (22), was discovered by Norris[5] in 1901.

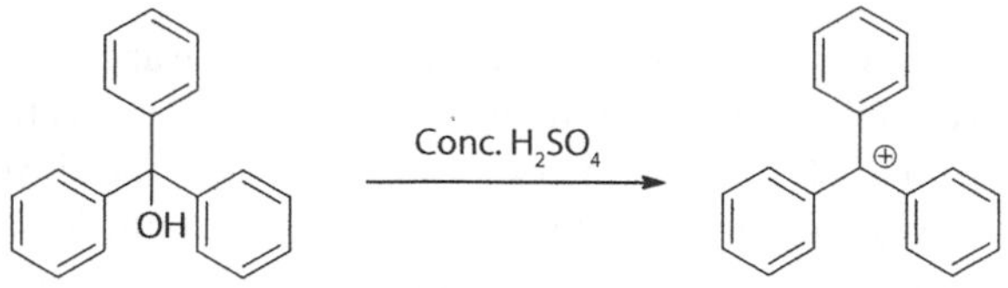

Triphenylmethanol (21)   Trityl cation (22)

Under the influence of a strong protic acid, trifluoroacetic acid (TFA), secondary alcohol 23 readily dehydrates to form a secondary carbocation 24.[6]

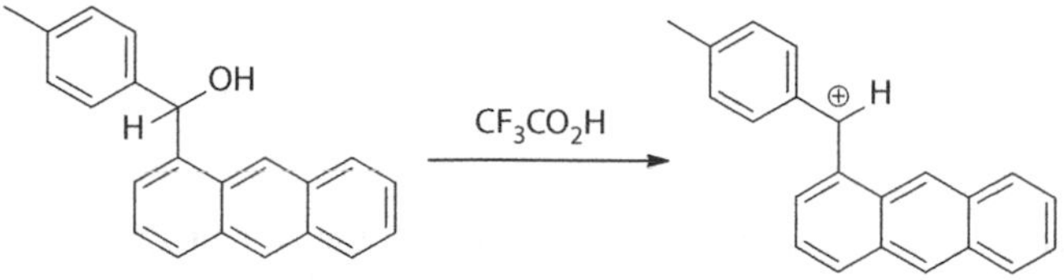

23   24

Dehydration of primary alcohols using strong acids to generate carbocations is complicated and beyond the scope of this chapter.

The second popular method of generating carbocations is through removal of an energy-poor anion from a neutral precursor using Lewis acids. For instance, tertiary chloride 25 was converted to tertiary carbocation 26 in the presence of Lewis acid pentafluoroantimony.[7] The concept can be extended. Chloride on 25 can be all the halogens and the methoxyl group and the Lewis acids could be $AlCl_3$, BF_3, $ZnCl_2$, $TiCl_4$, $SnCl_4$, $SnCl_5$, etc. for generating tertiary carbocation 26.

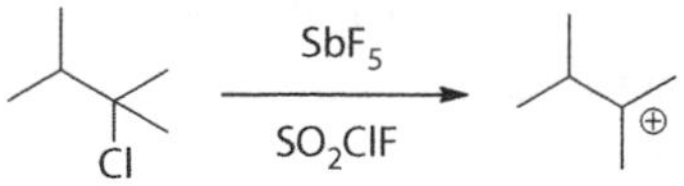

25   26

The third method of generating carbocations is the addition of electrophiles such as a proton to the π-systems such as double or triple bonds (C=C, C=O, C=N, etc.). As shown below, protonation of olefin 27 in strongly acidic condition would give rise to tertiary carbocation 28 as the intermediate. For example, 2,4-dimethylpent-1-ene (29) as a branched olefin undergoes a proton exchange in 73% H_2SO_4 to generate tertiary carbocation 30.[8]

The fourth category of generating carbocations is hydride abstraction from neutral precursors to give stable carbocation. In one example, the stable cation 1,1-dimethy-1*H*-azulenium cation (**32**) was obtained when 1,1-dimethyl-1,4-dihydroazulene (**31**) was treated with trityl perchlorate ($Ph_3C^{\oplus} ClO_4^{\ominus}$) in acetonitrile.[9] The Lewis acid used here is trityl perchlorate, but other Lewis acids such as trityl borontetrafluoride, borontrifluoride, and phosphorus pentachloride may be used for the purpose of hydride abstraction.

Finally, carbocations can be generated through the S_N1 or the E1 mechanisms, which will be the subjects of the later sections in this chapter.

1.3   THE NONCLASSICAL ION CONTROVERSY

The nonclassical ion controversy which took place during the second half of the last century was one of the more acrimonious and colorful events for organic chemistry. Despite hurt pride and feelings for the adversaries involved, organic chemistry is richer and we now have a better understanding of carbocations than arguably all other species.

The story goes back to the Wagner–Meerwein rearrangement (see also Section 7.6) where 2-norbornyl cation (**21**) was proposed as the intermediate. Since 1949, Winstein at UCLA (University of California, Los Angeles) investigated the solvolysis of *exo*-2-norbornyl brosylate (**22**) and its isomer *endo*-2-norbornyl brosylate (**22′**).[9–12] Brosylate is an analog of tosylate where the methyl group on the phenylsulfone is replaced by bromine. Winstein et al. observed that the reaction rate of solvolysis in acetic acid of the *exo*-isomer **22** was 350 times faster than that of the *endo*-isomer **22′**. Winstein proposed the reaction's cationic intermediate was, instead **21**, an σ-delocalized, symmetrically bridged norbornyl ion **23**. This "nonclassical" 2-norbornyl cation **23**, where two electrons delocalized over three carbon atoms, deviated from the accepted classical cation structure proposed by Meerwein as **21**, where the positive charge was considered to be localized on a single atom. It helped to explain that in the *exo*-isomer **22**, electrons from a C–C single bond near the brosylate help push off the leaving group, accelerating the reaction with the benefit of heightened levels of anchimeric assistance. But the geometry of the *endo*-isomer did not allow the anchimeric assistance by the two electrons on **22′**, therefore it had a slower reaction.

Brown: Winstein:

Equilibrating classical carbonium ion **21**   **22**   Nonclassical carbonium ion **23**

22′

In addition, regardless of which isomer (*endo/exo*) is utilized as the reactant, the *exo*-acetate is produced exclusively. Finally, optically active *exo*-starting materials react to give complete racemization, providing a 1:2 mixture of **24** and **25**. These data can be sufficiently explained through the intermediacy of a symmetrical and achiral structure **23**.[9–12]

Brown at Purdue was the future Nobel Laureate in chemistry in 1979 for organoboron chemistry. In the 1950s, he strongly disagreed with the existence of the "nonclassical" ion intermediate. He insisted on a rapid equilibrium between the two classical carbocation forms facilitated via the Wagner–Meerwein rearrangement.[13,14] The *exo*- and *endo*-rate ratios were attributed to steric effects, as strain caused *endo*-isomers to exhibit more hindrance to ionization. Finally, Brown criticized the bridged intermediate model for not providing sufficient electrons for all bonds.

The controversy became increasing vitriolic and personal. In 1954, Brown referred to Winstein as spicy peppers by writing "The Southwest must have a similar effect on the fauna of the region."[15] Winstein was less subtle, in reference to Brown in 1969, he said "That man is nothing but a shyster of a lawyer."[16]

In 1983, Olah, Saunders, and Schleyer were able to identify the intermediate as indeed the methylene-bridged nonclassical carbonium form of the norbornyl cation at low temperature. This was demonstrated through extensive spectroscopic analysis, including ^{1}H- and ^{13}C-NMR, Raman, ESCA (electron spectroscopy for chemical analysis), and further physical and kinetic studies.[17] Olah went on to win the 1994 Nobel Prize in chemistry for "his contributions to carbocation chemistry."

In 2013, Schleyer and colleagues obtained the long-sought x-ray crystallographic proof of the bridged nonclassical geometry of the 2-norbornyl carbonium ion **23** salt.[18]

The salt crystals were obtained by reacting norbornyl bromide with aluminum tribromide in CH_2B_2 at 86 K and then crystallization took place at 40 K.

Other important nonclassical ions are **26–29**.[19]

26	**27**	**28**	**29**

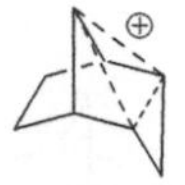

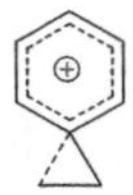

1.4   ELECTROPHILIC ADDITION TO ALKENES

We know more about nucleophilic aliphatic substitution than nearly all other classes of organic reactions. Ingold in 1928 defined S_N2 reaction as substitution *nucleophilic bimolecular* and S_N1 reaction as *substitution nucleophilic unimolecular*.[20] This section will focus on nucleophilic substitution, unimolecular (S_N1)[21] because the mechanism involves a formal carbocation intermediate. Nucleophilic substitution, bimolecular (S_N2) will be mentioned briefly because the reaction mechanism involves only partially charged carbocation as the intermediate.

In the 1930s, Hughes and Ingold proposed the "classic" mechanism for the S_N1 reaction where a formal carbocation is the intermediate.[22] In a typical S_N1 reaction, (*R*)-3-chloro-3-methylheptane (**30**) has a tertiary carbon at its reaction center thus *inversion* of configuration is completely out of question due to its steric hindrance. Instead, substrate **30** dissociates heterolytically to produce the chloride and carbocation **31** as the reaction intermediate. This step is slow and the rate-determining step (RDS). As a result, the reaction rate is proportional to sole the concentration of substrate **30**. Intermediate **31** is of sp^2 hybridization and trigonal planar configuration. A nucleophile (often a weak nucleophile) such as water may attack **31** from either side, giving rise to **32** as two possible enantiomers. Deprotonation of **32** then affords the final S_N1 product **32** as a mixture of two possible enantiomers.

30	**31**

A mixture of enantiomers

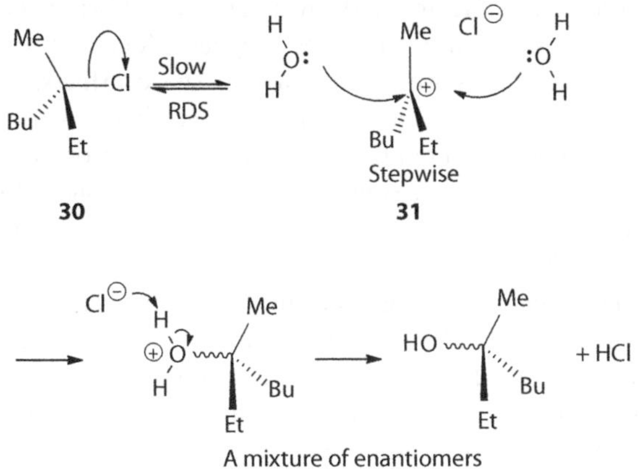

32	**33**

In the 1950s, Winstein et al.[23,24] discovered that the real S_N1 mechanism is more complex than the Hughes–Ingold mechanism shown above. They observed significant solvent effects during solvolysis of neutral substrates. It turns out that for substrate **34** to convert to a "free" carbocation **37**, **34** needs to form an "intimate ion pair" (IIP, **35**) first, which is the RDS. Intimate ion pair **35** is then transformed to a discreet intermediate **36** as a "solvent-separated ion pair" (SSIP). Since then on, more refined mechanisms have been proposed,[25] but the Winstein's mechanism seems to have withheld the test of time.

$$R–X \rightleftharpoons R^{\oplus} X^{\ominus} \longrightarrow R^{\oplus}/\!/X^{\ominus} \longrightarrow R^{\oplus} + X^{\ominus}$$

34   **35** (IIP)   **36** (SSIP)   **37**   **38**

Because the generation of carbocation as the reaction intermediate is the slowest step, the reactivity of the S_N1 reaction depends on the structure of the substrate. The more stable the carbocation intermediate, the faster the S_N1 reaction is. In case of haloalkanes, the rate of S_N1 reaction is fastest for the tertiary halogen, followed by secondary and primary halides. Methyl halides are the slowest. In fact, primary halides do not undergo through the S_N1 mechanism unless the substrate is a benzylic halide or a vinylic halide.

$$R_3CX > R_2CHX > RCH_2X > CH_3X$$

As far as the leaving group is concerned, as a rule of thumb, when all else is equal, a positively charged species is a better leaving group and a neutral species is a better leaving group than a negatively charged species. For instance, H_2O is a better leaving group than a $HO^{\ominus}$ group. For halides, $I^{\ominus}$ is the best leaving group because it has the largest size and the C–I is the easiest to break. Thus

$$I^{\ominus} > Br^{\ominus} > Cl^{\ominus} > F^{\ominus}$$

This section is divided into five subsections: addition of hydrohalides and acetic acid, hydration, addition of halogens, hydroboration, and oxymercuration.

1.4.1 ADDITION OF HYDROHALIDES AND ACETIC ACID

The mechanism of electrophilic addition to alkenes is shown below. Protonation of alkene **39** affords a carbocation intermediate **40**, which then captures a nucleophile to afford adduct **41**. The regiochemistry follows the Markovnikov rule because carbocation intermediate **40** is more stable than the intermediate formed when a proton is added to the less substituted carbon on the double bond.

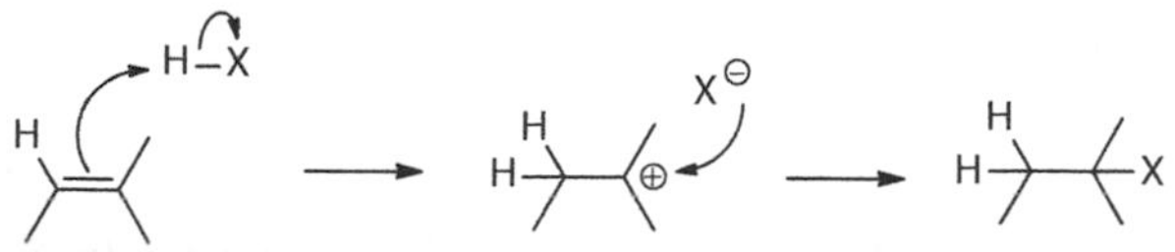

39   **40**   **41**

The aforementioned mechanism is also known as the Ad_E2 mechanism—addition-electrophilic bimolecular.[26] This reaction can be viewed as the reverse of the E1 reaction of **41**, which provide olefin **39** by eliminating HX, also via the intermediacy of **40**.

While the Ad_E2 mechanism readily explain the electrophilic addition to alkenes in polar solvents, the Ad_E3 (addition-electrophilic termolecular) mechanism[27] must be invoked when the electrophilic addition to alkenes takes place in nonpolar solvents. Under such circumstances, there is no distinctive carbocation intermediate: rather the reaction goes through a "concerted" mechanism with transition state **42**. This mechanism could be viewed as the reverse of the E2 mechanism from **41** to give olefin **39**.

The regiochemical outcome for the electrophilic addition of hydrohalides to olefins is more conspicuous if it is done using a cyclic olefin **43**. Treatment of **43** gave predominantly bromide **45**.[27] Evidently carbocation **44** is the major intermediate because it is more stable.

For alkynes, electrophilic addition works similarly except the intermediate is a vinylic cation in place of alkyl cation **40**.

Addition of acetic acid follows the same mechanism and stereochemical outcome as those for hydrohalides.

1.4.2 HYDRATION

The acid-catalyzed hydration of olefins gives rise to alcohols whereas hydration of alkynes gives rise to ketones.

In the presence of a catalytic amount of protic acid HX, hydration of olefin **39** begins with its protonation to afford carbocation intermediate **40**. This step is slow, thus the RDS. The combination of **40** with water as a weak nucleophile produces adduct **45**, which delivers the final product as alcohol **46** upon losing a proton. The last step for deprotonation–protonation is reversible. The regiochemistry follows the Markovnikov rule.

For hydration of terminal alkyne **47**, protonation with catalytic protic acid HX affords vinylic carbocation intermediate **48**. Similar to that of an alkene substrate, this step is also slow and the RDS. Combination with weak nucleophile water produces intermediate **49**, which readily loses a proton to afford enol **50**. Tautomerization of enol **50** takes place promptly to deliver the more stable tautomer ketone **51**. The regiochemistry generally follows the Markovnikov rule. Modern methodology affords many conditions that render functionalization of terminal alkenes and alkynes with anti-Markovnikov regiochemical outcome as well.[28]

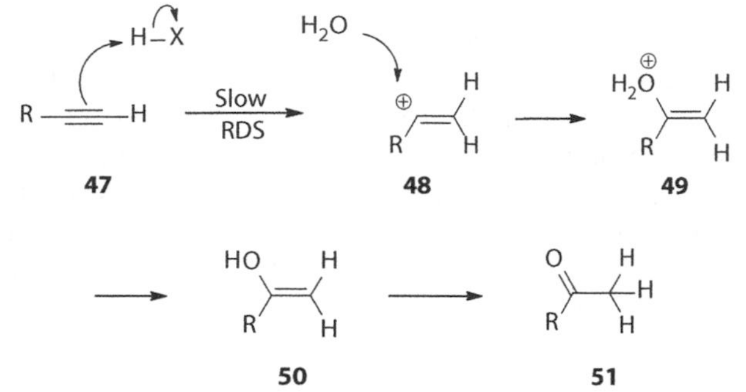

1.4.3 ADDITION OF HALOGENS

As we see at the beginning of this chapter concerning addition of bromine to an olefin **1**, bromonium ion **2** was invoked as the putative intermediate when explaining the mechanism of electrophilic addition of bromine to alkenes.[1] A generic scheme for addition of halogens to olefins may be depicted below: the π-electrons attacks the halogen molecule to form the π-complex **52** first. It is converted to the halonium ion intermediate **53**, which is in equilibrium with the nonbridged carbocation **54**. Nucleophilic attack of either halonium ion ring opening **53** or the nonbridged carbocation **54** with the available halide then delivered dihalide **54**. However, the outcome would be the same regardless which site the halide attacks the halonium ion intermediate **53**.[29] There are no regiochemical issues although the two halogens have an *anti* configuration.

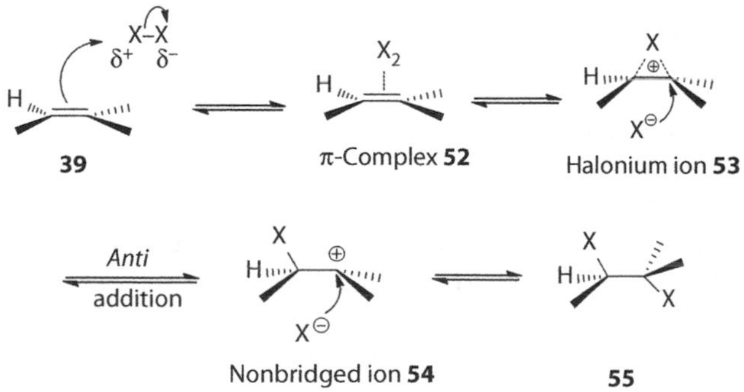

For the latest advances in catalytic, stereoselective dihalogenation of alkenes, see Demark's latest review in 2015.[30]

1.4.4 HYDROBORATION OF OLEFINS

After Brown's pioneering work, hydroboration flourished in organic synthesis. Hydroboration of olefins is now sometimes known as the Brown hydroboration as represented in the transformation of terminal olefin **56** to alcohol **57**.[31]

$$R \diagup\diagup \xrightarrow[\text{(2) } H_2O_2, \text{ NaOH}]{\text{(1) BH3 · THF}} R \diagup\diagdown \diagup OH$$

56   **57**

The regiochemical outcome follows the anti-Markovnikov rule and it is readily predicted using the concept of electronegativity. First, since R is an electron-donating group, therefore, the terminus of olefin **56** is partially negative. On the other hand, since boron's electronegativity is 2.04 and that of hydrogen (it is actually a hydride here) is 2.2, this makes the hydrogen atom partially negative. When they meet, it is not surprising that tetracyclic borane **58** is the predominant intermediate to afford the adduct **59**. Since there are still two hydrides left, borane 59 can consume two more molecules of olefin **56** to provide trialkyl borane **60**. Basic oxidative workup of **60** then delivered alcohol **57**.

56   **58**   **59**

60   **57**

Asymmetric hydroboration of olefins, especially challenging substrates such as 1,1-disubstituted alkenes, was reviewed in 2009.[32] In addition to olefins, hydroboration has found wide-spread application in the hydroboration of carbonyls, imines, and enamines.[33]

1.4.5 OXYMERCURATION OF OLEFINS

Oxymercuration is not widely used in real-world synthesis due to the use of toxic mercury reagents. However, it has been an excellent teaching tool with regard to the learning mechanism so a brief summary is presented here.

In addition to hydroboration, Brown also made significant contributions to the oxymercuation of olefins.[34] As shown in the scheme below, treatment of olefin **56** gives rise to organomercury compound **61**. The regiochemistry follows the Markovnikov rule. Therefore, this sequence is also called oxymercuation–demercuration (OM–DM), which is complimentary to the hydroboration in terms of the regiochemistry.

56   **61**   **62**

In terms of mechanism, the concept of electronegativity may be applied here as well. Since mercury's electronegativity is 2.00 and that of oxygen is 3.44, this makes the Hg atom partially positive. When olefin **56** is in contact with mercury acetate, the two π-electrons could attack the Hg atom and expel the acetate away. Meanwhile, two electrons on the Hg atom could migrate in between the C–C bond and the Hg atom to form a mercurium ion **63**, not unlike the halonium ion **53**. Because the R group is electron donating, a partial positive charge on the carbon α to R as shown in **64** is more stable than on its β position. Nucleophilic ring opening using water would give rise to oxonium ion **65**, which readily loses a proton to produce intermediate **61**. Demercuration of **61** under basic oxidative conditions is not discussed here but it most likely involves a reductive elimination step of the mercury hydride intermediate.

56	Mercurium ion **63**	**64**

65	**61**	**62**

1.5 ELECTROPHILIC AROMATIC SUBSTITUTION

1.5.1 MECHANISM AND ORIENTATION

The large majority of electrophilic aromatic substitution (EAR) reactions share a common mechanism: the arenium ion mechanism.[35] If we designate an electrophile as $E^\oplus$, then the first step of the electrophilic aromatic substitution is donation of a pair of electrons from benzene to $E^\oplus$. This is the slow, thus rate-limiting, step. The resulting cyclohexadienyl carbocation **1** exists in three resonance structures known as the Wheland intermediates, or σ-complex, or arenium ion. To simplify things, we generally draw resonance hybrid **67** rather than three σ-complexes **66-1**, **66-2**, and **66-3**. Once the cyclohexadienyl carbocation is formed, it loses a proton to regain the aromatic sextet to deliver the EAR product **68**. Since the second step restores the aromaticity, a thermodynamically favored process, it is a fast step.

When the electrophile is a dipole rather than a cation, the electrophilic aromatic substitution proceeds via a similar mechanism.

66-1	**66-2**	**66-3**

67	**68**

While the two-step mechanism involving the σ-complex has been widely accepted, a three-step mechanism involving a π-complex is also supported by experimental evidence.[36–38] The first step is a reversible process where benzene as a π-donor provides electrons from its π-bonds to the electrophile to form π-complex **69**. This step is the rate-limiting step. The second step is conversion of π-complex **69** to σ-complex **67**.

69	**67**	**68**
π-Complex	σ-Complex	

The regiochemical outcome of electrophilic aromatic substitution depends on the electronic nature of the substituents on the benzene ring. For deactivating groups (electron-withdrawing groups), monosubstituted benzene gives predominantly *meta*-substitutions. On the other hand, *ortho*- and *para*-directing groups are mostly activating groups (electron-donating groups), but some are deactivating groups.[39]

A deactivating group such as nitro on benzene is predominantly *meta*-directing. Other deactivating groups include halides, aldehyde, ketone, acid, ester, amide, sulfonic acid, nitrile, ammonium ion, etc. The halides are weak deactivating groups. The aldehyde, ketone, acid, ester, and amide are moderate deactivating groups. Finally, nitro, sulfonic acid, nitrile, and ammonium ion are strong deactivating groups.

Let us take a look of the simpler case first with regard to its orientation of electrophilic aromatic substitution reactions.

The nitro group is one of the most powerful deactivating groups. When nitrobenzene attacks an electrophile $E^{\oplus}$, there are three possible outcomes. One is the *ortho*-substitution to give product **70**. As shown by the three resonance structures of the intermediates, the carbocation is highly destabilized because the nitro group is electron withdrawing. The same scenario is encountered for the intermediates to produce the *para*-substitution product **71**. In stark contrast, the resonance structures to generate the *meta*-substitution product **72** does not place the positive charge right next to the nitro group, rendering them the most stable among all three possible outcomes. Therefore, the activation energy (Ea) for *meta*-substitution is smaller than that for *ortho*- and *para*-substitution, and *meta*-product **72** is the predominant product. However, the Ea leading to *meta*-substitution product **72** is still higher than that of benzene because the nitro group is a deactivating group.

The outcome of electrophilic aromatic substitution reactions for a benzene ring with an electron-donating group is exactly the opposite.

For *ortho*-substitution to give product **73**, the methoxyl group sits right adjacent to the cation on the intermediate. The same is true for the *para*-substitution to give product **74**. This makes the Ea for both *ortho*- and *para*-substitutions not only lower than that for *meta*-substitution to give product **75**, it is also lower than that of benzene because the methoxyl group is an activating group. The final outcome is that electron-donating groups such as methoxybenzene favor *ortho*- and *para*-substitutions.

The energy diagram of the electrophilic aromatic substitution of nitrobenzene is shown below.

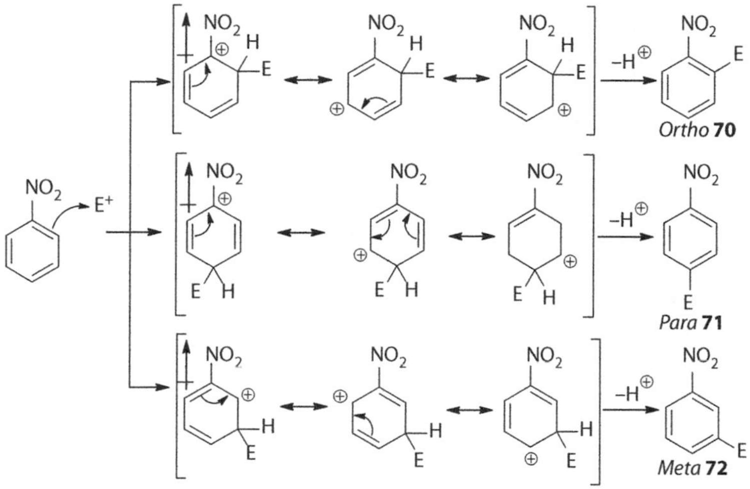

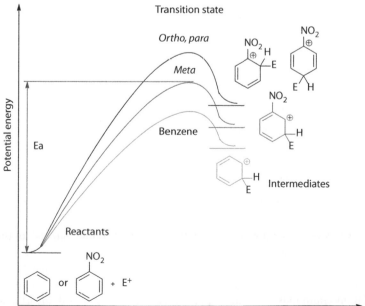

The energy diagram of the electrophilic aromatic substitution of methoxybenzene is shown below.

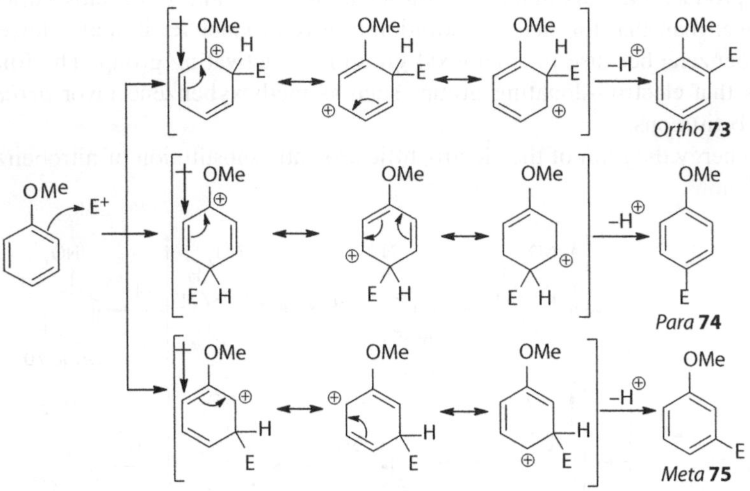

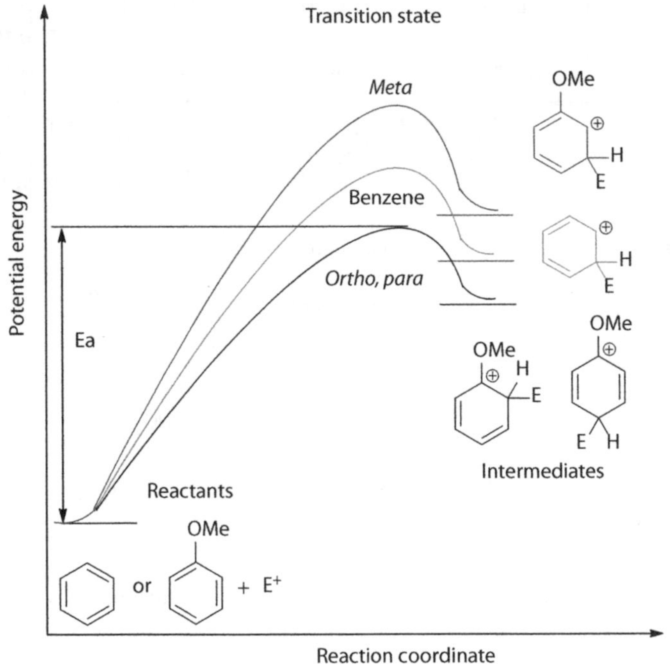

As far as halobenzenes are concerned, they are the "trouble makers" of the rules. On the one hand, halogens are electron-withdrawing groups, thus they are weak deactivating groups. On the other, they possess three lone pairs of electrons that are capable of halonium ions **52-2** and **53-2**. While they are not as stable as oxonium ions and immonium ions, they are important enough to make both *ortho-* and *para-* substitution more predominant than the corresponding *meta*-substitution.

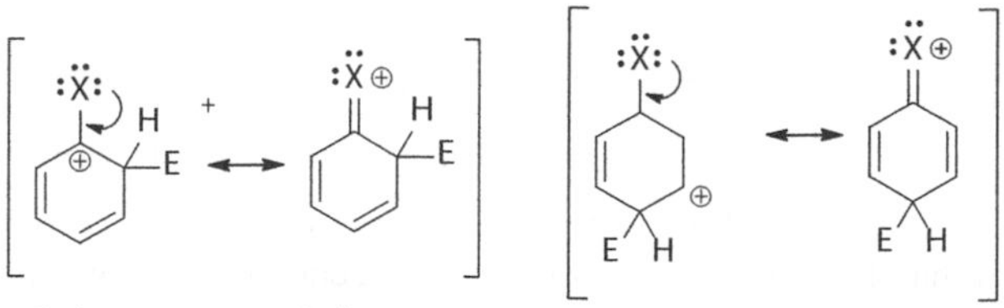

Ortho-**76-1**   Ortho-**76-2**   Para-**77-1**   Para-**77-2**

Halonium ion   Halonium ion

Three major types of electrophilic aromatic substitution reactions are reviewed here: nitration, halogenation, and Friedel–Crafts alkylation and acylation reactions. Protonation, sulfonation, diazonium ion formation, etc. are not covered in this chapter.

1.5.2 NITRATION

Benzene is readily nitrated in a mixture of nitric acid in concentrated sulfuric acid, where a powerful electrophile nitronium ion $NO_2^{\oplus}$ is generated.[40–42]

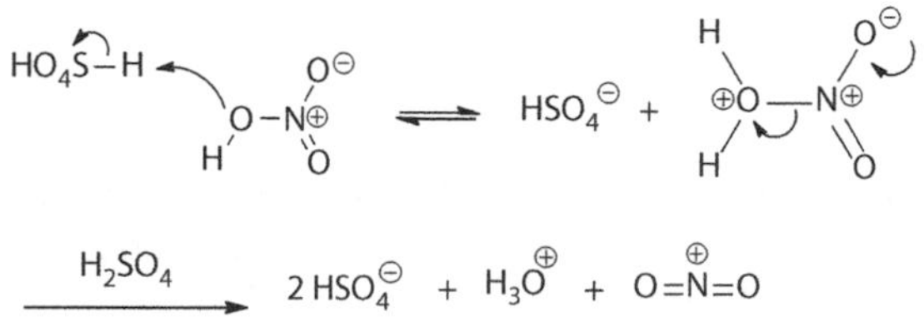

Nitronium ion

According to a mechanism advanced by Ingold and Hughes,[43] the electron pairs on the benzene ring attacks nitronium ion, producing the intermediate as σ-complex **78**, which rapidly loses a proton to rearomatize to nitrobenzene **79**.

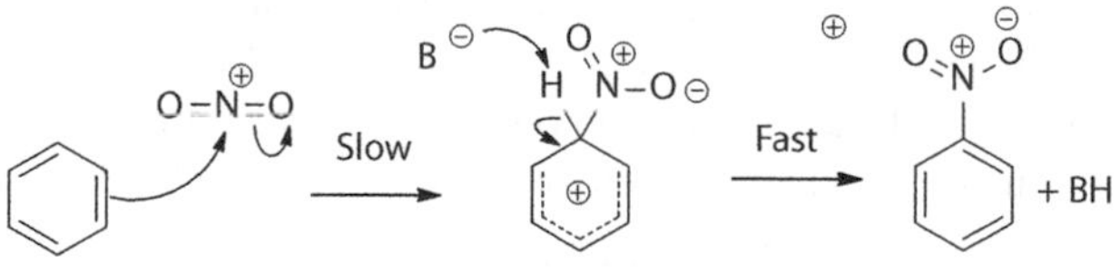

This classical mechanism, now known as the Ingold–Hughes mechanism, has largely withstood the test of time and many experimental data.[44]

Olah suggested a modified mechanism with an additional step of π-complex **80** formation.[45] Formation of the π-complex in a RDS explains the low substrate selectivity observed upon nitration of activated aromatics while retaining positional selectivity.

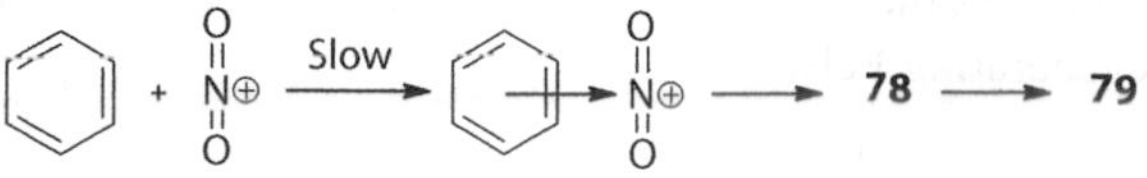

80, π-Complex   α-Complex   Nitrobenzene

Both the Ingold and Olah mechanisms involve transfer of two electrons. Another mechanism invoking the single-electron transfer (SET) with **81** as the counter ion was proposed and has gained much credence.[46,47]

$$\text{(benzene)} + \underset{\overset{\parallel}{O}}{\overset{\overset{O}{\parallel}}{N^{\oplus}}} \rightleftharpoons \left[PhH \cdot NO_2^{\oplus}\right] \rightleftharpoons 78 \longrightarrow 79$$

81, Encounter pair α-complex nitrobenzene

For nitration of monosubstituted benzenes, *ipso*-substitution, in which X is displaced by nitro, has been known for over half a century.[48,49] Normally, X = alkyl, acyl, –SiRr, –SO₃H, and –N₂Ar.[50] The mechanism of *ipso*-nitration is summarized below.[51,52] The *ipso*-intermediate then loses X to provide the corresponding nitrobenzene.

Ipso   *Ortho*   *Meta*   *Para*

Side reactions[53] and some unconventional pathways[54] for aromatic nitration have been reviewed in the literature.

1.5.3 HALOGENATION

In the presence of Lewis acid catalysts, electrophilic aromatic bromination and chlorination readily take place. Iodine is the least reaction electrophiles among halogens. Electrophilic aromatic iodination requires the aid of an oxidizing agent even for activating substrates so that iodine is converted to a more powerful electrophile.[55] Generally speaking, the power of electrophilicity follows the trend below

$$Cl_2 > BrCl > Br_2 > ICl > I_2$$

Recall that bromination of benzene with bromine alone is very slow. Adding an iron nail drastically accelerates the bromination reaction, by reacting with bromine to form the ferric bromide salt, which serves as a Lewis acid to react with another molecule of bromine to produce the bromonium ion.

$$2\,Fe + 3Br_2 \longrightarrow 2\,FeBr_3$$
Ferric bromide

$$FeBr_3 + Br_2 \longrightarrow Br^{\oplus} + FeBr_4^{\ominus}$$
Bromonium ion

The bromonium ion is a much more powerful electrophile than bromine. Even nonactivated benzene may be readily brominated to provide carbonium ion **82** as a α-complex. Intermediate **82** then loses a proton to the base FeBr₄⁻, affording bromobenzene (**83**) and regenerating FeBr₃.

An electron transfer (ET) mechanism was invoked for aromatic iodination using I–Cl in hexafluoropropa-2-ol.[56] Spectral and kinetics data suggested the existence of the cation radical intermediates.[57] The mechanism below is capable of explaining the fact that there are three products for aromatic iodination using I–Cl: ArI, ArCl and mixed iodination/chlorination on the same molecule.

$$ArH + ICl \longrightarrow ArH^{\cdot \oplus} + I^{\cdot} + Cl^{\ominus}$$

$$ArH^{\cdot \oplus} + I^{\cdot} \longrightarrow Ar\!-\!I + H^{\ominus}$$

$$ArH^{\cdot \oplus} + Cl^{\ominus} \longrightarrow Ar\!-\!Cl + H^{\oplus} + e^{\ominus}$$

A review appeared in 2009 to summarize oxidative halogenation using green oxidants such as oxygen and hydrogen peroxide.[58] Halogenation via C–H activation was also reviewed in 2010.[59]

Fluorination has become a more and more popular research area because of the important role fluorine plays in medicine.[60] Electrophilic aromatic fluorination[61,62] using fluorine in inert liquid is not operationally convenient. Therefore, electrophilic fluorinating reagents are used as the fluoronium ion ($F^{\oplus}$) source. In addition to *N*-fluorobenzenesulfonimide (NFSI, **84**), there are two other classes of fluorinating reagents:[63,64] *N*-fluoropyridinium salts (FP salts) as represented by [pyF]BF$_4$ (**85**) and *N*-fluorotriethyldiamine salts (F-TEDA salts) as represented by Selectfluor (**86**).

NFSI, **84**   [pyF]BF$_4$, **85**   F-TEDA-BF$_4$ (Selectfluor, **86**)

When hydroxyl-tetrahydronaphthalene **87** was treated with Selectfluor (**86**), cyclohexa-dienone **88** was the predominant product.[65]

87   **88**

Recently, fluorination via C–H activation is intensively being investigated.[66–69] Last year, Xu developed a selective *ortho*-fluorination by installing aryl-*N*-heterocycles such as quinoxaline, pyrazole, benzo[*d*]oxazole such as **89**, and pyrazine as the directing groups.[70] While those heteroarene directing groups survived the strongly acidic Pd(OAc)$_2$–NFSI–TFA system, they are not removable.

89   **90**

Also in the last year, Daugulis accomplished a copper-catalyzed fluorination of arene and heteroarene C—H bonds.[71] Daugulis used his signature auxiliary 8-aminoquinoline such as **91** as the directing group. In contrast to Xu's fluoronium ion, Daugulis used a nucleophilic F$^{\ominus}$ in the form of AgF as the fluorinating agent. While the auxiliary is readily removed via hydrolysis to expose the carboxylic acid functionality, the atom economy is not ideal because the auxiliary has similar mass as the substrate.

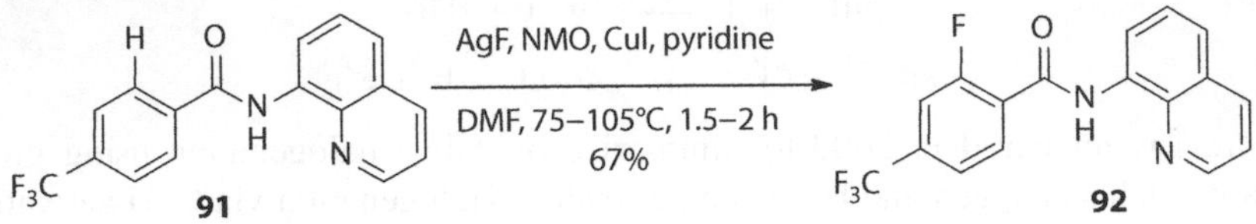

1.5.4 FRIEDEL–CRAFTS ALKYLATION

The Friedel–Crafts reaction has been a workhorse for organic synthesis since its discovery in 1877. A number of books[72–76] and many reviews have appeared in the literature.

Friedel–Crafts alkylation involves an electrophilic addition of a benzene ring to a carbocation intermediate generated from an alkyl halide or an alkyl alcohol.[77] For example, ethylation of a benzene takes place between ethyl chloride and benzene under the catalysis of AlCl$_3$. Lewis acid AlCl$_3$ interacts with ethyl chloride to produce ethyl cation and AlCl$_4{}^{\ominus}$. Electrophilic attack by the benzene ring to the ethyl cation leads to α-complex **93**, which loses a proton readily to afford ethylbenzene (**94**).

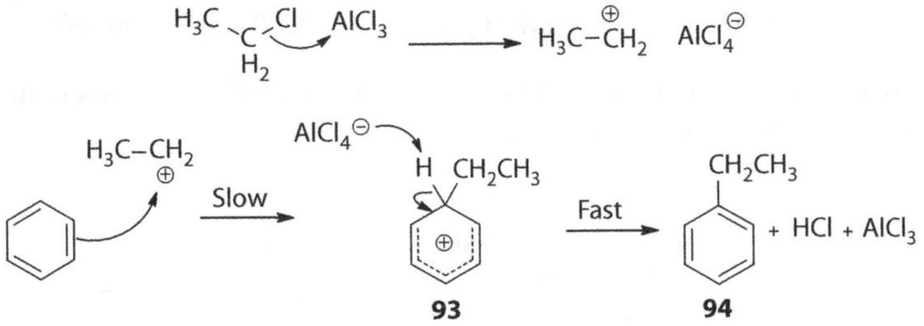

A more nuanced mechanism for Friedel–Crafts alkylation is shown below[78,79]

Generalized Friedel-Crafts intermolecular RX/ROH alkylation mechanism

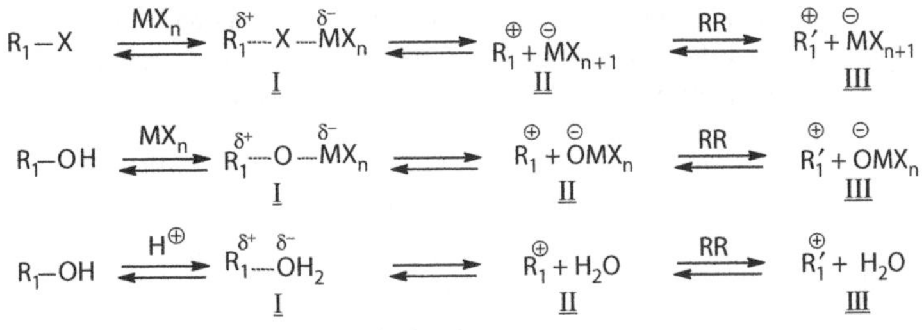

(RR denotes rearrangement took place)

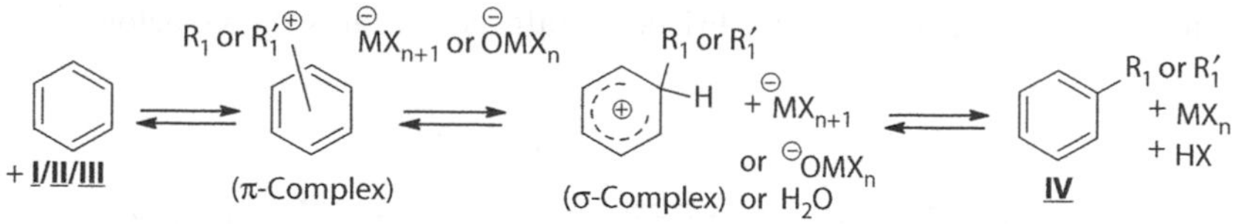

Two points need to be added to the simplified mechanism above. One, the carbocation intermediate could also be generated from the reaction of alcohols and Lewis acids. For instance, see below.

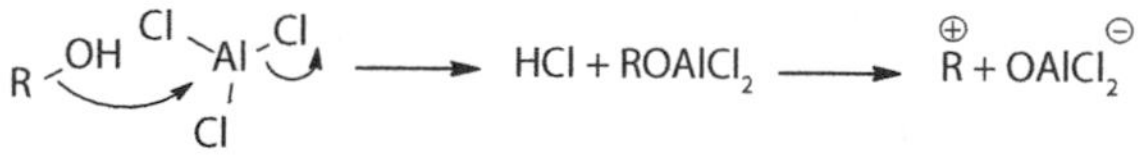

The second point to add to the simplified mechanism above is cationic rearrangement. The Friedel–Crafts alkylation of benzene with *n*-butyl chloride and $AlCl_3$ actually gives the rearrangement product 2-phenylbutane (**96**) in addition to **95**. This is a clear evidence of cationic rearrangement of the *n*-butyl cation via *hydride shift.*

95, 35% **96**, 35%

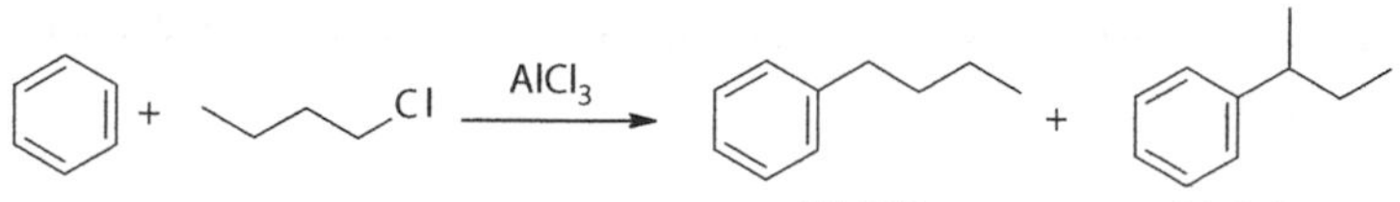

Since one of the resulting sp^3 carbon could be chiral such as on **96**, much has been achieved in the field of *asymmetric* Friedel–Crafts alkylation.[79,80]

1.5.5 Friedel–Crafts Acylation

Friedel–Crafts acylation cannot be asymmetric because the resulting carbonyl compounds are sp^2 hybridized.[81] In case of acid chloride **97**, it complexes with Lewis acid catalyst $AlCl_3$ to form donor–acceptor complex **98**, which undergoes a heterolytic C–Cl bond cleavage to provide acylium ion **99** as the key intermediate. The highly reactive electrophile **99** and benzene undergoes the electrophilic aromatic substitution to produce the a-complex **100**, which readily rearomatizes to deliver the carbonyl product **101**.

97 Donor–acceptor complex **98** Acylium ion **99**

100 **101**

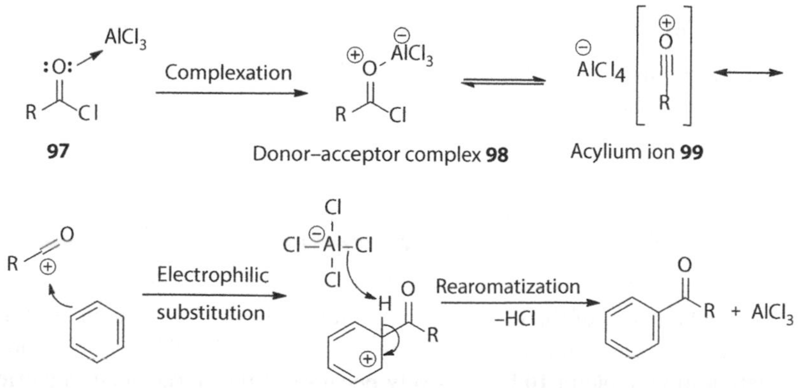

A more nuanced mechanism for Friedel–Crafts acylation is shown below[77,78]

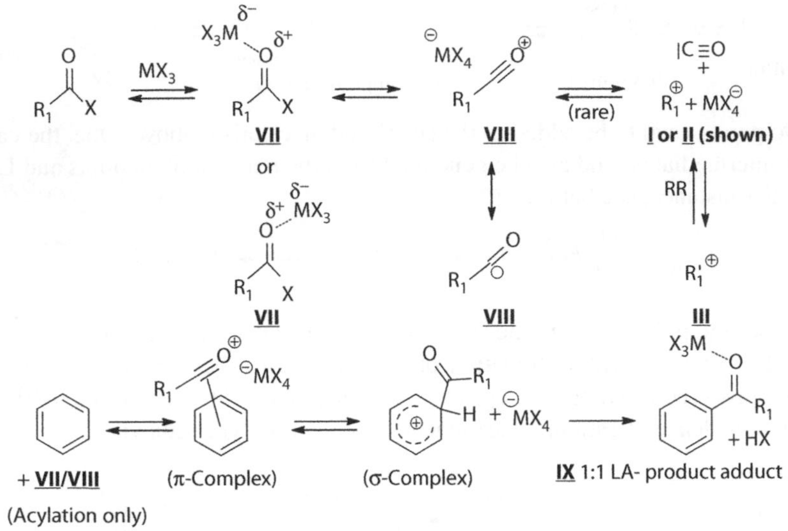

(Acylation only)

1.6 β-ELIMINATION REACTIONS

There are three common mechanisms for elimination: elimination, unimolecular (E1); elimination, bimolecular (E2); and elimination, unimolecular, conjugate base (E1cb). Since the E1, E2, and E1cb mechanisms all involve elimination of the hydrogen atom β- to the leaving group X, we group them together and call all three mechanisms β-elimination reactions.

A typical E1 mechanism involves generation of a carbocation, which is the RDS followed by expulsion of a proton to yield an olefin, as exemplified below. Both E1 and S_N1 are two-step process and they may be considered as the reverse of each other.

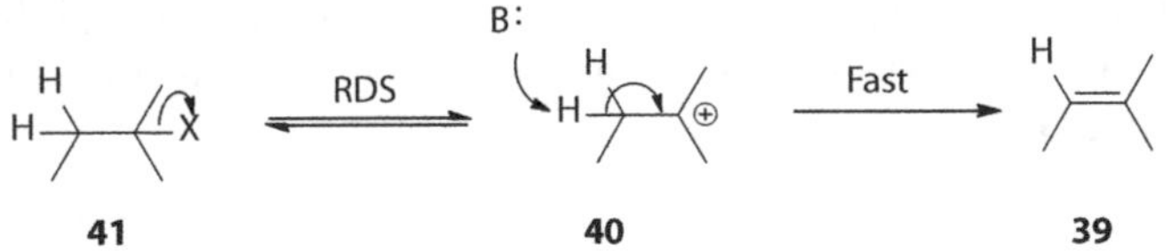

The E2 mechanism is a concerted process. Both E2 and S_N2 are one-step concerted process and they may be considered as the reversal of each other.

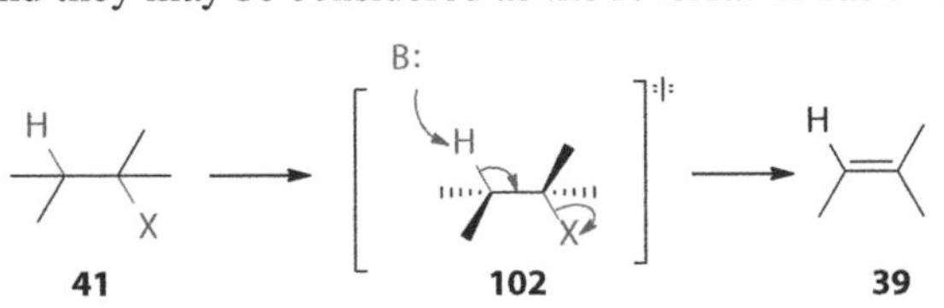

The stereochemistry of the E2 mechanism prefers the *anti*-periplanar conformation as shown on transition state **102**. For instance, bromide **103** can only undergo β-elimination to give olefin **104** exclusively because of the stereoelectronic effects. In contrast, olefin **106** is the sole product from the E2 elimination of bromide **105**.

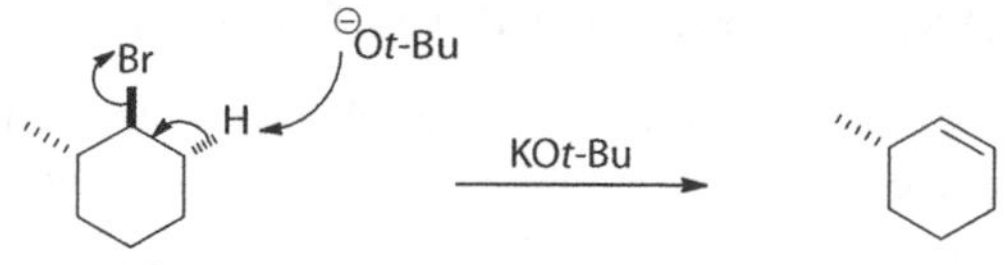

103 **104**

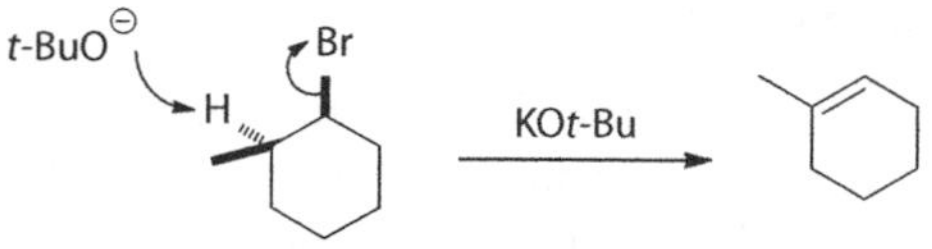

105 **106**

The E1cb mechanism is a two-step process. The first step is to generate the conjugate base **107**. Usually, there is an electron-withdrawing group α to the carbanion on **107**, often in the form of a carbonyl group, which is the stabilizing force here.

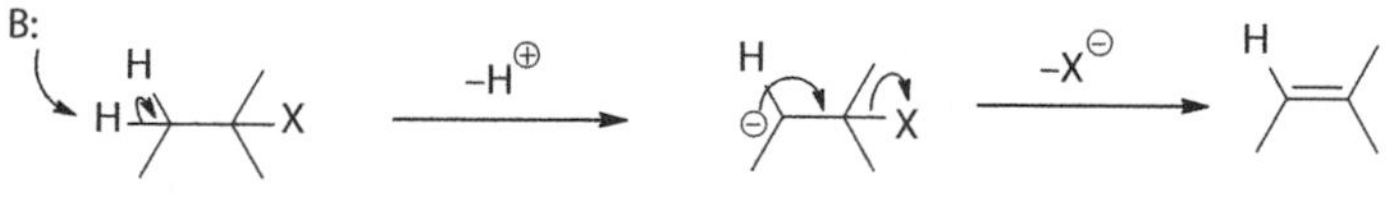

41 **107** **39**

One of the important steps for the mechanisms of organometallic reaction is β-hydride elimination. It is the *syn*-elimination of hydrogen and Pd(II) from a palladium alkyl complex with no change in oxidation state.

β-Hydride elimination

For instance, oxidative addition of the heteroaryl chloride **108** to Pd(0) would provide Pd(II) intermediate **109**, which subsequently inserts into benzoxazole to form the arylpalladium(II) complex **110**.[82] β-Hydride elimination of **110** would afford **111** with concomitant regeneration of Pd(0) for the next catalytic cycle. However, alternative mechanisms for product formation via C–H activation processes involving the C$_2$–H bond on benzoxazole are also conceivable.

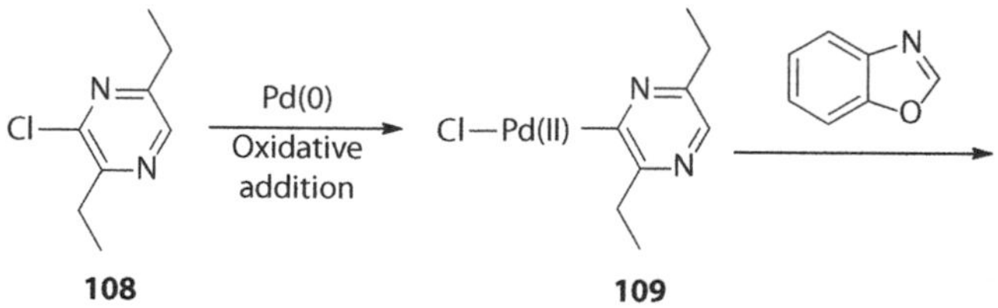

108 **109**

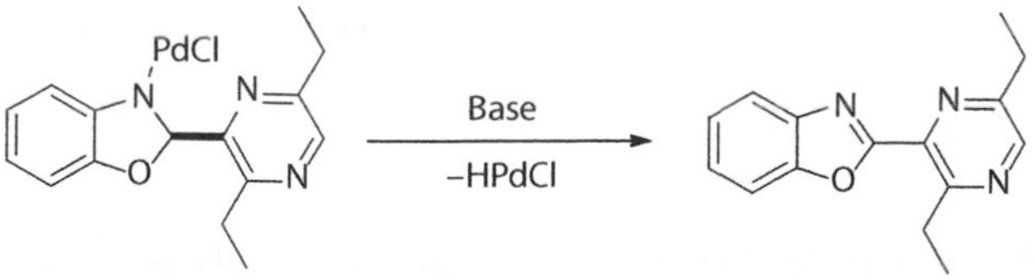

110 **111**

1.7 REARRANGEMENT REACTIONS OF CARBOCATIONS

One of the most unique aspects of carbocation chemistry is the propensity for the carbocation to rearrange to a more stable carbocation. This ability is what affords carbocation chemistry many diverse fascinating rearrangement reactions.

1.7.1 PINACOL REARRANGEMENT

The carbocation rearrangement most familiar to organic chemists is the pinacol rearrangement.[83] This is one of the first carbocation rearrangement reactions we encounter when we traverse the road of learning organic chemistry. When *gem*-diol (pinacol) **112** was treated with a catalytic amount of acid, the rearrangement product was obtained as pinacolone **113**.

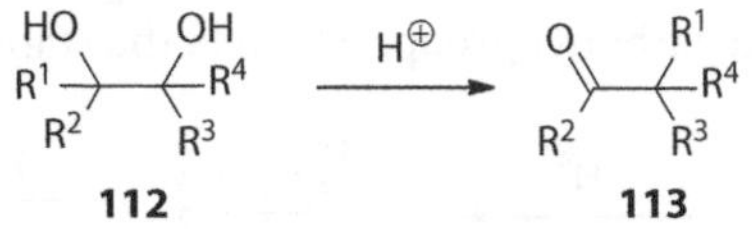

The mechanism of pinacol rearrangement commences with protonation of pinacol (**112**) with a proton to provide intermediate **114**. Departure of a water molecule from **114** affords the key carbocation intermediate **115**. Experimental evidence supporting carbocation intermediates in pinacol rearrangements has been furnished in the literature.[84] The crucial [1,2]-migration of the adjacent alkyl, aryl, or hydride group transforms carbocation intermediate **115** to another carbocation intermediate **116**, which readily loses a proton to furnish pinacolone **113**.

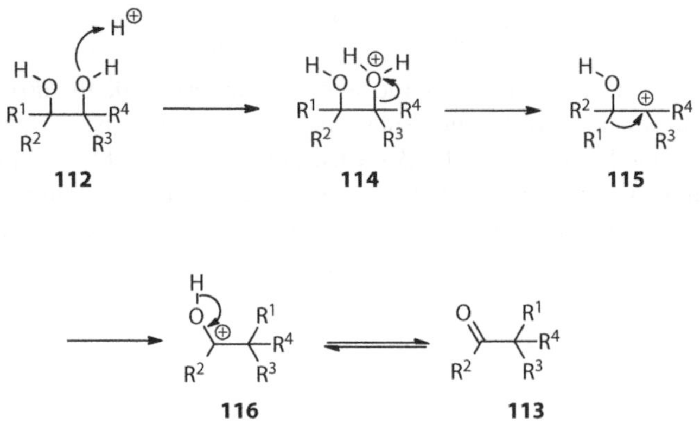

1.7.2 BECKMANN REARRANGEMENT

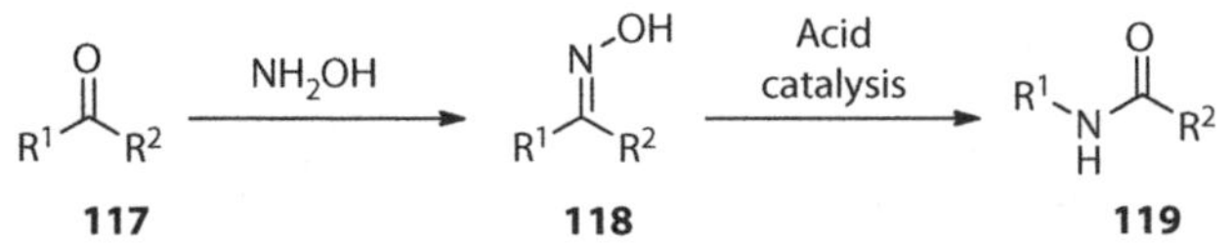

The acid-catalyzed rearrangement of an oxime **118**, easily prepared from ketone **117** and hydroxylamine, to an amide **119**.[85] The intramolecular version using cyclic oxime such as **120** results in ring expansion affording lactam **121**, which is a reaction performed on an industrial scale since the product is used as a monomer for manufacturing synthetic fibers.

120 **121**

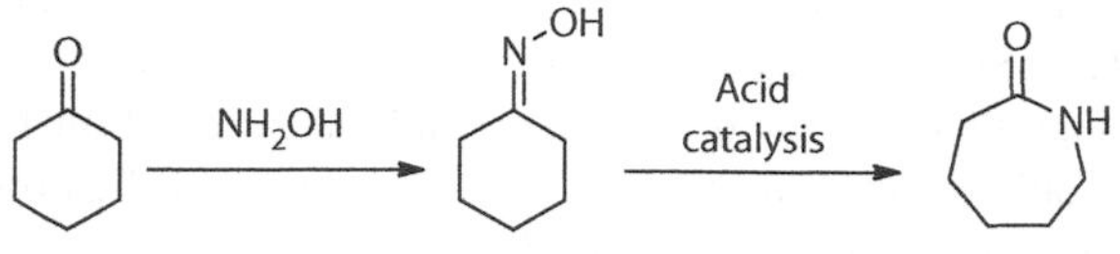

Under acidic conditions, the hydroxyl group of the oxime **118** is protonated to give the oxonium ion **122**. The migration of R^1 and loss of a water molecule occur concurrently to afford the cation **123**. Water addition to iminyl cation **123** then gives rise to oxinium ion **124**, which loses a proton to afford iminol **125**. In general, the substituent *trans* to either the hydroxyl or the leaving group migrates. Tautomerization of **115** results in the more stable amide **119**.

118 **122** **123**

124 **125** **119**

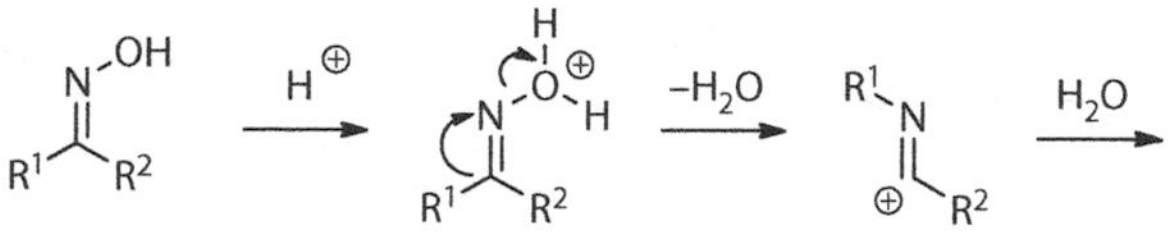

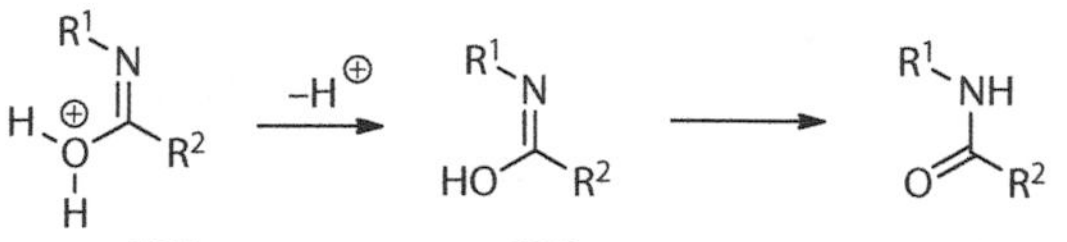

1.7.3 Demjanov and Tiffeneau–Demjanov Rearrangement

The Demjanov rearrangement involves the reaction of a primary alkyl amine **126** with a diazotizing reagent to form the diazoalkane **127**. Alkyl azide **127** losses of N$_2$, forms the primary cation which rearranges normally by ring expansion to afford the corresponding secondary cation **128**. A solvent such as water then acts to trap the rearranged cation typically leading to mixtures of isomeric alcohols **129**.

126 **127**

128 **129**

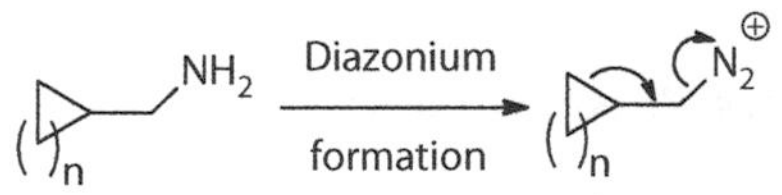

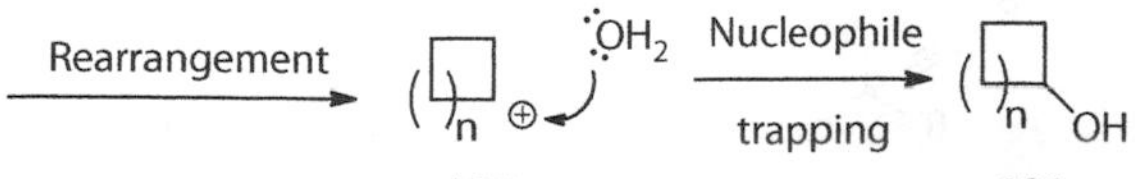

Mechanistically,[86] reaction of the amine **126** with activated O=N–X (**130**, X is thought to be ONO)[87] delivers the *N*-nitroso compound **131**, which undergoes rearrangement to provide hydroxyl-diazo material **132**. Protonation of **132** gives rise to oxonium **133** and its loss of water provides the diazonium **127**, which losses N_2 to provide cation **134**. Since carbocation **134** is an unstable primary carbocation, it undergoes rearrangement to provide cyclic secondary carbocation **128**. Cation **128** then traps a nucleophile (typically water sometimes nucleophilic cosolvents or acid counterions like acetate) to provide alcohol **129**.

126 **130** **131** **132**

133 **127**

134 **128** **129**

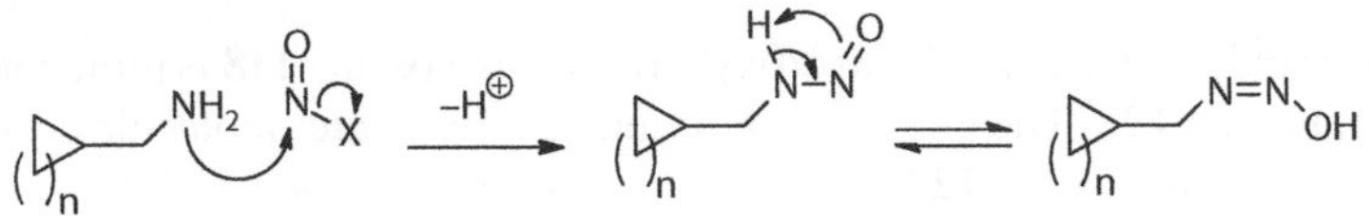

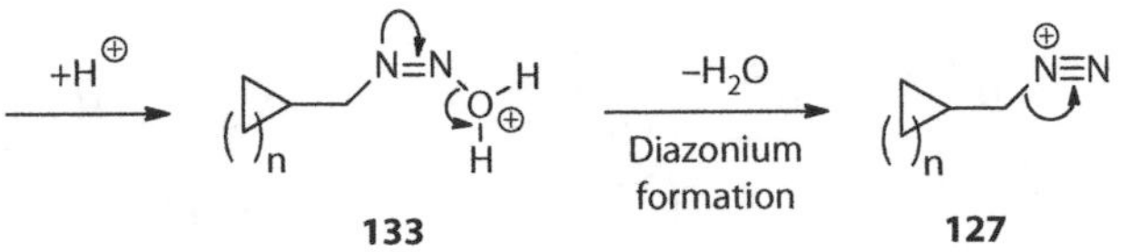

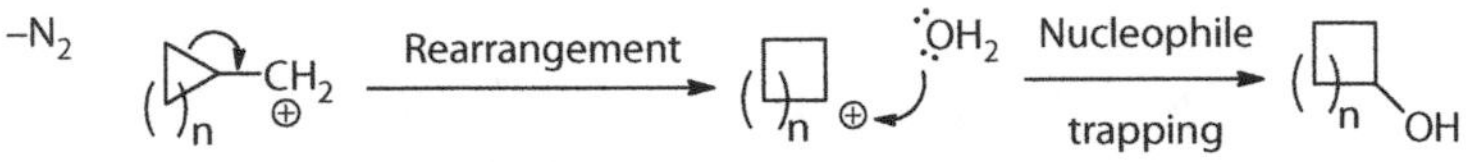

A variation of the Demjanov rearrangement is the Tiffeneau–Demjanov rearrangement that concerns the formation of diazo-alcohol **136** from amino-alcohol **135** and the subsequent rearrangement of **136** to cyclic ketone **137**. The mechanism of the Tiffeneau–Demjanov rearrangement is similar to that of the Demjanov rearrangement.

135 **136** **137**

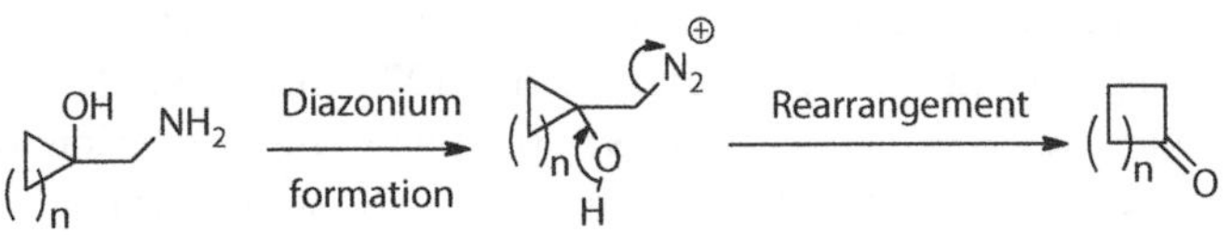

1.7.4 Meyer–Schuster Rearrangement

The Meyer–Schuster rearrangement refers to the isomerization of secondary or tertiary propargyl alcohols such as **138** to α,β-unsaturated carbonyl compounds such as **139** under acidic conditions. It involves a propargyl carbocation.

138 **139**

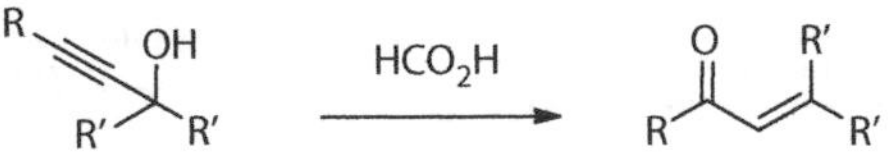

In terms of the mechanism,[88] propargyl alcohol **138** first gets protonated under acidic conditions to afford oxonium ion **140**, which loses a molecule of water to produce propargyl cation **141**. Rearrangement of propargyl cation **141** gives rise to allenyl cation **142**, which then traps water as a nucleophile to produce allenol **143** after loss of a proton. Tautomerization of allenol **143** then delivers α,β-unsaturated carbonyl compound **139**.

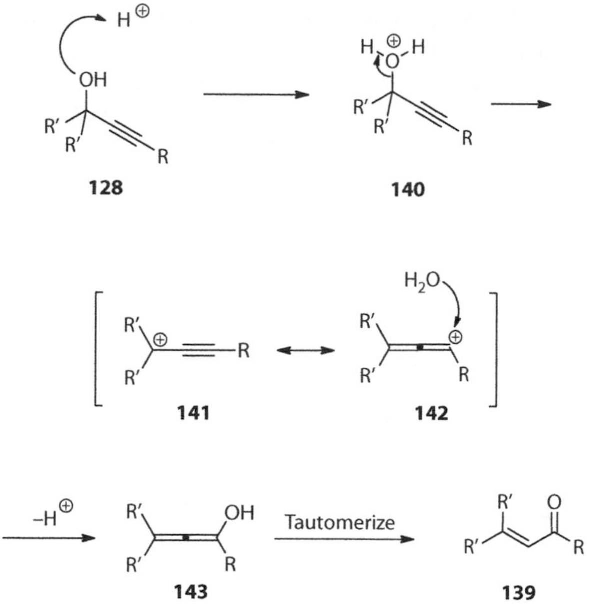

1.7.5 SCHMIDT REACTIONS

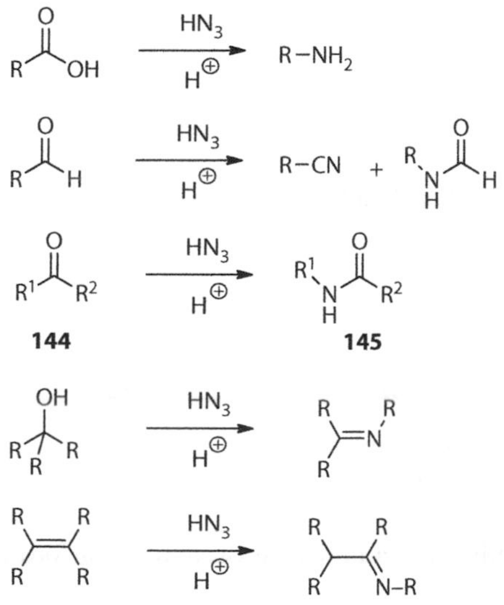

Many variation of the Schimdt reaction exist.[89] Some examples are listed in the box below. However, despite the existence of many different substrates, they all involve using hydrozoic acid and they follow a similar mechanistic pathway involving carbocations.

The conversion of ketone **144** to amide **145** is the quintessential example to illustrate the mechanism. Protonation of ketone **144** gives rise to oxonium ion **146**, which is considerably more electrophilic than its un-protonated cousin **144**. Nucleophilic addition of azide to **146** then affords azido-alcohol **147**, which is protonated to produce oxonium intermediate **148**. Loss of water from **148** provides a tertiary carbocation but a more stable resonance structure as **149** is presented as the major species. Migration of R^1 on **149** takes place with concurrent loss of N_2 leads to nitrilium ion **150**, which traps water as a nucleophile to offer hydroxyl-imine **151**, which readily tautomerizes to the final product as amide **145**.

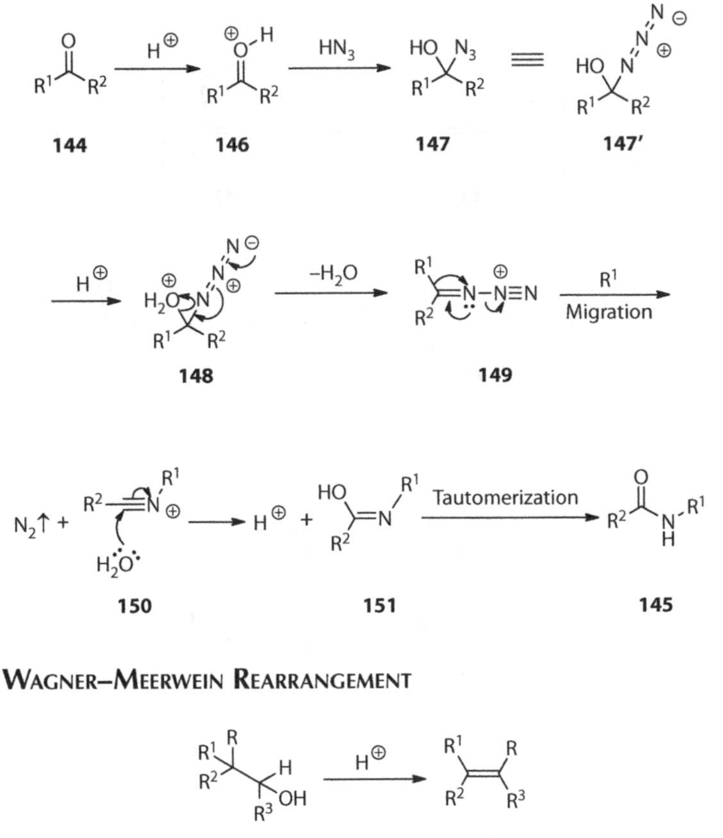

1.7.6   WAGNER–MEERWEIN REARRANGEMENT

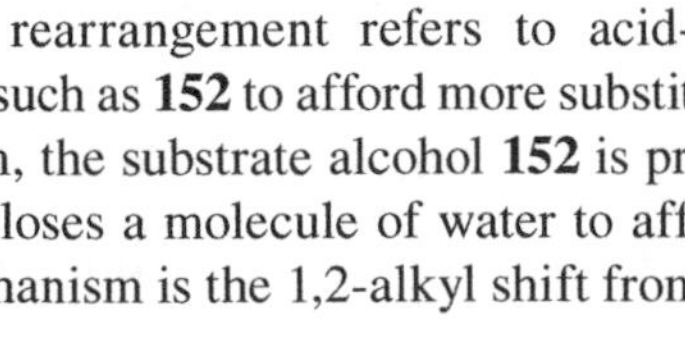

The Wagner–Meerwein rearrangement refers to acid-catalyzed alkyl group's 1,2-migration of alcohols such as **152** to afford more substituted olefins such as **153**.[90]

In terms of mechanism, the substrate alcohol **152** is protonated to give oxonium ion **154**, which promptly loses a molecule of water to afford carbocation **155**. The key operation of this mechanism is the 1,2-alkyl shift from a secondary carbocation

155 to produce the more stable tertiary carbocation **156**. It rapidly loses a proton to deliver olefin **153**.

In summary, I have gone over the fundamental mechanisms of several salient classes of reactions involving the intermediacy of carbocations. In the ensuing chapters, their applications in synthesis will be highlighted with an emphasis on the advances in the last decade.

REFERENCES

1. Slebocka-Tilk, H.; Ball, R. G.; Brown, R. S. *J. Am. Chem. Soc.* **1985**, *107*, 4504–4508.
2. Olah, G. A. *J. Org. Chem.* **2001**, *66*, 5943–5957.
3. Schultz, J. C.; Houle, F. A.; Beauchamp, J. L. *J. Am. Chem. Soc.* **1984**, *106*, 3917–3927.
4. Lossing, F. P.; Holmes, J. L. *J. Am. Chem. Soc.* **1984**, *106*, 6917–6920.
5. Norris, J. F. *Am. Chem. J.* **1901**, *25*, 117.
6. Fujimaro Ogata, F.; Takagi, M.; Nojima, M.; Kusabayashi, S. *J. Am. Chem. Soc.* **1981**, *103*, 1145–1153.
7. Krishnamurthy, V. V.; Prakash, G. K. S.; Iyer, P. S.; Olah, G. A. *J. Am. Chem. Soc.* **1985**, *107*, 5015–5016.
8. Kramer, G. M. *J. Am. Chem. Soc.* **1968**, *33*, 3453–3457.
9. Oda, M.; Nakajima, N.; Thanh, N. C.; Kajioka, T.; Kuroda, S. *Tetrahedron* **2006**, *62*, 8177–8183.
10. Winstein, S.; Trifan, D. S. *J. Am. Chem. Soc.* **1949**, *71*, 2953.
11. Ibid. **1952**, *74*, 1147–1154.
12. Ibid. **1952**, *74*, 1154–1160.
13. Brown, H. C. *Top. Curr. Chem.* **1979**, *80*, 1–18.
14. Brown, H. C. *Acc. Chem. Res.* **1983**, *16*, 432–440.
15. Torrice, M. *Chem. Eng. News* **2012**, *90*, 44–45.
16. Weininger, S. J. *Bull. Hist. Chem.* **2012**, *25*(2), 130.
17. Olah, G. A.; Prakash, G. K. S.; Saunders, M. *Acc. Chem. Res.* **1983**, *16*, 440–448.
18. Scholz, F.; Himmel, D.; Heinemann, F. W.; Schleyer, P. v. R.; Meyer, K.; Krossing, I. *Science* **2013**, *341*, 62–64.
19. Olah, G. A.; Klopman, G.; Schlesberg, R. H. *J. Am. Chem. Soc.* **1969**, *91*, 3261–3268.
20. Ingold, C. K. *J. Chem. Soc.* **1928**, 1217–1221.
21. Bateman, L. C.; Church, M. G.; Huges, E. D.; Ingold, C. K.; Taher, N. A. *J. Chem. Soc.* **1940**, 979.
22. Bateman, L. C.; Hughes, E. D.; Ingold, C. K. *J. Chem. Soc.* **1940**, 1017.

23. Winstein, S.; Clippinger, E.; Fainberg, A. H.; Robinson, G. C. *J. Am. Chem. Soc.* **1954**, *76*, 2597.
24. Winstein, S.; Robinson, G. C. *J. Am. Chem. Soc.* **1958**, *80*, 169–181.
25. Katritzky, A. R.; Brycki, B. E. *Chem. Soc. Rev.* **1990**, *19*, 83–105.
26. Harris, J. M.; Wamser, C. C. *Fundamentals of Organic Reaction Mechanisms*, Wiley: New York, New York, 1976, 191pp.
27. Schneider, H. J.; Philippi, K. *J. Chem. Res. S* **1984**, 104–105.
28. Beller, M.; Seayad, J.; Tillack, A.; Jiao, H. *Angew. Chem. Int. Ed.* **2004**, *43*, 3368–3398.
29. Herges, R. *Angew. Chem. Int. Ed.* **1995**, *34*, 51–53.
30. Cresswell, A. J.; Eey, S. T.-C.; Denmark, S. E. *Angew. Chem. Int. Ed.* **2015**, *54*, 15642–15682.
31. Clay, J. M. In *Name Reactions for Functional Group Transformations*, Eds., Li, J. J.; Corey, E. J., Wiley: Hoboken, New Jersey, 2007, pp. 183–188.
32. Thomas, S. P.; Aggarwal, V. K. J. *Angew. Chem. Int. Ed.* **2009**, *48*, 1896–1898.
33. Chong, C. C.; Kinjo, R. *ACS Catal.* **2015**, *5*, 3238–3259.
34. Brown, H. C.; Geoghegan, P. J., Jr.; Kurek, J. K. *J. Org. Chem.* **1981**, *46*, 3810–3812.
35. Harris, J. M.; Wamser, C. C. *Fundamentals of Organic Reaction Mechanisms*, Wiley: New York, New York, 1976, 192pp.
36. Taylor, R. *Electrophilic Aromatic Substitution*, Wiley: New York, 1990.
37. Olah, G. A. *Acc. Chem. Res.* **1971**, *4*, 240–248.
38. (a) Hubig, S. M.; Kochi, J. K. *J. Org. Chem.* **2000**, *65*, 6807–6818. (b) Koleva, G.; Galabov, B.; Hadjieva, B.; Schaefer, H. F., III; Schleyer, P. R. *Angew. Chem. Int. Ed.* **2015**, *54*, 14123–14127.
39. Rosokha, S. V.; Kochi, J. K. *J. Org. Chem.* **2002**, *67*, 1727–1737.
40. Hoggett, J. G.; Moodie, R. B.; Penton, J. R.; Schofield, K. *Nitration and Aromatic Reactivity*, Cambridge University Press: Cambridge, 1971.
41. Schofield, K. *Aromatic Nitration,* Cambridge University Press: Cambridge, 1980.
42. Olah, G. A.; Malhotra, S. C. *Nitration: Methods and Mechanisms,* VCH: New York, 1989.
43. Hughes, E. D.; Ingold, C. K.; Reed, R. I. *J. Chem. Soc.* **1950**, 2400–2440.
44. Eberson, L.; Hartshorn, M. P.; Radner, F. *Acta Chem. Scand.* **1994**, *48*, 937–950.
45. Olah, G. A.; Kuhn, S.; Flood, S. H. *J. Am. Chem. Soc.* **1961**, *83*, 4571–4580.
46. Kochi, J. K. *Acc. Chem. Res.* **1992**, *25*, 39–47.
47. Eberson, L.; Radner, F. *Acc. Chem. Res.* **1987**, *20*, 53–59.
48. Nightingale, D. V. *Chem. Rev.* **1947**, *40*, 117–140.
49. Hartshorn, S. R. *Chem. Soc. Rev.* **1974**, *3*, 167–192.
50. Bunce, N. J. *J. Chem. Soc., Perkin Trans.* **1974**, *2*, 942–944.
51. Moodie, R. B.; Schofield, K. *Acc. Chem. Res.* **1976**, *9*, 287–292.
52. Prakash, G. K. S.; Mathew, T. *Angew. Chem. Int. Ed.* **2010**, *49*, 1726–1728.
53. Suzuki, H. *Synthesis* **1977**, 217–238.
54. Ridd, J. H. *Acta Chem. Scand.* **1998**, *52*, 11–22.
55. Butler, A. R. *J. Chem. Educ.* **1971**, *48*, 508.
56. Fabbrini, M.; Galli, C.; Gentili, P.; Macchitella, D.; Petride, H. *J. Chem. Soc., Perkin Trans.* **2001**, *2*, 1516–1521.
57. Hubig, S. M.; Jung, W.; Kochi, J. K. *J. Org. Chem.* **1994**, *59*, 6233–6244.
58. Podgorsek, A.; Zupan, M.; Iskra, J. *Angew. Chem. Int. Ed.* **2009**, *48*, 8424–8450.
59. Bedford, R. B.; Engelhart, J. U.; Haddow, M. F.; Mitchell, C. J.; Webster, R. L. *Dalton Trans.* **2010**, *39*, 10464–10472.
60. Ojima, I., Ed., *Fluorine in Medicinal Chemistry and Chemical Biology*, Wiley: Hoboken, New Jersey, 2009.
61. Taylor, S. D.; Kotoris, C. C.; Hum, G. *Tetrahedron* **1999**, *55*, 12431–12477.
62. Borodkin, G. I.; Shubin, V. G. *Russ. Chem. Rev.* **2010**, *79*, 259–283.

63. German, L. S.; Zemskov, S. V. *New Fluorinating Agents in Organic Synthesis*, Wiley: New York, New Jersey, 1989.
64. Jerome; B.; Dominique, C. *Org. React.* **2007**, *69*, 347–672.
65. Stavber, G.; Zupan, M.; Jereb, M.; Stavber, S. *Org. Lett.* **2004**, *6*, 4873–4976.
66. Lin, A.; Huehls, C. B.; Yang, J. *Org. Chem. Front.* **2014**, *1*, 434–438.
67. Li, Y.; Wu, Y.; Li, G.-S.; Wang, X.-S. *Adv. Synth. Catal.* **2014**, *356*, 1412–1418.
68. Liang, T.; Neumann, C. N.; Ritter, T. *Angew. Chem. Int. Ed.* **2013**, *52*, 8414–8264.
69. Campbell, M. G.; Ritter, T. *Chem. Rev.* **2015**, *115*, 612–633.
70. Lou, S.-J.; Xu, D.-Q.; Xia, A.-B.; Wang, Y.-F.; Liu, Y.-K.; Du, X.-H.; Xu, Z.-Y. *Chem. Commun.* **2013**, *49*, 6218–6220.
71. Truong, T.; Klimovica, K.; Daugulis, O. *J. Am. Chem. Soc.* **2013**, *135*, 9342–9345.
72. Olah, G. A.; Krishnamurti, R.; Prakash, G. K. S. In *Friedel–Crafts Alkylation in Comprehensive Organic Synthesis*, Vol. 3, Eds., Trost B. M., Fleming I., Pergamon Press: Oxford, 1991, 293.
73. Olah G. A.; Dear R. R. *Friedel–Crafts and Related Reactions*, Wiley-Interscience: New York, 1963–1965.
74. Olah G. A. *Friedel–Crafts Chemistry*, Wiley-Interscience: New York, 1973.
75. Olah G. A.; Krishnamurthy, R.; Prakash, G. K. S. In *Kirk-Othmer Encyclopedia of Chemical Technology*, 5th Ed., Vol. 12, Wiley: New York, 2005, 159pp.
76. Sheldon, R. A.; Bekkum, H., Eds., *Friedel–Crafts Reaction*, Wiley-VCH: New York, 2001.
77. Roberts, R. M.; Khalaf, A. *Friedel–Crafts Alkylation Chemistry: A Century of Discovery,* Marcel Dekker: New York, 1984.
78. Barton, D.; Ollis, W. D. *Comprehensive Organic Chemistry*, Pergamon Press: New York, **1979**, Vol. 1, pp. 268–269.
79. Campbell, J. A. Asymmetric Friedel–Crafts reactions: Past to present, In *Name Reaction in Carbocyclic Ring Formations*, Ed., Li, J. J., Wiley: Hoboken, New Jersey, 2010.
80. Bandini, M.; Umani-Ronchi, A., Eds., *Catalytic Asymmetric Friedel–Crafts Alkylations*, Wiley-VCH: Weinheim, 2009.
81. Sartori, G.; Maggi, R. *Advances in Friedel–Crafts Acylation Reactions: Catalytic and Green Processes,* CRC Press: Boca Raton, Florida, 2009.
82. Aoyagi, Y.; Inoue, A.; Koizumi, I.; Hashimoto, R.; Tokunaga, K.; Gohma, K.; Komatsu, J. et al. *Heterocycles* **1992**, *33*, 257–272.
83. Goess, B. In *Name Reactions for Homologations—Part II*, Ed., Li, J. J., Wiley: Hoboken, New Jersey, 2009, pp. 319–333.
84. Bunton, C. A.; Carr, M. D. *J. Chem. Soc.* **1963**, 5854–5861.
85. Kumar, R. R.; Vanitha, K. A.; Balasubramanian, M. In *Name Reactions for Homologations–Part II*, Ed., Li, J. J., Wiley: Hoboken, New Jersey, 2009, pp. 274–292.
86. Thomson, T.; Stevens, T. S. *J. Chem. Soc.* **1932**, 55–69.
87. Curran, T. T. In *Name Reactions for Homologations—Part II*, Ed., Li, J. J., Wiley: Hoboken, New Jersey, 2009, pp. 293–304.
88. Mullins, R. J.; Collins, N. R. In *Name Reactions for Homologations—Part II*, Ed., Li, J. J., Wiley: Hoboken, New Jersey, 2009, pp. 305–318.
89. Wu, Y.-J. In *Name Reactions for Homologations—Part II*, Ed., Li, J. J., Wiley: Hoboken, New Jersey, 2009, pp. 353–373.
90. Mullins, R. J.; Grote, A. L. In *Name Reactions for Homologations—Part II*, Ed., Li, J. J., Wiley: Hoboken, New Jersey, 2009, pp. 373–394.

2 Nucleophilic Aliphatic Substitution

S_N1

Yu Feng, Safiyyah Forbes, and Jie Jack Li

CONTENTS

2.1 INTRODUCTION

In organic chemistry, nucleophilic substitution is a fundamental catalog of reactions involving the addition of a nucleophile to an electrophilic atom or ion. When such substitution takes place at a tetrahedral or sp^3 carbon, it is termed as a nucleophilic aliphatic substitution. There are two major mechanisms for nucleophilic substitutions: S_N1 reaction and S_N2 reaction. In such, S stands for substitution, N stands for nucleophilic, and the number indicates the kinetic order of the reaction. So S_N1 reaction is a first-order nucleophilic substitution which means the rate of the reaction depends only on the concentration of substrate (Equation 2.1).

$$Rate = k[S] \qquad (2.1)$$

The mechanism for the reaction is described in Scheme 2.1, and the corresponding energy level changes have been depicted in Figure 2.1. This pathway is a multistep process: (1) slow loss of the leaving group (LG) to generate a carbocation intermediate and (2) rapid attack of a nucleophile on the electrophilic intermediate to form a new δ bond. In S_N1 reaction, the rate-determining step is the first elementary step (loss of the LG to form the intermediate carbocation), which means the more reactive the LG, the faster reaction rate. The carbocation is a real-life positive ion with a limited lifetime and thus is not just a transition state. Bulky groups attached help stabilize the charge on the carbocation via

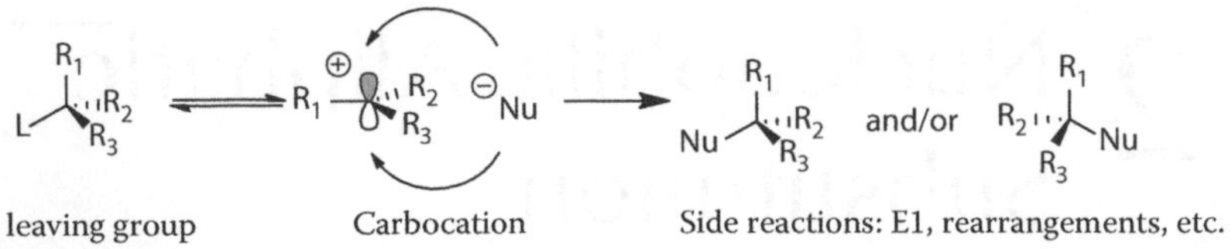

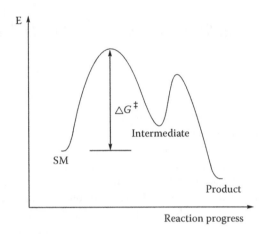

L: leaving group   Carbocation   Side reactions: E1, rearrangements, etc.

SCHEME 2.1   General mechanism for an S_N1 reaction.

FIGURE 2.1   Reaction energy profile of S_N1 reaction.

distribution of charge and resonance. Polar solvents which stabilize carbocation can also favor the S_N1 reaction. The common side reactions include elimination (E1) and rearrangements.

2.2 Π-ACTIVATED ALCOHOLS: BRØNSTED ACIDS

Usually, π-activated alcohols can generate allylic, benzylic, and propargylic cations. Direct nucleophilic substitution reactions[1,2] of π-activated alcohols are considered as green chemistry processes, allowing new allylic compound formations and generating only water as a by-product.

Benzofuran-3(2*H*)-one (**1**) reacted with Michler's hydrol (**2**) under the influence of 20 mol% TFA (trifluoroacetic acid) catalyst to form S_N1 reaction adduct **3** in 64% yield.[3] The methodology was also applied to the alkylation of many oxazolones and related heterocycles through an S_N1 reaction pathway.

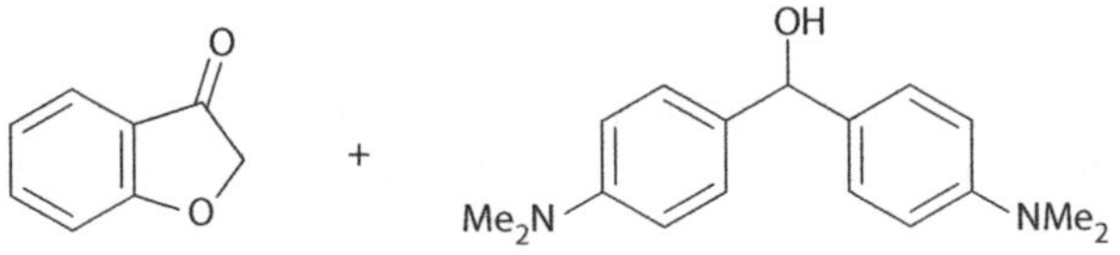

1   +   **2**

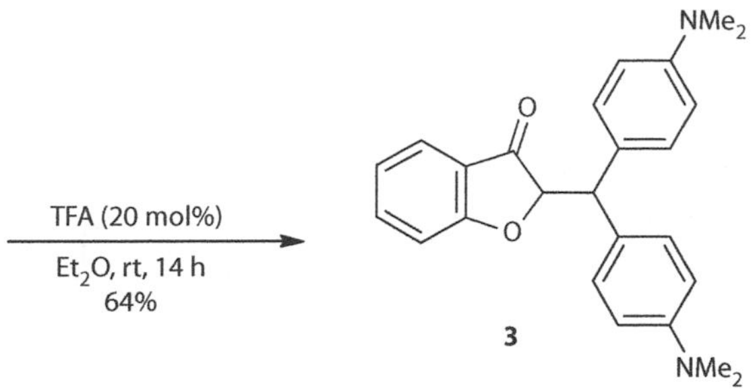

3

Enantioselective α-alkylation of ketones was a challenging endeavor until the emergence of chiral organocatalysts.[4] One of them, pyrrolidinethioxotetrahydro-pyrimidinone (**5**), proved to be an efficient catalyst for enantioselective α-alkylation of ketones. For example, with the aid of chiral organocatalyst **5**,[5] and using 4-nitro-benzoic acid (4-NBA) as the Brønsted acid, ketone **4** was alkylated with activated alcohol Michler's hydrol (**2**) in a direct, S_N1, and diastereoselective fashion to provide adduct **6** in 62% yield and 44% ee.

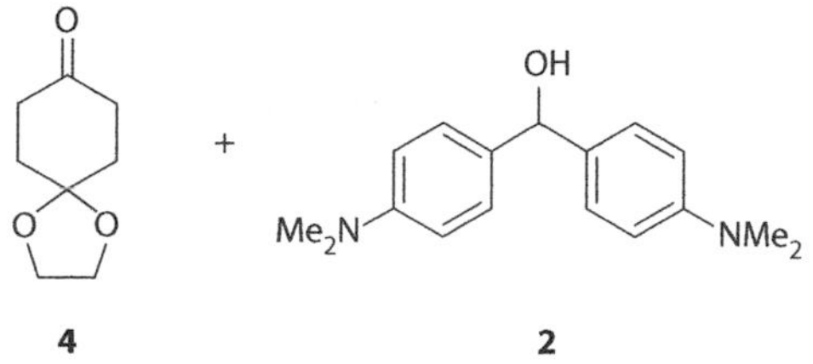

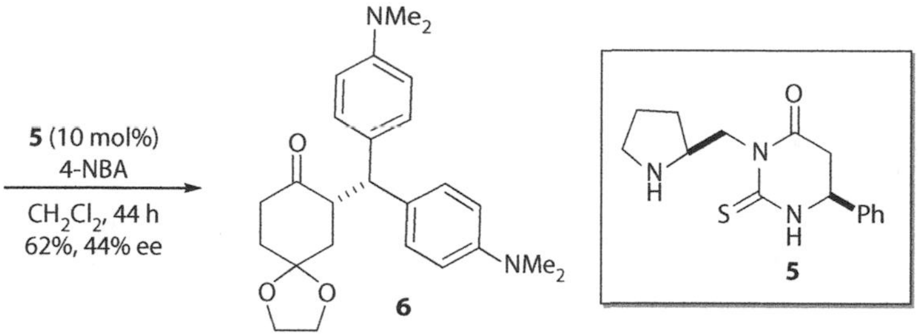

4   **2**

6   **5**

Fluorinated solvents such as trifluoroethanol (TFE) and hexafluoroisopropanol (HFIP) are unique solvents with high ionizing power (acidic) and yet low nucleo-philicity. When TFE was used as the solvent, nitro compound **7** was alkylated with activated alcohol Michler's hydrol (**2**) in a direct, S_N1, and diastereoselective fashion to provide adduct **8** in nearly quantitative yield and excellent diastereoselectivity.[6]

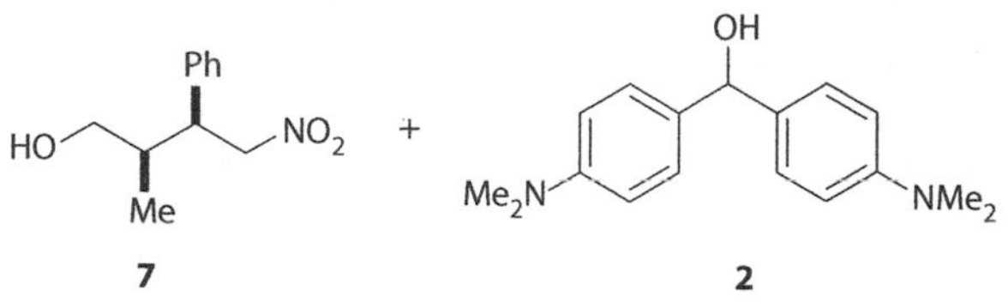

7   **2**

$$\xrightarrow[\text{99\%, 10:1:1}]{\text{0.2 M CF}_3\text{CH}_2\text{OH}}$$

8

Tris-propargyl alcohol **9** was treated with benzyl azide to synthesize *tris*-"benzyl" alcohol **10** using the copper-promoted alkyne-azide [2 + 3] cycloaddition click chemistry (Huisgen reaction). Under the influence of trifluoroacetic anhydride, alcohol **9** was converted to tri(triazolyl)carbenium ion **11**, which may be captured by a variety of nucleophiles including *O*-, *N*-, *S*-, and *C*-nucleophiles to give the corresponding triazolylmethane derivatives **12** in moderate-to-high yields.[7]

$$\xrightarrow[\text{EtOH/H}_2\text{O}]{\substack{\text{3 equiv. PhCH}_2\text{N}_3 \\ \text{CuSO}_4 \\ \text{sodium ascorbate}}}$$

9   **10**

$$\xrightarrow[\text{CH}_2\text{Cl}_2, \, 0°\text{C}]{(\text{CF}_3\text{CO})_2\text{O}} \qquad \xrightarrow{\text{Nu}-\text{H}}$$

11   **12**

Sometimes, for asymmetric synthesis, *bis*-cinchona alkaloids work better than *mono*-cinchona alkaloids by providing stronger interactions. One *bis*-cinchona alkaloid (DHQ)$_2$-PHAL [1,4-*bis*(9-*O*-dihydroquinine)phthalazine, **13**] is most famous for its application in Sharpless asymmetric dihydroxylation. When **13** was used as a cocatalyst with a Brønsted acid such as mesylic acid, direct α-alkylation of 2-oxindole **14** with Michler's hydrol (**2**) generated the C-2 symmetrical cation **15** via an S$_N$1-type pathway in the noncovalent activation mode. The direct α-alkylation product **16** with a tertiary chiral carbon center was obtained in good yield and ee.[8]

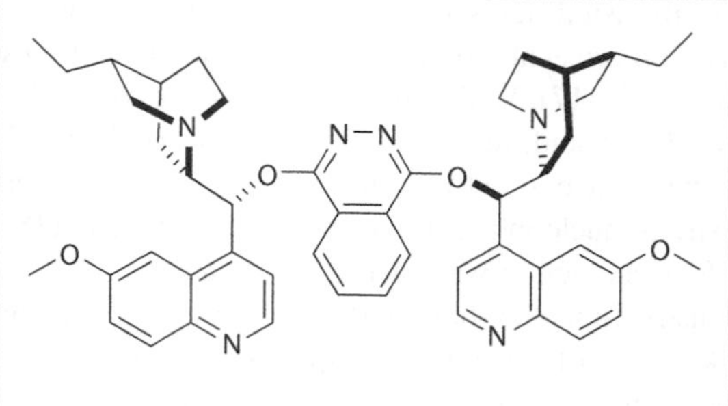

13, $(DHQ)_2$-PHAL

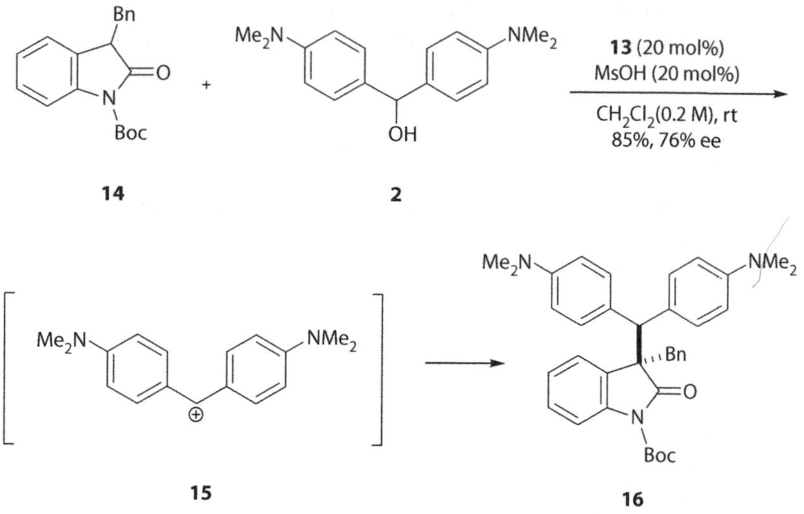
Bn
O
N
Boc
+
Me2N
NMe2
OH
13 (20 mol%)
MsOH (20 mol%)
CH2Cl2(0.2 M), rt
85%, 76% ee
14
2
Me2N
NMe2
15
Me2N
NMe2
Bn
O
N
Boc
16

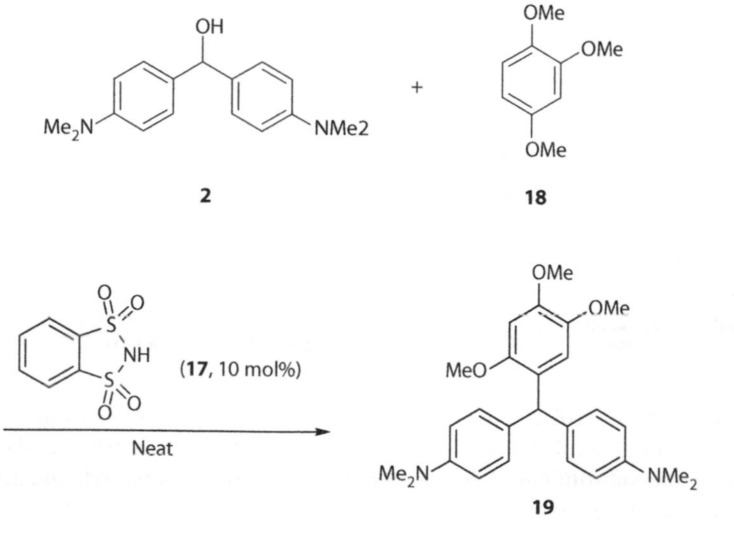
OH
Me2N
NMe2
2
+
OMe
OMe
OMe
18
(17, 10 mol%)
Neat
OMe
OMe
MeO
Me2N
NMe2
19

Like Lewis acids, Brønsted acids have been used to generate benzylic cations from benzyl alcohols. Barbero et al.[9] took advantage of an interesting strong Brønsted acid, *o*-benzenesulfonimide (**17**). With a pK_a value of approximately 1.0, compound **17** is as acidic as phosphoric acid. In an S_N1 fashion, benzyl alcohol **2** was converted to the corresponding benzylic cation by catalytic strong Brønsted acid **17**, which was then captured by a strong nucleophile 1,2,4-trimethoxybenzene (**18**), giving rise to alkylation product **19**. The reaction was carried out neat.

Stereoselective construction of carbon–carbon bonds can be accomplished through asymmetric *C*-α-alkylation of carbonyl compounds. Chiral *N*-acyl thiazolidinethiones can undergo highly stereoselective S_N1 direct-type alkylation catalyzed by structurally simple, commercially available, and easy to handle nickel(II) complexes. (*S*)-4-isopropyl-*N*-propanoyl-1,3-thiazolidine-2-thione (**20**), with 4,4′-dimethoxybenzhydryl methyl ether (**21**) is catalyzed by 5 mol% of $(Me_3P)_2NiCl_2$ to produce adduct **22** as a single diastereomer in 92%–94% yield in a simple and very efficient manner.[10]

20 **21** **22**

23 **24**

23a, X = O, 99%
23b, X = N–Ar, 99%
23c, X = S, 95%

24a, X = O, chirality 91%
24b, X = N–Ar, chirality 96%
24c, X = S, chirality 91%

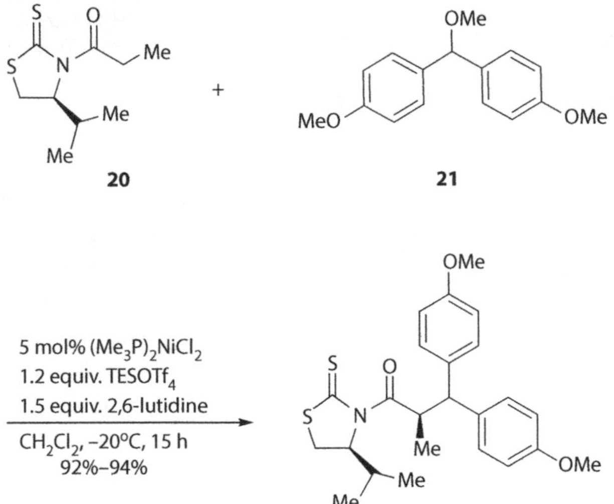

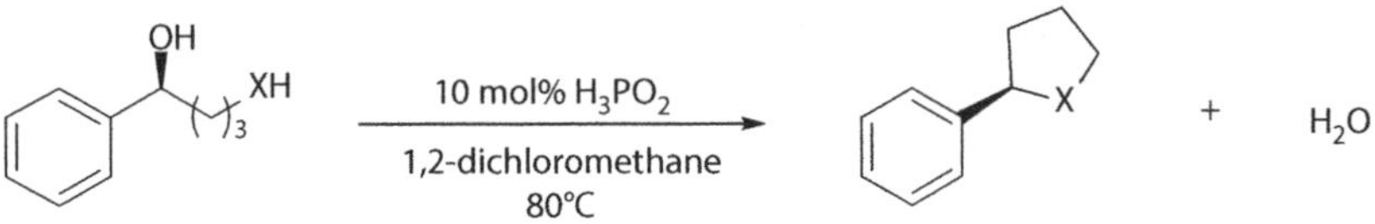

Direct substitution of hydroxyl groups with transfer of chirality is a major challenge in *C*-heteroatom bond-forming reactions. Samec et al. used Brønsted acid catalyzed intramolecular substitution of enantioenriched aryl, allyl, propargyl, and alkyl alcohols by *O*-, *N*-, *S*-centered nucleophiles. The reaction of (*S*)-1-phenylbutane-1,4-diol

(**23a**) with phosphinic acid catalysis generated (*R*)-2-phenyltetrahydrofuran (**24a**) with inversion of configuration with excellent yields and chirality transfers.[11]

2.3 Π-ACTIVATED ALCOHOLS: LEWIS ACIDS

Highly stable chiral benzylic cations reacting with weak nucleophiles can produce adducts with highly diastereoselectivity. The reaction of acetate **25** with the acetophenone-derived silylenolether **26** was conducted in the presence of $ZnCl_2$. The carbenium ion occurred smoothly with high level of stereocontrol and only a single diastereomer **28** was obtained.[12]

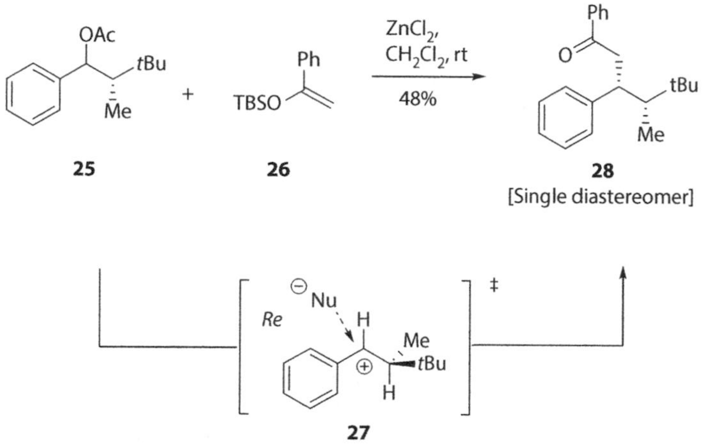

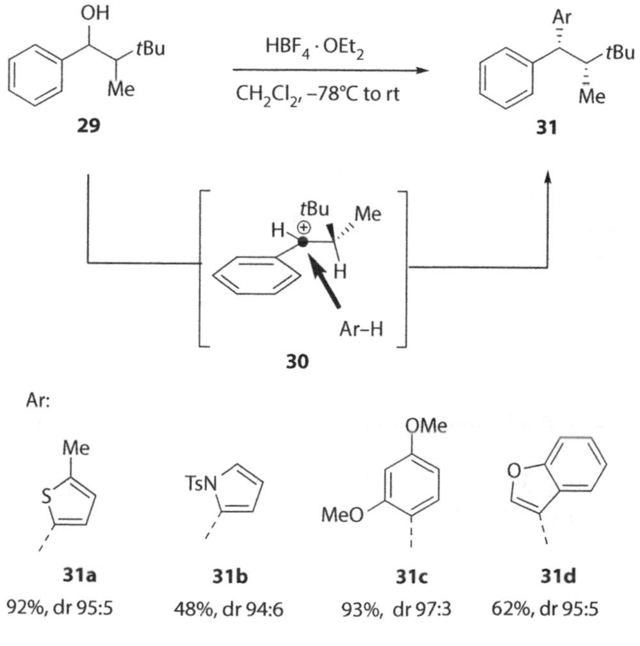

Benzylic carbocations can be produced under optimized conditions using Lewis acid $HBF_4 \cdot OEt_2$ in the nonpolar solvent methylene chloride. The reaction of alcohol **29** with several arene nucleophiles generated diastereomeric ratio exceeding 94/6 and satisfactory to excellent product yields.[13] It should be pointed out that the epimeric composition of **29** (no matter syn- or anti-) did not influence the stereo-outcomes.

While π-activated alcohol derivatives such as allylic esters, carbonates, carbamates, phosphates, halides, etc. have been extensively used for the synthesis of allylic amines, π-activated alcohol itself per se has not been used as frequently. Using a rhenium catalyst Re_2O_7, allylic alcohol **32** reacted with electron-deficient amines such as oxazolidin-2-one (**33**) to furnish the direct substitution adduct **34** via the S_N1 mechanism.[14] Benzylic and propargylic alcohols also work for this methodology.

32   **33**   **34**

35a, R = t-Bu   **37a**, 98%, dr > 95/5
35b, R = i-Pr   **37b**, 98%, dr > 81/19
35c, R = Et   **37c**, 82%, dr > 61/39

Chiral propargylic cations may be generated under the influence of Lewis acid $Bi(OTf)_3$ in polar solvent nitromethane. For the reaction between propargyl acetate **35** with silylated nucleophiles such as (1-t-butylvinyloxy)trimethylsilane (**36**) resulted in diastereoselective product **37**. While the yields were excellent regardless of the size of the α-substituent R, the diastereoselectivity decreased as the size of R became smaller: R = t-Bu > i-Pr > Et.[15]

The dichotomy between reactivity and chemoselectivity in direct S_N1 reactions of alcohols in the presence of acid-sensitive alkenes and protecting groups can be accomplished using $B(C_6F_5)_3$, a strong nonhydrolyzable neutral Lewis acid whose hydrates undergo rapid ligand exchange. When comparing $B(C_6F_5)_3$ to establish mild boronic acid cyclodehydration catalyst, 1 mol% $B(C_6F_5)_3$ is able to cyclize isomer **38** in just 2 h producing **39** in good yields. Likewise, secondary propargylic alcohol **40** is cyclized to 2-alkynyl tetrahydrofuran (**41**) in excellent yield without the additional preactivation step.[16]

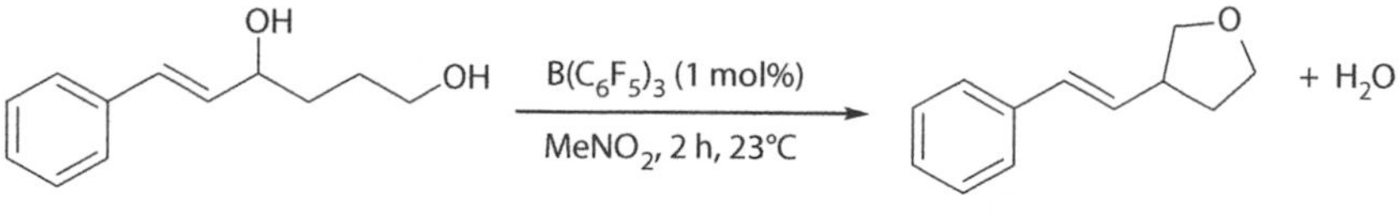
OH
OH
B(C6F5)3 (1 mol%)
MeNO2, 2 h, 23°C
O
+ H2O

38 **39**

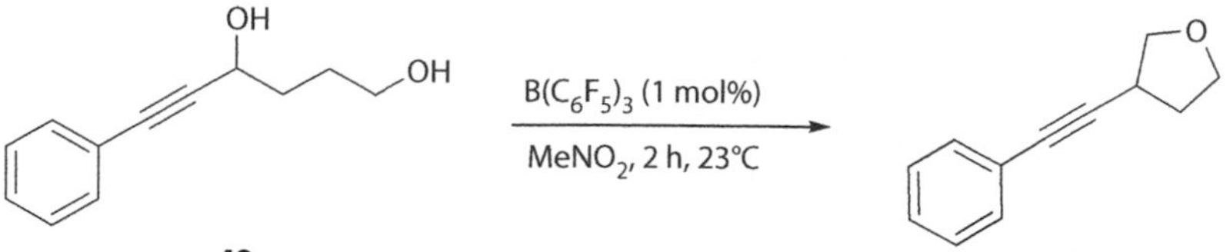
OH
OH
B(C6F5)3 (1 mol%)
MeNO2, 2 h, 23°C
O

40 **41**

Unsymmetrical diarylmethanes and other 1,1-diarylalkane may be generated using a superior air and moisture-tolerant catalyst ferroceniumboronic acid hexafluoroantimonate salt (**42**) in hexafluoroisopropanol as cosolvent. The reaction of *m*-xylene (**43**) with benzylic alcohols (**44**) containing an electronically neutral arene core or others with slightly deactivating substituents such as fluoride or bromide resulted in high yield with good to high regioselectivity products (**45**).[17]

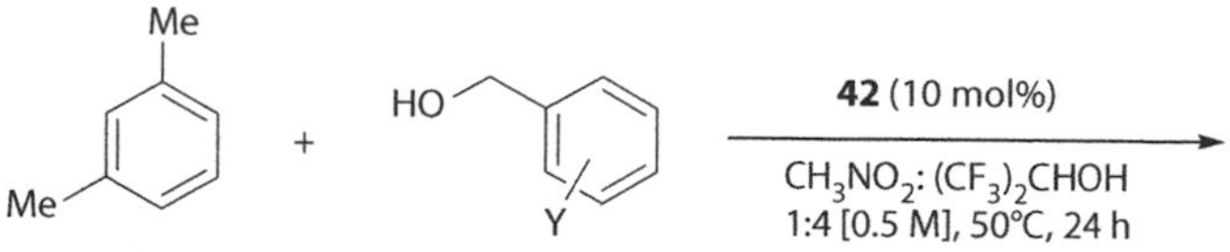
Me
Me
+
HO
Y
42 (10 mol%)
CH3NO2: (CF3)2CHOH
1:4 [0.5 M], 50°C, 24 h

43 **44**

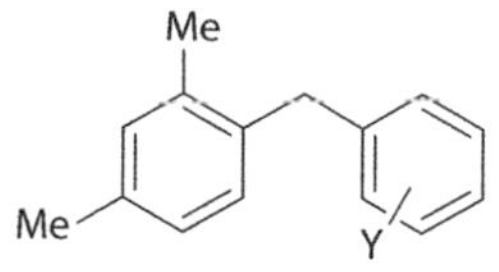
Me
Me
Y

45

45a, Y = F (99%, a:b = 78:22)
45b, Y = CF_3(99%, a:b = 81:19, 80°C)
45c, Y = Br (87%, a:b = 78:22)

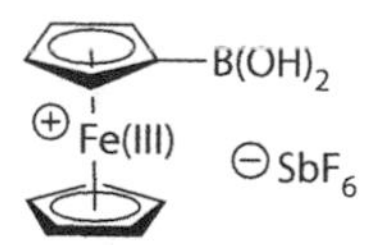
B(OH)2
Fe(III)
SbF6

42

2.4 ALKYLATION OF ALDEHYDES AND KETONES

The asymmetric organocatalyzed α-alkylation of aldehydes under galvanostatic conditions may result in versatile aldehydes in a stereoselective manner. An achiral sec-amine catalyst, pyrrolidine was used to promote the coupling of hydrocinnamaldehyde (**46**) with xanthene (**47**) producing **49** in reasonable yield.[18]

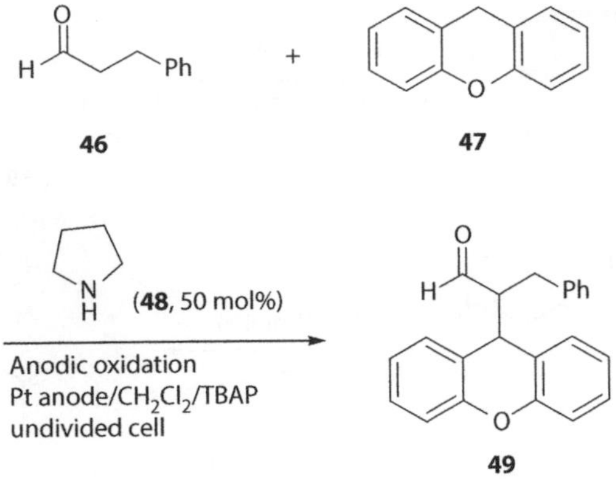
H
Ph
+
O
46
47
(48, 50 mol%)
Anodic oxidation
Pt anode/CH2Cl2/TBAP
undivided cell
49

The catalytic enantioselective intermolecular α-alkylation of aldehydes is one of the most challenging fields in asymmetric catalysis. An organocatalytic asymmetric oxidative hydrogenative α-alkylation of aldehydes may be generated via benzylic C–H bond activation using molecular oxygen as the oxidant. Jiao et al. demonstrated that the reaction of commercially available xanthene (**47**) with hexanal (**50**) in the presence of MacMillan catalyst **51** in nitromethane under 1 atm of oxygen produced the adduct **52** in relatively good enantiomeric excess and yields.[19]

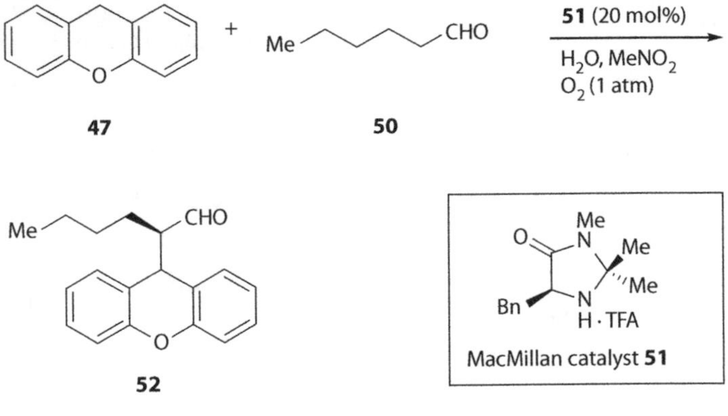
+
Me
CHO
51 (20 mol%)
H2O, MeNO2
O2 (1 atm)
47
50
52
Bn
H · TFA
MacMillan catalyst 51

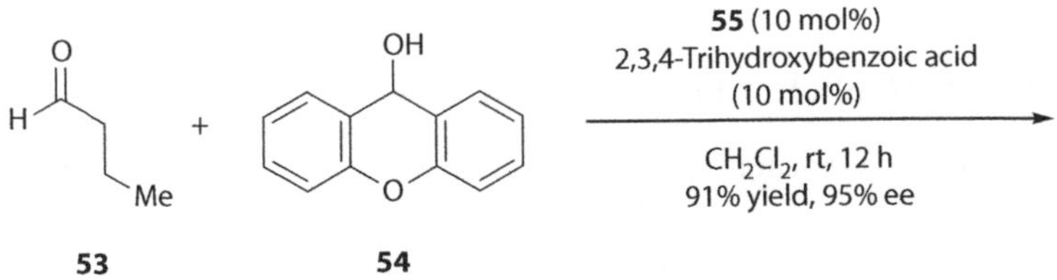
OH
+
Me
55 (10 mol%)
2,3,4-Trihydroxybenzoic acid
(10 mol%)
CH2Cl2, rt, 12 h
91% yield, 95% ee
53
54

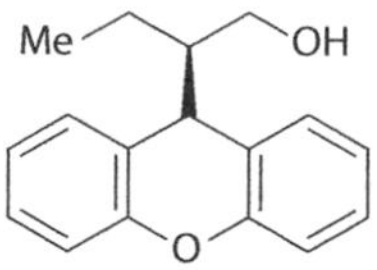

56

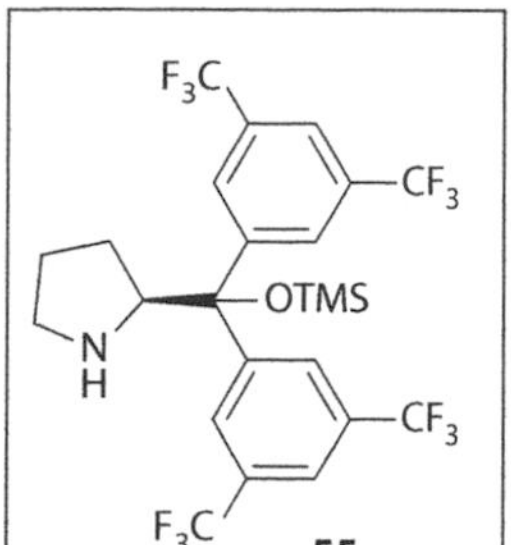

55

Similarly, the butyraldehyde (**53**) and xanthydrol (**54**) was carried out in the presence of a catalytic amount of **55** and trisubstituted benzoic acid. The desired enantioenriched adduct **56** was obtained in good yield and excellent ee.[20]

In the case below, indium(III) triflate worked in tandem with a chiral organocatalyst **57** (MacMillan catalyst) to promote a synergistic stereoselective S_N1 reaction. Thus, under the influence of both In(OTf)$_3$ and **57**, benzyl alcohol **58** was converted to the corresponding carbenium ion, which was then captured by propionaldehyde **59** to offer adduct **60**.[21]

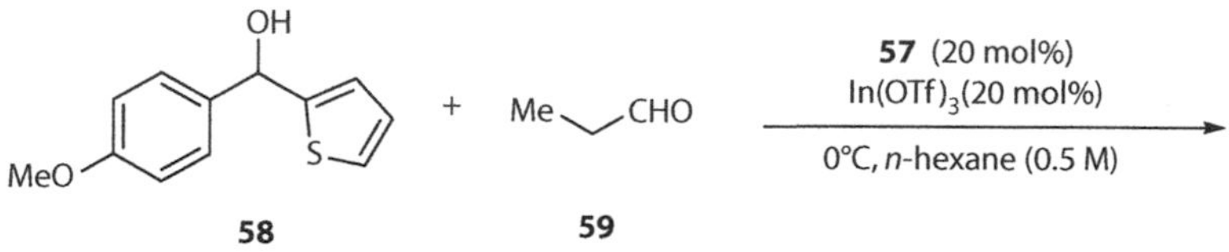

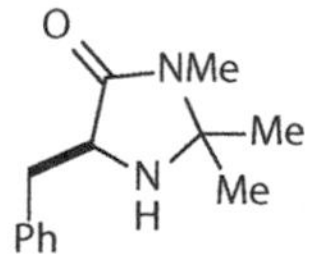

58   **59**

57 (20 mol%)
In(OTf)$_3$(20 mol%)

0°C, *n*-hexane (0.5 M)

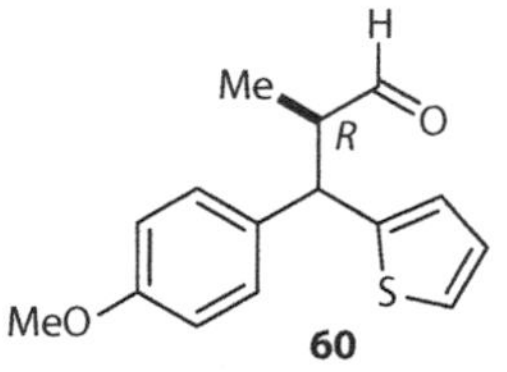

60

MacMillan catalyst **57**

MacMillan catalyst (**57**) also found utility in another S_N1 alkylation. Under the influence of catalytic amount **57**, stereoselective organocatalytic alkylation of aldehyde **62** took place with benzodithiolylium tetrafluoroborate (**61**) to offer adduct benzothiol **63** in 62% yield and 92% ee. Here **61** is a commercially available carbenium ion and benzothiol **63** may serve as a useful synthon for further manipulations.[22]

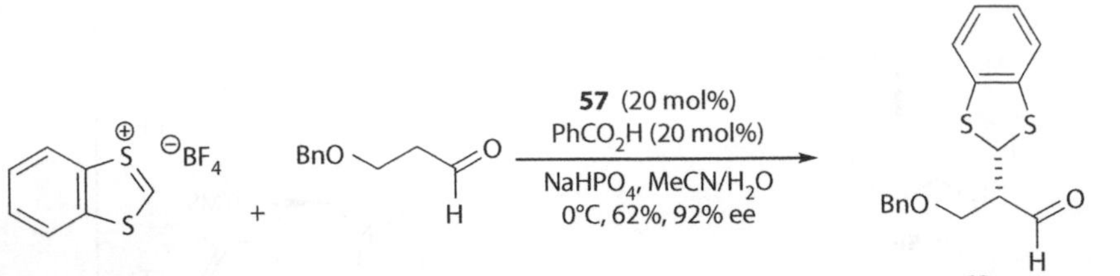

62 61 63

A direct enantioselective nucleophilic substitution using less reactive carbocations (from alcohols) has been achieved by the Cozzi group.[23] In the presence of MacMillan imidazolidinonium catalyst (**64**), ferrocenyl alcohol **65** smoothly reacted under mild reaction conditions with propionaldehyde to form the desired (2*S*, 3*R*)-*syn*-product **66** in 48%, a 3:1 dr and 90% ee. This remarkable methodology revealed the entries for domino, consecutive, and multicomponent organocatalytic reactions.

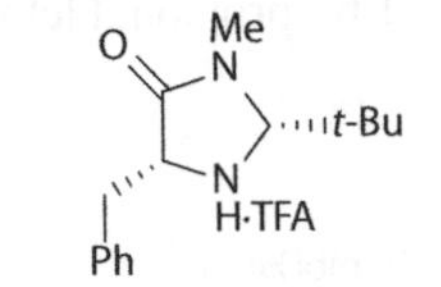

MacMillan catalyst **64**

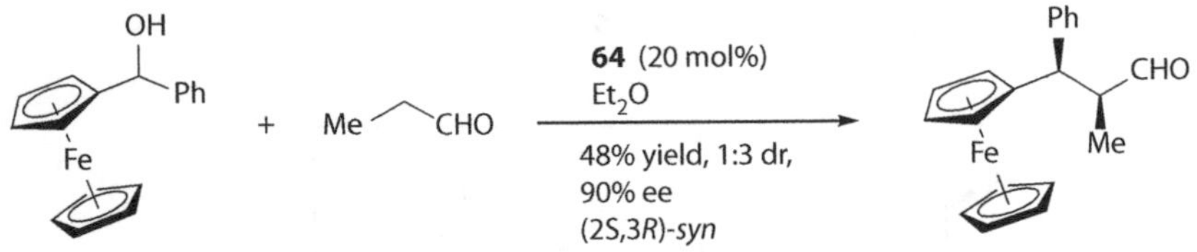

65 59 66

Subsequently, the same group demonstrated that MacMillan organocatalyst (**64**) was also able to catalyze the α-alkylation of aldehyde.[24] As shown below, bis(4-dimethylamino-phenyl)methylium tetrafluoroborate (**67**) reacted with propionaldehyde **59** in the presence of (*S*,*S*)-MacMillan catalyst *ent*-**64** to give the adduct **68** in 75% yield and 65% ee. The temperature is crucial for the stereoselection, and better results were obtained at lower temperature.

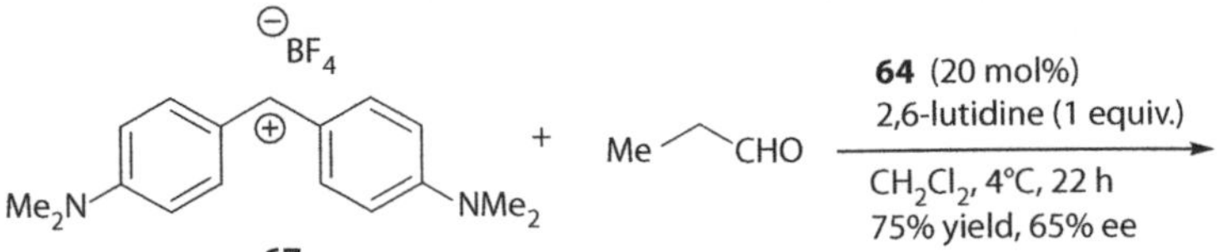

67

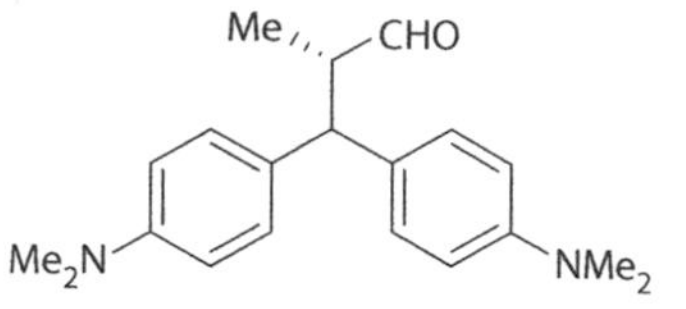

68

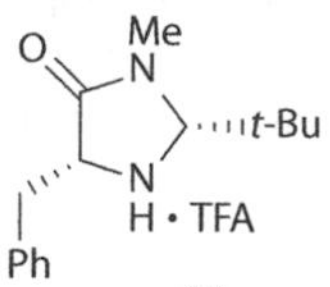

64

Alkylation of a variety of 2-arylpropionaldehydes proceeded in moderate-to-good yields and high enantioselectivities in the presence of thiourea-derived catalyst (**69**).[25] The α-alkylation of 2-arylpropionaldehyde **70** with bromodiphenylmethane (benzhydryl bromide **71**) was shown to produce compound **72** with excellent enantioselectivity and relatively good yield.

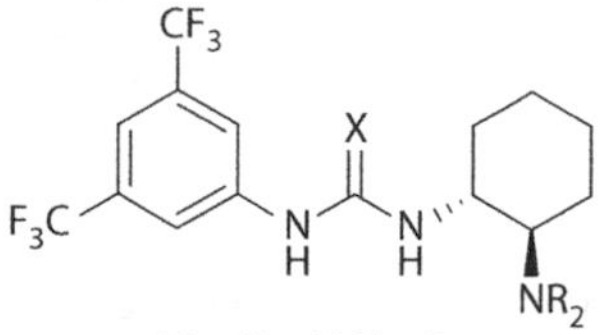

69a: R = H; X = S
69b: R = H; X = O
69c: R = Me; X = S

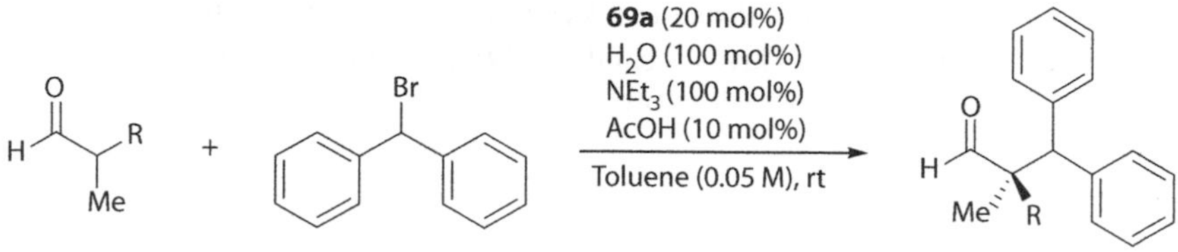

70 71 72

Using Brønsted acids such as *p*-toluenesulfonic acid (*p*-TSA), α-alkylation of aldehydes such as **73** with allylic/benzylic alcohol **74** gave the S_N1 alkylation adducts such as **75**.[26]

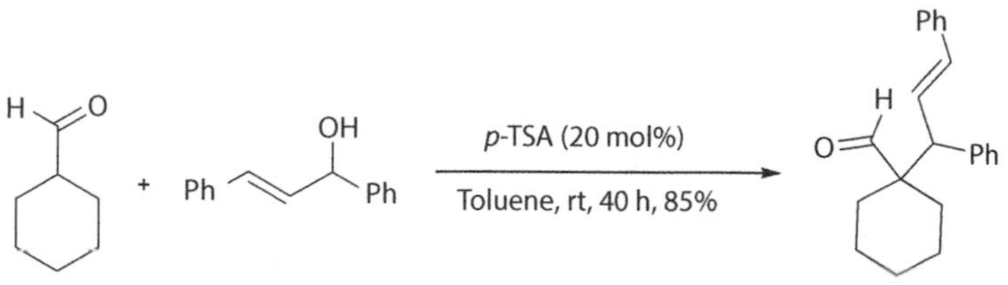

73 74 75

In order to generate benzylic cations, benzyl alcohols have been often used as the precursors by the breaking C–O bond with the help of Brønsted or Lewis acid. *N*-benzylic sulfonamides may be also employed to produce benzylic cations using TFA by breaking the C–N bond.[27] Using a chiral organocatalyst **76**, *N*-benzylic sulfonamide **77** reacted with propionaldehyde to prepare the α-alkylation product **78** in 83% yield and 65% ee.

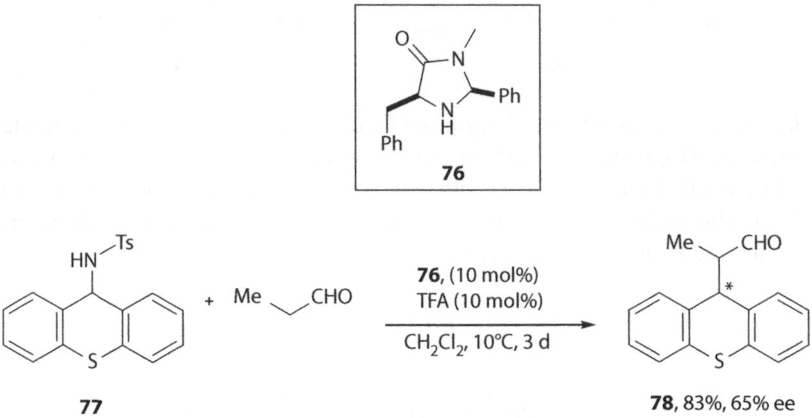

O
N
N
H
Ph
Ph
76
HN
Ts
S
+ Me CHO
76, (10 mol%)
TFA (10 mol%)
CH2Cl2, 10°C, 3 d
Me
CHO
*
S
77
78, 83%, 65% ee

Jørgensen–Hayashi organocatalyst (**79**) was employed for an enantioselective tandem reaction at elevated temperature for a one-pot hydroformylation followed by an S_N1 alkylation of the resulting aldehyde.[28] Taking diene **80** as an example, hydroformylation was accomplished using both Rh and Ru catalysts and 1,4-*bis*(dip henylphosphinobutane) (DPPB) as the bidentate ligand. *In situ*, the aldehyde was not isolated but was subjected to a direct α-alkylation using Michler's hydrol (**2**) as the cation source, the Jørgensen–Hayashi catalyst (**79**), and valeric acid as the promoter to assemble adduct **81** in 76% yield, 80% ee and 3:1 dr.

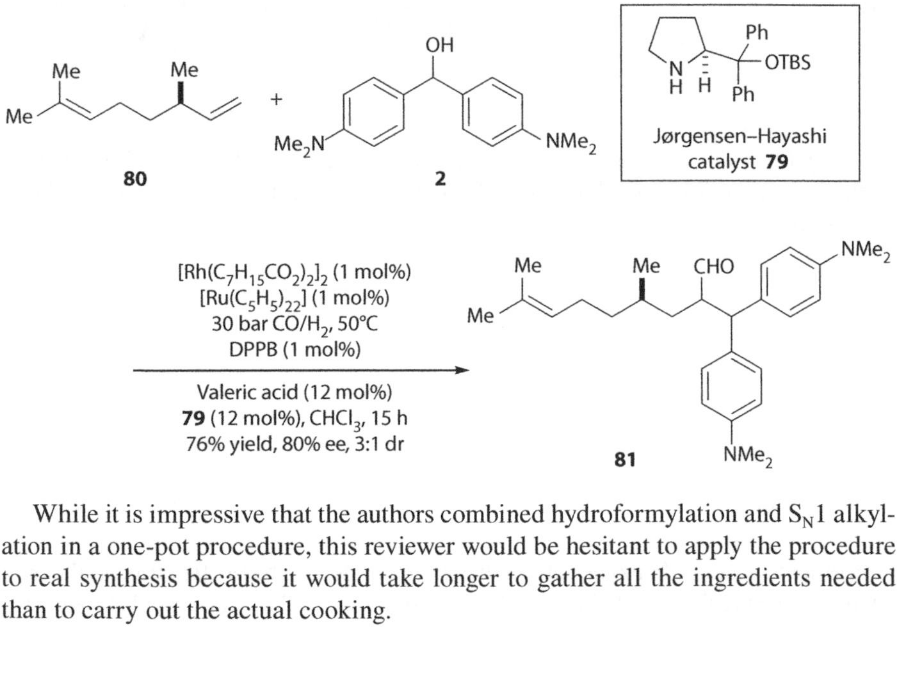

Me
Me
Me
80
+
OH
Me2N
NMe2
2
N
H H
Ph
OTBS
Ph
Jørgensen–Hayashi
catalyst 79
[Rh(C7H15CO2)2]2 (1 mol%)
[Ru(C5H5)22] (1 mol%)
30 bar CO/H2, 50°C
DPPB (1 mol%)
Valeric acid (12 mol%)
79 (12 mol%), CHCl3, 15 h
76% yield, 80% ee, 3:1 dr
Me
Me
Me
CHO
NMe2
NMe2
81

While it is impressive that the authors combined hydroformylation and S_N1 alkylation in a one-pot procedure, this reviewer would be hesitant to apply the procedure to real synthesis because it would take longer to gather all the ingredients needed than to carry out the actual cooking.

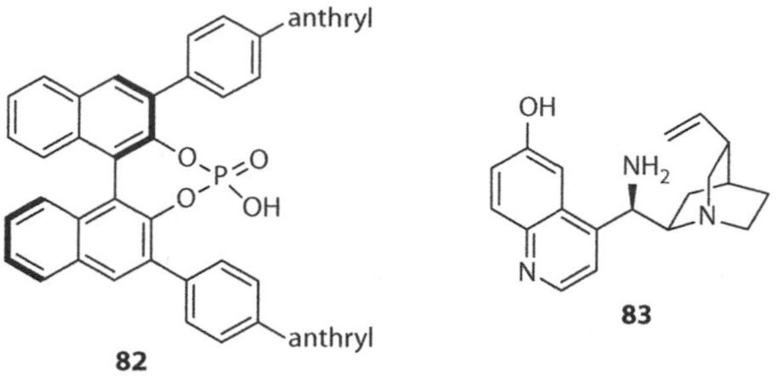

(+)-Gliocladin C (86)

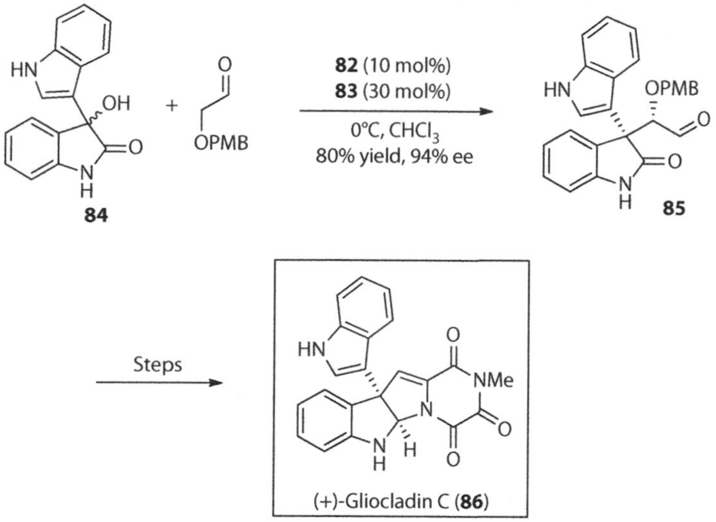

The combined catalyst of a chiral primary amine and phosphoric acid **82** has been demonstrated to be an efficient system to α-alkylation of aldehydes with 3-hydroxyoxindole **84**. The approach allowed to construct functionalized 3,3′-*di*-substituted oxindole **85**, which led to a concise total synthesis of (+)-glicladin C (**86**) in 12 steps from 3-hydroxyoxindole **84** in 19% overall yield.[29]

Indium(III) salts are unique because, as a Lewis acid, it is compatible with water and amines. Therefore, indium(III) Lewis acid has been widely used in aqueous reactions. As shown below, α-alkylation of aldehyde **88** with propargyl alcohol **87** was carried out in water using indium(III) triflate as the Lewis acid catalyst.[30] With the aid of the MacMillan's catalysts **57** as the organocatalyst, the resulting propargylated aldehyde **89** and **90** were obtained with high enantioselectivity.

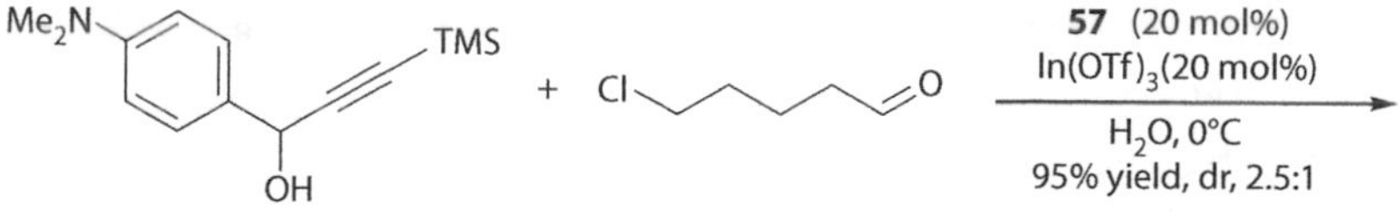

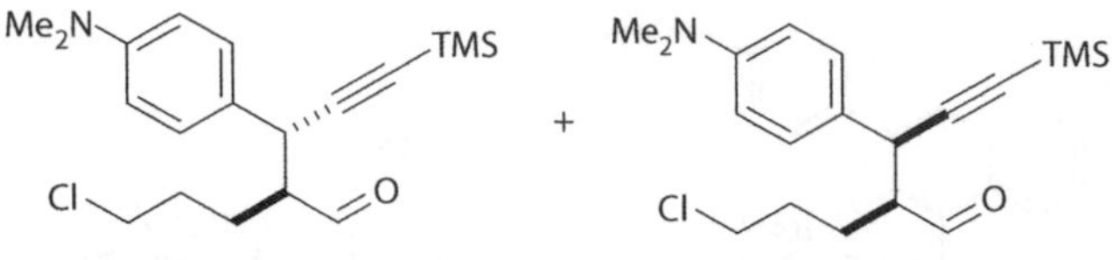

89, anti, 94% ee   **90**, syn, 93% ee

The α-alkylation of ketones with primary alcohols may also be accomplished using the highly effective and versatile Cp*Ir (**91**) catalyst, functionalized with bipyridonate ligand. Under the influence of 1 mol% catalyst, the reaction of aceto-phenone (**92**) with benzyl alcohol (**93**) was carried in *tert*-amyl alcohol as solvent generating the α-alkylated product (**94**) in exceptional good yield.[31]

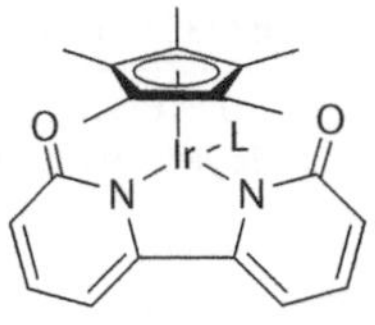

91 L = H$_2$O

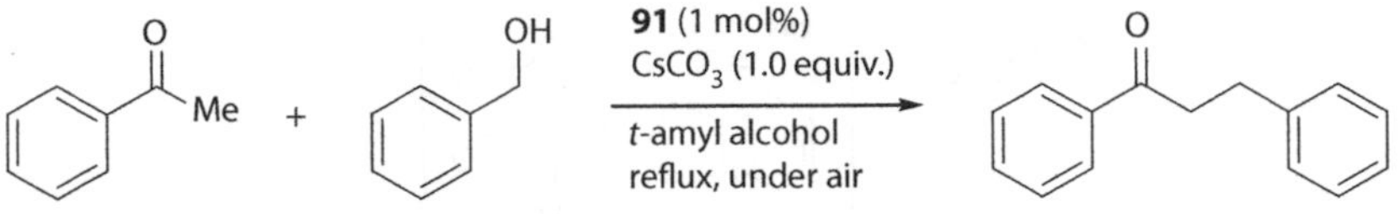

| **92** | **93** | | **94** |

The copper(II) triflate-*tert*-butyl-bisoxazoline [Cu(OTf)$_2$-*t*-Bu-BOX, **95**]-cata-lyzed asymmetric alkylation of β-keto esters **96** using free benzylic alcohols such as xanthydrols (**54**), as alkylating agents, is herein described for the first time. This green protocol renders in general the corresponding products **97** with good results in terms of both yields and enantioselectivities using different keto esters, even when quaternary stereocenters were created.[32] The scope, limitations, and mechanistic aspects of the process are also discussed.

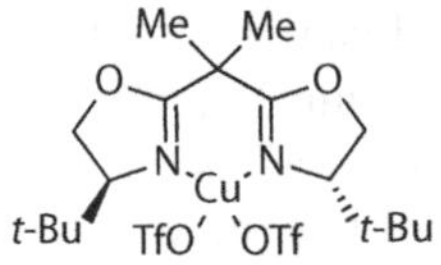

95

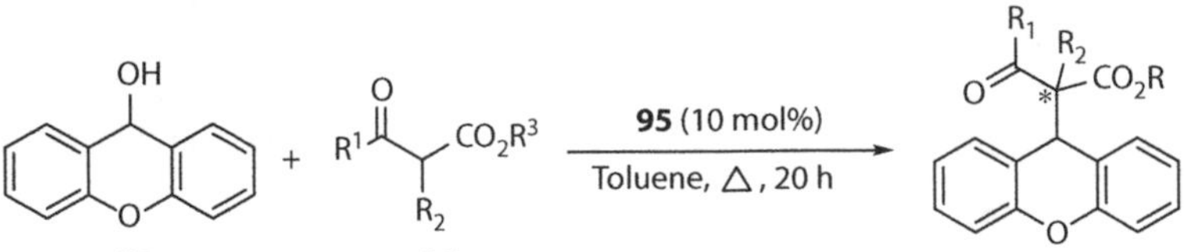

54   **96**   **97**

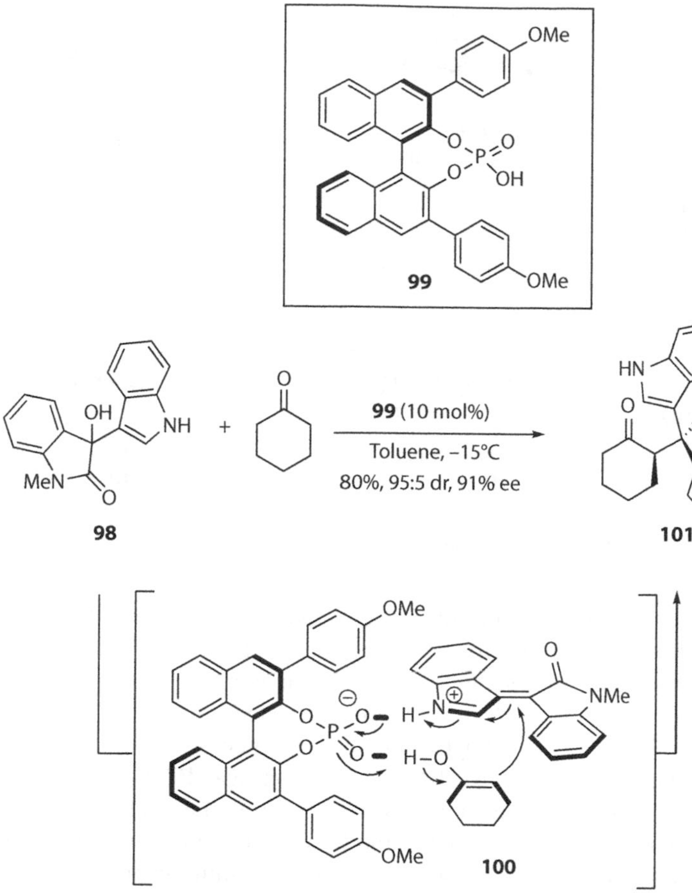

99

98   +   **99** (10 mol%)

Toluene, −15°C
80%, 95:5 dr, 91% ee

101

100

The alkylation of unmodified ketones with alcohols and using Brønsted acid catalysis has resulted in high yields and high enantioselectivities. The reaction of isatin-derived 3-hydroxy-3-indolyoxindole **98** with cyclohexanone using the chiral phosphoric acid **99** as a catalyst could generate chiral 3-indolyl-oxindole **101**.[33]

2.5   GLYCOSYLATION

Glycosylation is one of the most classical and essential reactions in the field of carbohydrate chemistry. Crich and coworkers have extensively studied the mechanism for the glycosylation.[34]

A general S_N1-type glycosylation mechanism is depicted in Scheme 2.2. Usually, it involves a reaction between an activated glycosyl donor (**103**) and glycosyl acceptor (nucleophile) to form a glycosidic bond. In the presence of an activator or promoter, the departure of the LG leads to the formation of an oxocarbenium ion intermediate (**104**, glycosyl cation). The real existence of the glycosyl cations have been demonstrated by Crich through spectroscopic methods. The nucleophile then attack either from the top or the bottom face of the flattened ring. The stereochemistry would be

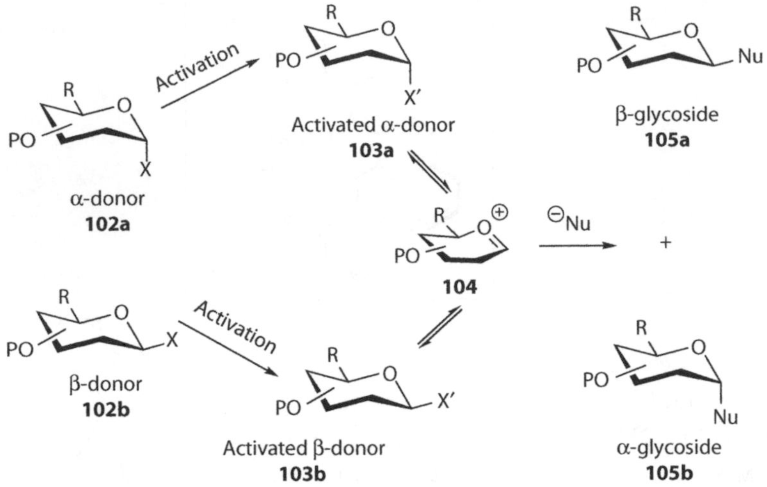

SCHEME 2.2 A general S_N1-glycosylation mechanism.

controlled by the neighboring substituent at C-2 and uncontrolled reactions may lead to a mixture of α-glycoside (**105b**) and β-glycoside (**105a**).[35]

Eukaryotic enzymes are able to use 2-*C*-acetylmethylsugars as replacement of 2-*N*-acetamidosugars. When 1,2-cyclopropaneacetylated sugar **106** was treated with strong Lewis acid such as trimethylsilyl trifluoromethanesulfonate (TMSOTf), an S_N1 displacement took place to give 2-*C*-acetylmethylsugar **107** with predominantly β-substitution product **107** when alcohols were used as the nucleophiles.[1] On the other hand, when $BF_3 \cdot OEt_2$ was used as the Lewis acid promoter, alcohols reacted with glycosyl donor **106**, the major product is the β-substitution product (not shown here) via the S_N2 displacement pathway.[36]

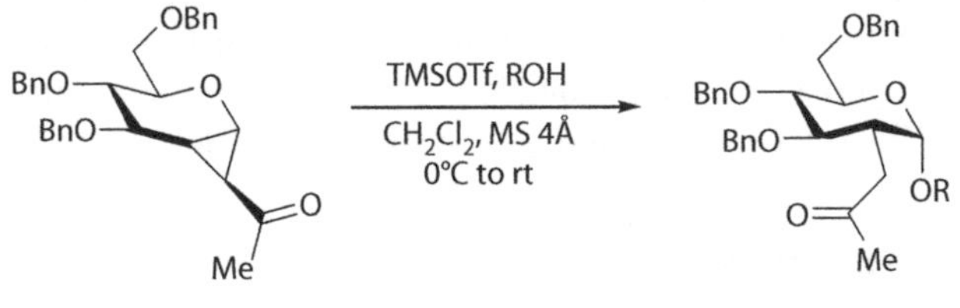

For glycosylation, the widely invoked mechanism involves the intermediate of glycosyl oxocarbenium ion, a species that eluded chemists. In 2012, Crich et al. reported evidence that they obtained through a relative kinetics study using a competing cyclization reaction as a clock.[37] Chiral sulfoxide **108** was activated by triflic anhydride in the presence of 2,4,6-tri-*tert*-butylpyrimidine (TTBP) to give two cyclization products after quenching. The expected *cis*-fused system **109** was the major product and the unexpected *trans*-fused system **110** with a twisted boat conformation in the middle.

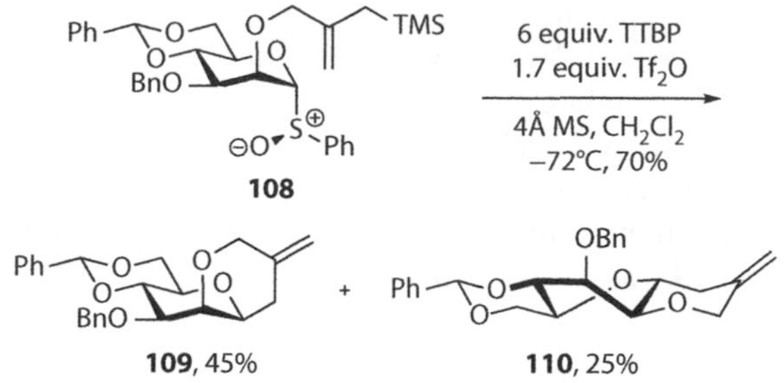

The results may be rationalized by invoking the intermediacy of a mannosyl oxocarbenium ion **111**, which exists with α-glycosyl triflate **112**. The intramolecular Sakurai reaction would then give the major product from β-face attack and the minor product from α-face attack. Competition reactions of sulfoxide **108** with isopropanol and trimethyl(methallyl)silane indicated that β-*O*-mannosylation proceeded via an associative S$_N$2-like mechanism. Meanwhile, α-*O*-mannosylation and β-*C*-mannosylation are dissociative and S$_N$1-like.

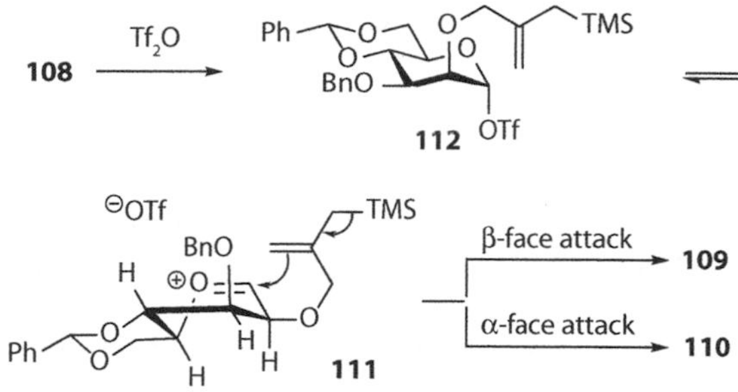

Woerpel et al. discovered profound solvent effects when they carried out their stereoselective *C*-glycosylation.[38] When tetrahydropyran acetal **113** was treated with silyl ketene acetal **114** using Me$_3$SiOTf as the catalyst and nonpolar solvents, the S$_N$2 product **115a** was isolated as the major product. In contract, the S$_N$1 product **115b** prevailed, presumably through the intermediacy of oxocarbenium ion **116**.

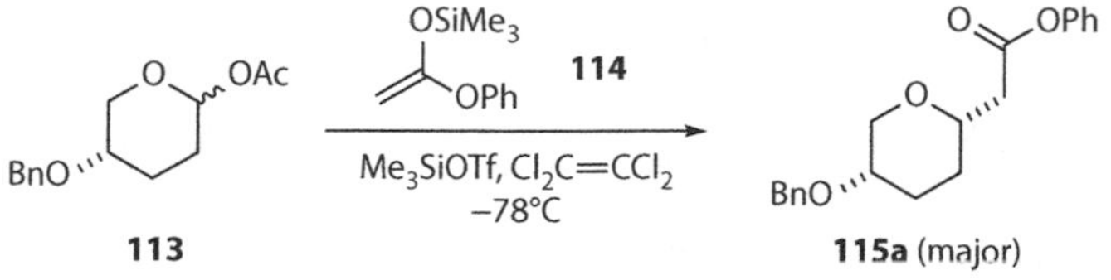

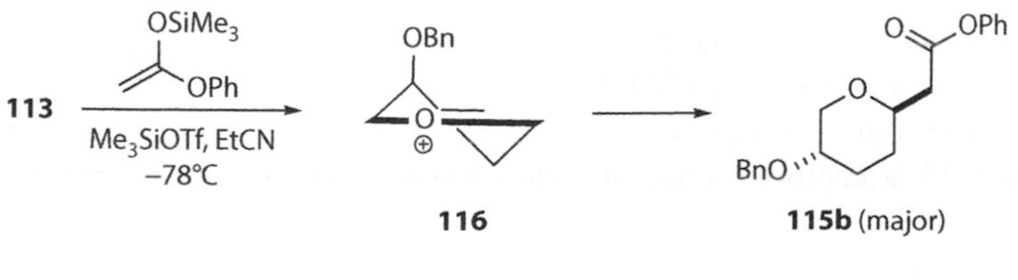

To avoid formation of potential bicyclic ring products (via intramolecular addition), De Brabander et al. developed smooth conditions for introduction of a cyano group.[39] In a one-pot procedure, trimethylsilyl cyanide was added to alcohol **117** (neat) in the absence of Lewis acid to allow protection of the secondary alcohol as a silyl ether, followed by addition of a solution of zinc iodide in acetonitrile to initiate oxonium **118** formation, followed by axial cyanide attack. After acidic aqueous workup, compound **119** was obtained in 91%.

Per-*O*-acetylation is one of the most common reactions in carbohydrates research. Hung et al. found that Cu(OTf)$_2$ is an extremely efficient catalyst (lowest loading: 0.03 mol%) for per-*O*-acetylation of hexoses. They developed an efficient one-pot cascade per-*O*-acetylation-anomeric substitution of hexoses to thioglycosides.[40] Initially, solvent-free per-*O*-acetylation of hexoses **120** with a stoichiometric amount of acetic anhydride in the presence of Cu(OTf)$_2$ proceeded at room temperature to afford pyranosyl products as an anomeric mixture (structure not shown), the α/β ratio of which was dependent on the temperature and amount of catalyst used. Subsequently, anomeric substitution with *p*-thiocresol employing of BF$_3$ etherate gave the β-isomer **121** as the single product in a 73% overall yield.

Gervay-Hague and coworkers developed an efficient, one-pot syntheses of biologically active α-GalCer and BbGL-II analogs via glycosylation between a TMSI (trimethylsilyl iodide)-protected glycosyl iodide (**122**, donor) and an unprotected ceramide (**123**, acceptor).[41] Tetrabutylammonium iodide served as a promoter that accelerated the reaction and induced α-stereoselectivity via *in situ* anomerization. This method has been demonstrated that tolerates unsaturation and free hydroxyl group in the fatty acid side chain.

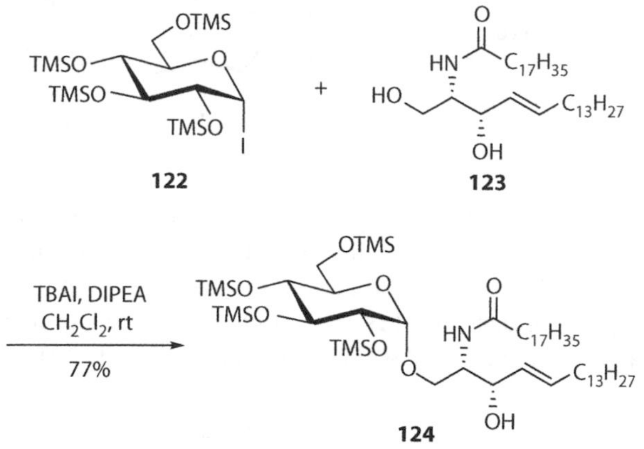

To gain an insight for the mechanism underlying the exceptional 1,2-*cis* selectivity, Codee and coworkers explored the use of a 1-thio mannosaziduronate for glycosylation.[42] Employing the Ph_2SO–Tf_2O activator system, oxocarbenium ion **125** was formed, which can lead to the β-linked product via an S_N2-like displacement. However, the transformation can also proceed via an S_N1-like pathway in this case. As a result, it allowed the nucleophile **126** preferentially to attack the oxocarbenium ion (**125b/c**) intermediate from the β-face to produce disaccharide **128** with complete 1,2-*cis* selectivity in 85% yield.

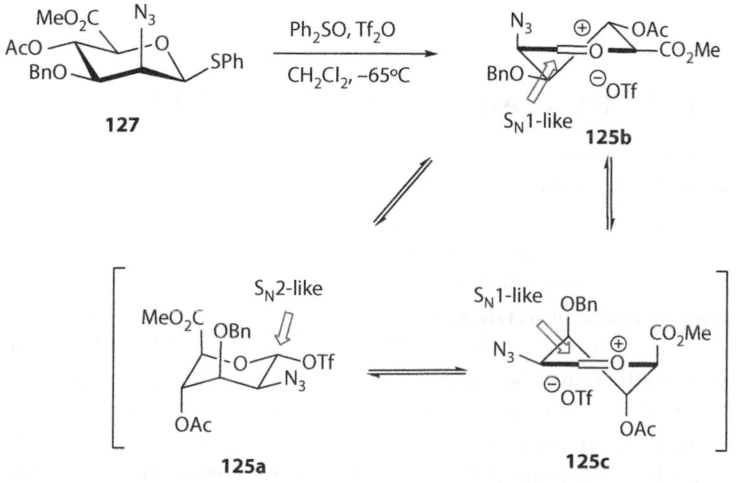

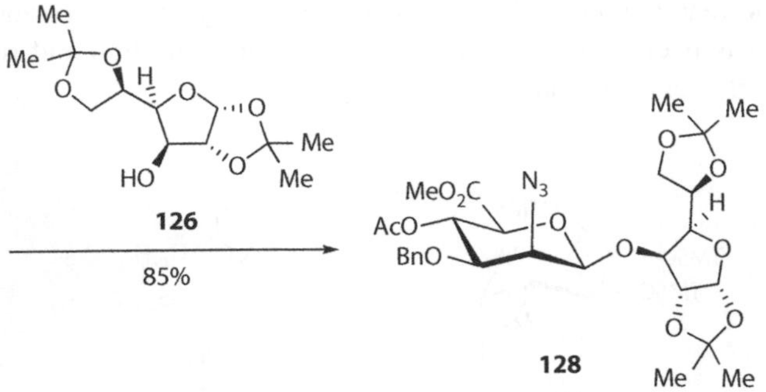

126

85%

128

Shiga toxins 1 and 2, expressed by genes are two major virulence factors of *Escherichia coli* (*E. coli*) O157:H7. To differentiate between structurally homologous Shiga toxins 1 and 2, Weiss and coworkers synthesized a series of glycoconjugates.[43] Initially, the coupling of acceptor **129** with trichloroacetimidate donor **130** in the presence of catalytic amount of TMSOTf afforded disaccharide **131** in 63% yield.

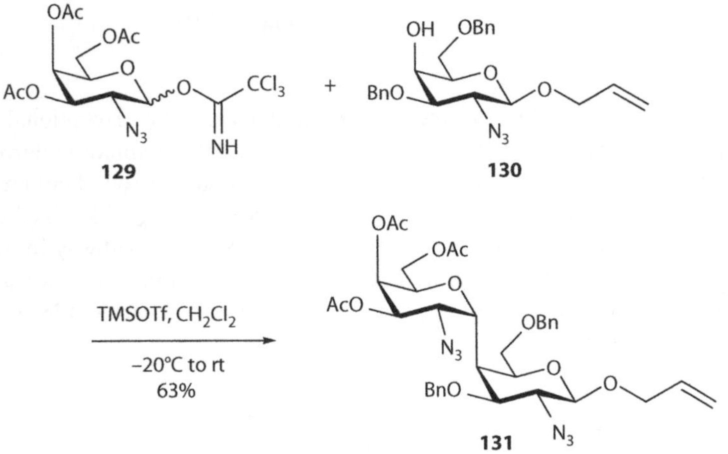

129   **130**

TMSOTf, CH$_2$Cl$_2$

−20°C to rt
63%

131

2.6   FRIEDEL–CRAFTS ALKYLATION AND ACYLATION

Both Friedel–Crafts alkylation and acylation are typical S$_N$1 reactions because they all involve the generation of carbocation and acylium ion intermediates.

For some S$_N$1 reactions such as dehydrative transformations of alcohols, Lewis acid catalysts are accompanied by undesirable Brønsted acid-catalyzed reactions when water or other protic functional groups are present. As a consequence, one has to choose between powerful but harsh catalysts or poor but mild ones. A commercially available strong Lewis acid BPh$_3$ is nonhydrolyzable and neutral. It may be the answer to the prayer because its hydrates also undergo rapid ligand exchange. Taking the Friedel–Crafts alkylation reaction as an example, a catalytic amount of BPh$_3$ promoted conversion of allylic alcohol **132** to allylic cation **134**,[16] which was then captured by mesitylene (**133**) to furnish the Friedel–Crafts alkylation adduct **135**. Unlike most Lewis acids, there was little (<5%) of olefin isomerization from **135** to the more stable isomer **136** when BPh$_3$ was used as the strong Lewis acid catalyst, which inevitably took place when other Lewis acid catalysts and Brønsted acids were used as catalysts.

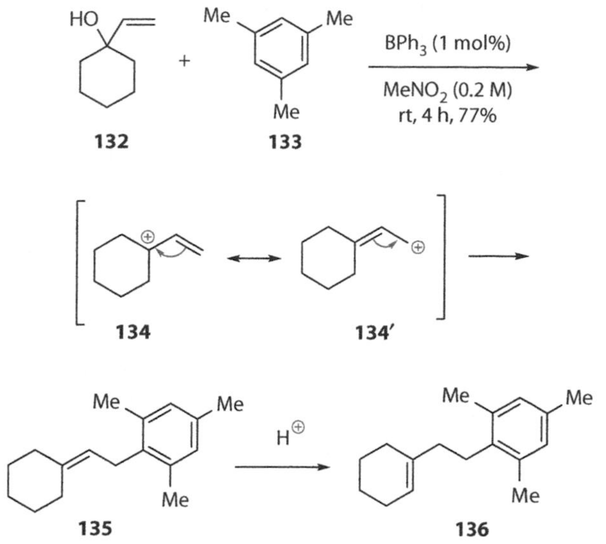

A first catalyzed enantioselective propargylation of aromatic compounds with propargylic alcohols via allenylidene intermediates was developed by Nishibayashi and coworkers.[44] The chirality was controlled by an *in situ* prepared thiolate-bridged diruthenium complex. Although the reactivity of *N,N*-dimethylaniline is less than that of 2-alkylfuran in the propargylation of aromatic molecules, under this condition that in the presence of a large excess of *N,N*-dimethylaniline, such transformation was completed in moderate-to-good yields and high enantioselectivities. They proposed that an allenylidene intermediate **140** was generated and then its resonance structure alkynyl complex **140'** was attacked by nucleophiles from the *Si* face. Employing the propargylic alcohol **141** as electrophile, this article recorded a novel protocol for the catalytic asymmetric Friedel–Crafts alkylation of aromatic compounds.

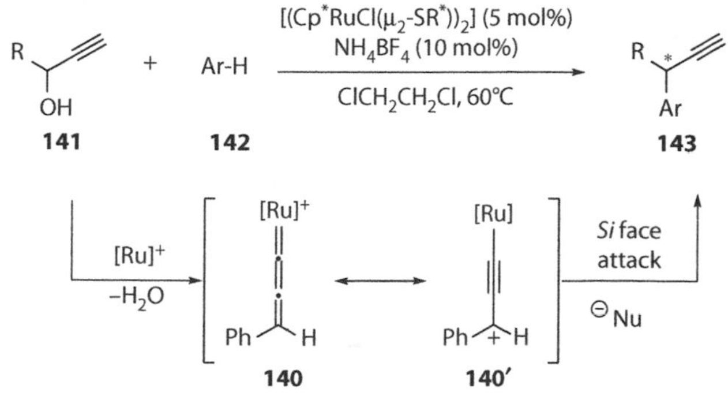

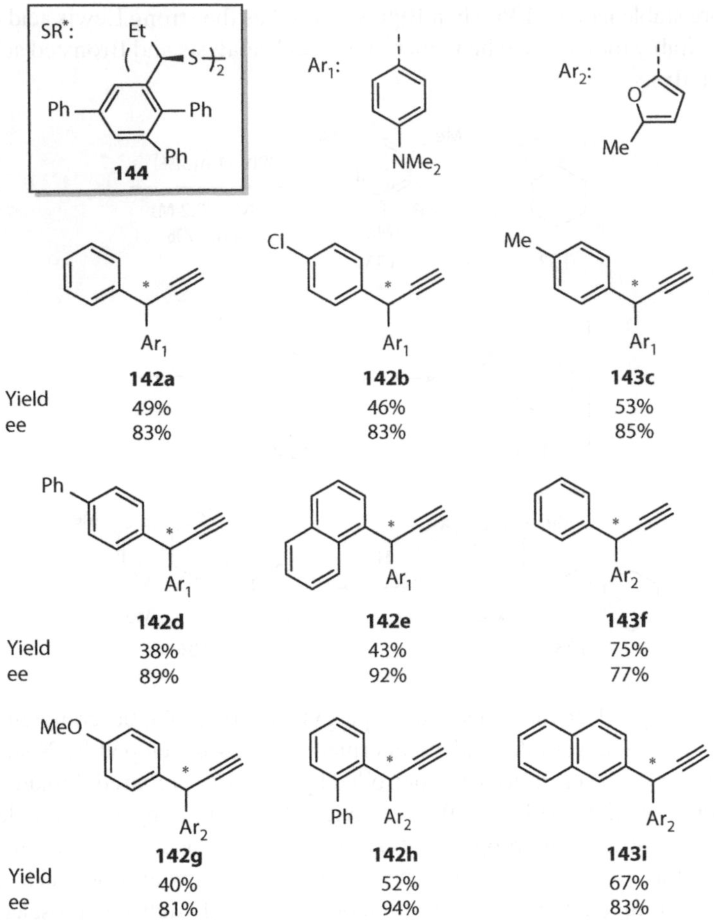

SR*: Et S 2
Ph Ph Ph
144
Ar1: NMe2
Ar2: O Me
Cl Me
Ar1 Ar1 Ar1
142a 142b 143c
Yield 49% 46% 53%
ee 83% 83% 85%
Ph
Ar1 Ar1 Ar2
142d 142e 143f
Yield 38% 43% 75%
ee 89% 92% 77%
MeO
Ar2 Ph Ar2 Ar2
142g 142h 143i
Yield 40% 52% 67%
ee 81% 94% 83%

In place of Lewis acids, organocatalysts have been applied to promote Friedel–Crafts alkylation. Kotsuki et al. developed a series of thioureas such as **145** as halophilic catalyst because these thioureas can bind anions through double hydrogen bonding.[45] These thiourea catalysts have been able to effectively induce ionization of alkyl halides as electrophiles to generate carbocation intermediates in an S_N1-type Friedel–Crafts alkylation. Since **145** is only weakly basic, with a pK_a of 8.5, it is a very mild catalyst.

CF3
CF3
S
F3C
N H
N H
CF3

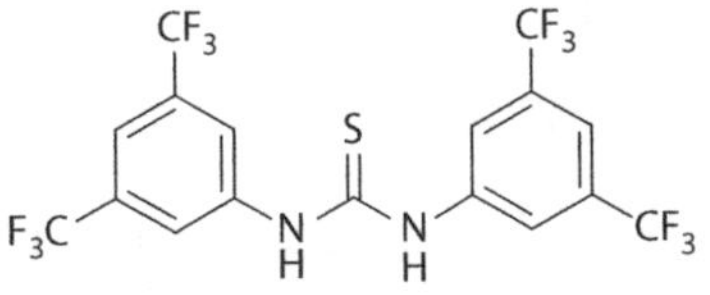

145

For example, treatment of a π-excessive heteroaromatic **146** with benzylic chloride **147** gave rise to the Friedel–Crafts benzylation using organocatalyst **145**.[45] Other π-excessive heteroaromatics such as furans, thiophenes, pyrroles, benzofurans, and indoles also underwent similar Friedel–Crafts benzylation in good yields although bis-benzylation often competed with mono-benzylation for furans, thiophenes, and pyrroles.

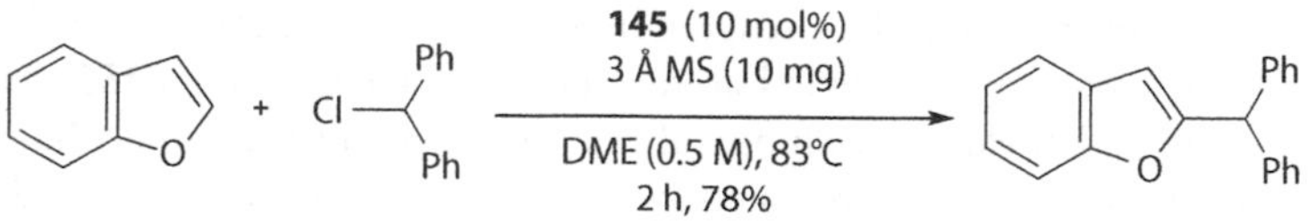

Fluorinated alcohols feature unique properties, including strong H-bonding donor ability, high ionizing power, and low nucleophilicity, enabling them to easily generate cationic intermediates through hydrogen bonding interactions. As a result, reactions can be launched in the absence of Lewis acid or Brønsted acid catalysts. Recently, Xiao and coworkers developed the first fluorinated alcohol-mediated S_N1 reaction of indolyl alcohols with diverse nucleophiles. For example, 2-phenyl-3-indolylmethanol **149** was selected to form the alkylideneindoleninium ion **150** using TFE as solvent. Treatment with *N*-methyl indole, it gave adduct **151** in excellent yield.[46]

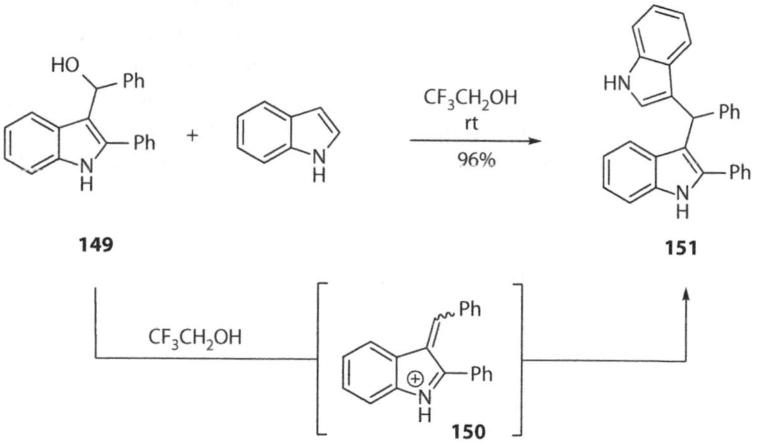

For Friedel–Crafts benzylation of *p*-xylene, it was discovered that XtalFluor-E ([Et$_2$NSF$_2$]BF$_4$), a relatively more thermally stable (than DAST) deoxofluorinating agent, may activate benzyl alcohols *in situ*.[1] For instance, benzyl alcohol **152** reacted with 5 equiv. of *p*-xylene **153** to afford 1,1-diarylmethane **154** in 61% yield with the aid of XtalFluor-E. Here cosolvent HFIP stands for 1,1,1,3,3,3-hexafluoro-2-propanol [HOCH(CF$_3$)$_2$].[47]

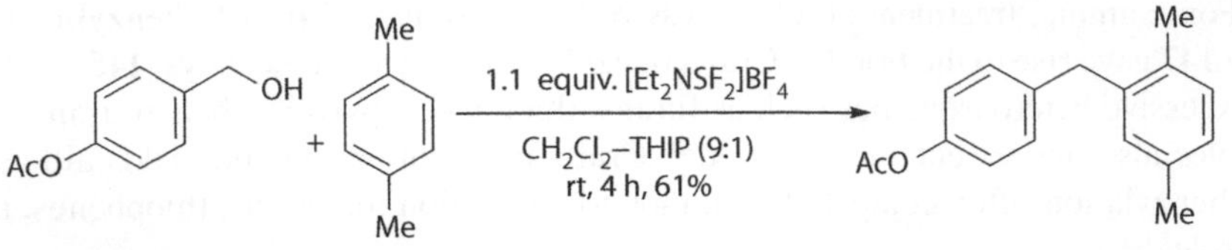

152   **153**   **154**

Mechanistically, XtalFluor-E, a reagent that requires an external fluoride source to induce the deoxofluorination reaction, reacted with benzyl alcohol **152** to an activated alkoxy-*N,N*-diethylaminodifluorosulfane intermediate **155** rather than benzyl fluoride. Ionization of **155** then produced a stabilized benzylic carbocation **157** along with diethylaminosulfinyl fluoride (**156**) after loss of a fluoride. Cation **157** then could react with nucleophile *p*-xylene to deliver 1,1-diarylmethane **154**.

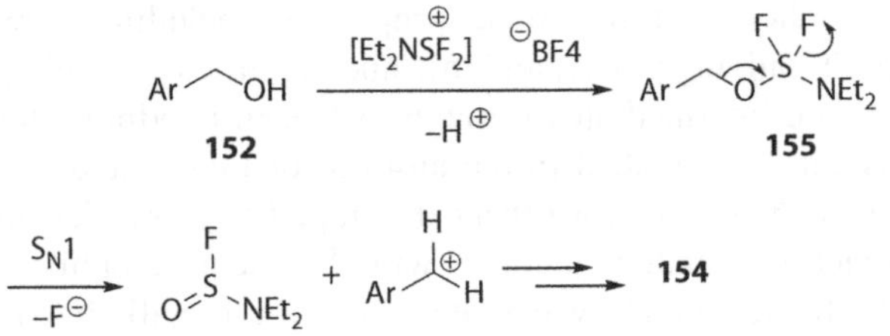

Mirroring the Bi(OTf)$_3$-catalyzed method, Hua et al. developed a BiCl$_3$-catalyzed synthesis of 1,1-diarylalkanes starting from electron-rich arenes and styrenes. Additionally, they found that heating of styrene **158** in the presence of catalytic amounts of BiCl$_3$ yielded substituted dihydroindenes **159** as a result of styrene dimerization.[48] This reaction may proceed via an intermolecular ene reaction between styrene **158** and the carbocationic intermediate **160** followed by an intramolecular Friedel–Crafts alkylation of the resulting carbocationic 1,3-diarylpropane **161**.

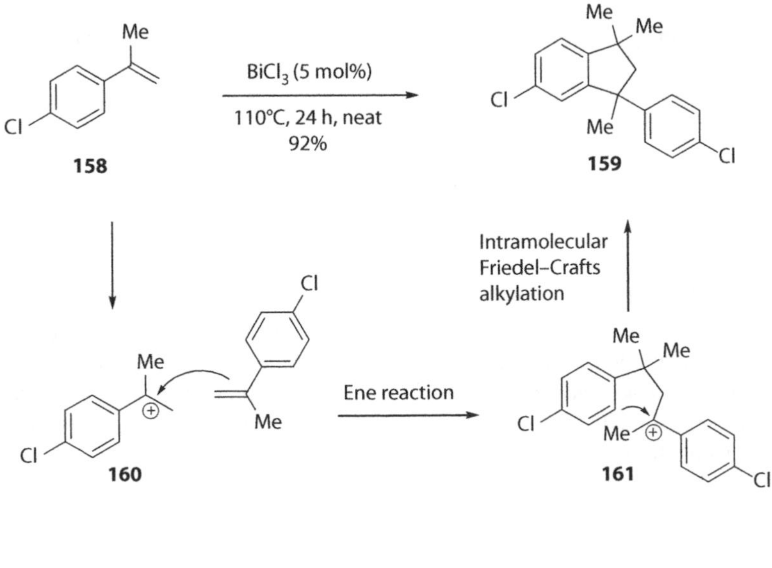

A dehydrative nucleophilic substitution of benzyl alcohols in water using a dodecylbenzenesulfonic acid (DBSA) as a surfactant-type Brønsted acid catalyst has recently been developed by Kobayashi and coworkers.[49] With this environmental-friendly methodology, a variety of aromatic compounds were employed as nucleophiles to yield corresponding diarylmethanes and 3-substituted indoles. This reaction system can particularly be extended to the stereoselective C-glycosylation of 1-hydroxysugar **162** yielding the products **164** in good yields and with remarkable anomeric ratios.

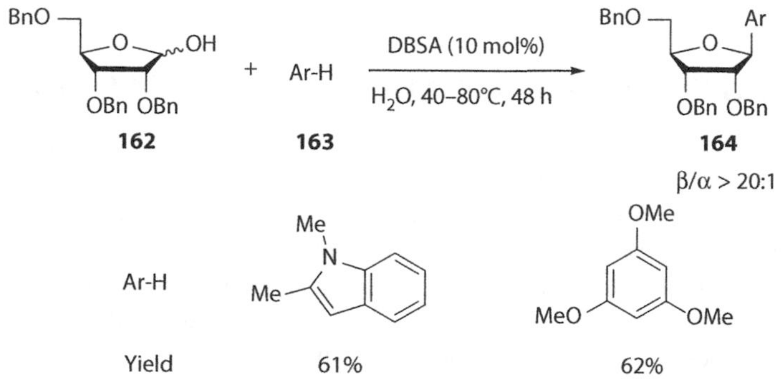

Toste and coworkers developed a mild aromatic propargylation reaction, using an air- and moisture-tolerant rhenium oxo complex as a catalyst and a propargyl alcohol **166** as the electrophile.[50] With this methodology in hand, the authors achieved the synthesis of a couple of bioactive molecules, including β-apopicropodophyllin (**168**) and podophyllin (**169**). The synthesis was starting from the rhenium-catalyzed Friedel–Crafts alkylation of ethyl propiolate **166** with safrole (**165**).

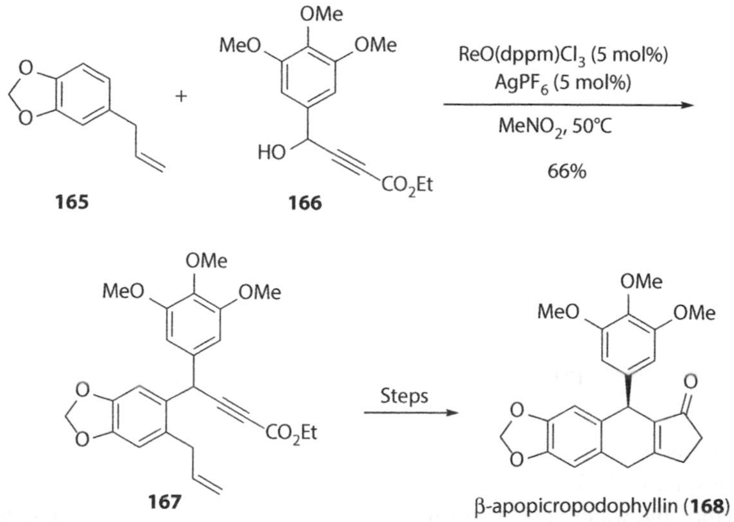

Podophyllotoxin (**169**)

2.7 ELECTROPHILIC FLUORINATION USING FLUORONIUM ION

Electrophilic fluorination is an electrophilic source of fluorine to react with carbon nucleophiles to afford C–F bond. The common electrophilic fluorinating agents include *N*-fluoropyridinium salts, *N*-fluorosulfonamide and its derivatives, and Selectfluor and its derivatives.[51] Although elemental fluorine and reagents incorporating an oxygen–fluorine bond can be used for this purpose, they have largely been replaced by reagents containing a nitrogen–fluorine bond. Specially, in this section, we only introduce the electrophilic fluorination using fluoronium ion.

Halonium ion **170** is well known in organic chemistry. Any sophomore organic chemistry course would cover it extensively to provide mechanistic insight for fundamental reactions. Iodonium ions are stable and some have been isolated. Bromonium ions are less stable, but some bromonium ions such as **172** have been isolated when caged in a crowded molecule such as adamantylideneadamantane (**171**).[52] Chloronium ions are less documented, but some stable chloronium ions have been synthesized.[53] However, the existence of the fluoronium ion has been doubted because fluorine is small in size yet it is the most electronegative element, making it difficult for fluorine's electron pairs to engage in hypervalent bonding, like other halogens can.[54]

170, halonium ion

X = I, Br, Cl
X = F?

171   Br$_2$ **172**

In 2013, Lectka et al. obtained strong evidence of the existence of the fluoronium ion (F$^{\oplus}$) in solution.[55] Solvolysis of isotope-labeled fluoro-triflate **173** with 2,2,2,-trifluoroethanol and water at 60°C furnished a 1:1 mixture of two isomers **174** and **175**. The outcome strongly suggests the existence of a symmetrical fluoronium ion **176**.[56]

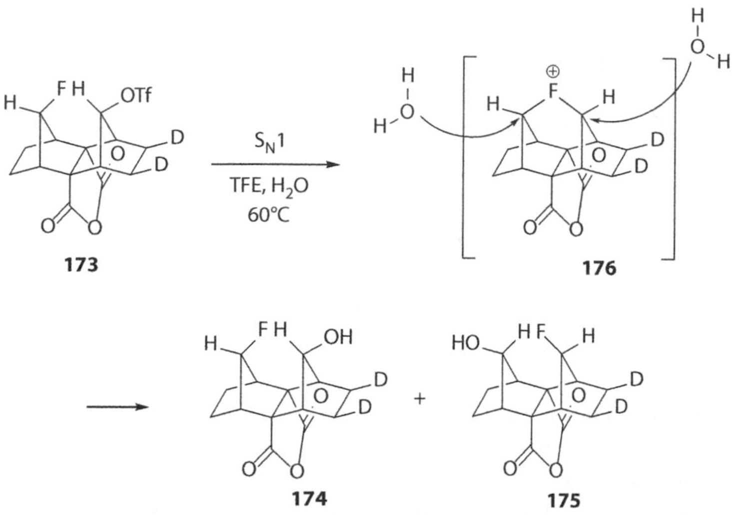

173   S$_N$1
TFE, H$_2$O
60°C

176

174   +   **175**

2.8 MISCELLANEOUS S$_N$1-RELATED REACTIONS

Using ionic liquid BmimBF$_4$ as the medium, the S$_N$1 reaction between 9H-xanthen-9-ol (**54**) and indole **177** took place at room temperature. Presumably, carbocation **178** was generated from **54** under the influence of BmimBF$_4$.[57] The S$_N$1 reaction of **178** with indole **177** occurred at the C-3 position of the indole ring to produce adduct **179**.

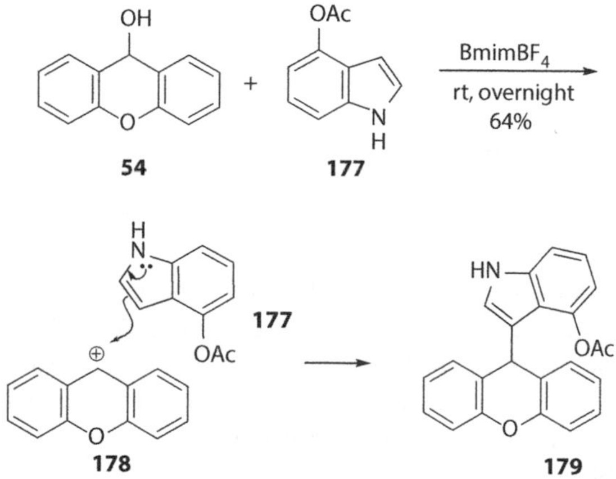

54   **177**   BmimBF$_4$
rt, overnight
64%

178   **179**

Azabicyclo[5.3.0]alkanone amino acids have been modified to prepare mimics of peptide β-turn backbone and side-chain geometry.[58] In particular, 6-iodo-pyrroloazepine-2-one amino ester **180** has served as a common precursor in reactions with a variety of nucleophiles to provide an array of constrained peptide mimics. The nucleophile included alcohols, phenols, nitrates, and azides. Nitrate **182** and azido-compound **183** were isolated in 89% and 76% yield, respectively, presumably via the intermediacy of secondary carbenium ion **181**.

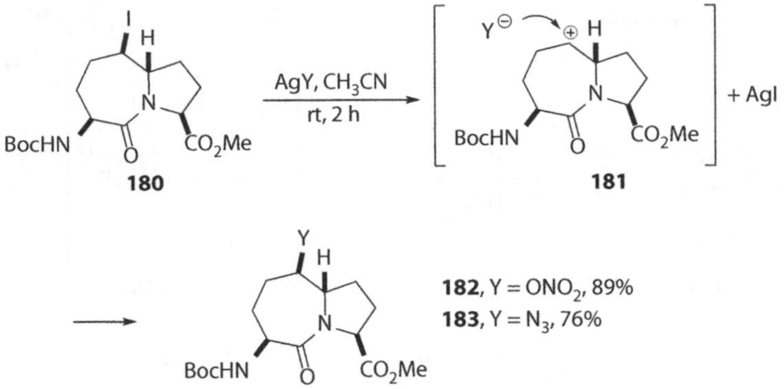

180 → **181** + AgI

Y = ONO$_2$, AgY, CH$_3$CN, rt, 2 h

182, Y = ONO$_2$, 89%
183, Y = N$_3$, 76%

Organophosphonates are a class of important compounds. One of their utilities is to serve as one of the two coupling partners in the Horner–Wadworth–Emmons reaction with aldehydes. The Michaelis–Arbuzov reaction, in turn, is the most frequently used method to prepare organophosphonates. Interestingly, a Lewis acid-mediated Michaelis–Arbuzov reaction at room temperature undergoes through the S$_N$1 mechanism. For instance, chiral benzyl alcohol **184** and chiral benzyl bromide **185** were treated with triethylphosphite and catalyzed by ZnBr$_2$ to give the corresponding phosphonates **186** and **187** as 1:1 mixtures of enantiomers.[59] This is in accord with an S$_N$1 pathway.

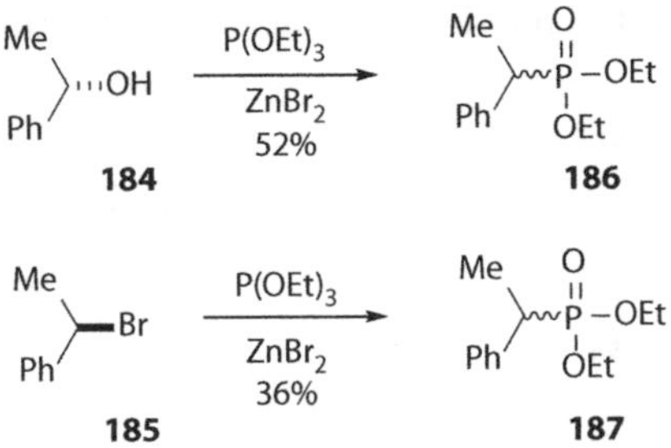

Multimetallic complexes are supposed to be superior sometimes in terms of reactivity and selectivity over mono-metal catalysts. A multimetallic complex [Cp*Ir-(SnCl$_3$)$_2${SnCl$_2$(H$_2$O)$_2$}, **188**] was able to promote γ-hydroxylactams via the intermediacy of N-acyliminium ions.[1] For instance, under the influence of **188**, serving as a Lewis acid, γ-hydroxylactam **189** was converted to α-amidoalkylation product **190**, presumably via the N-acyliminium ion intermediate **191** in an S$_N$1 fashion, being captured by a strong nucleophile, 2,5-dimethylfuran. The solvent for the reaction was dichloroethane (DCE).[60]

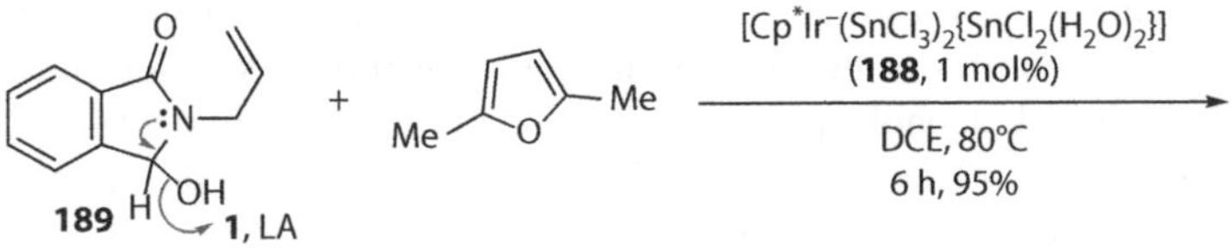

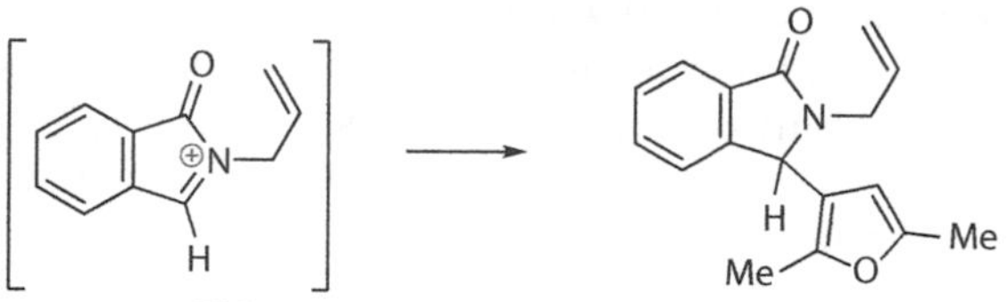

191 190

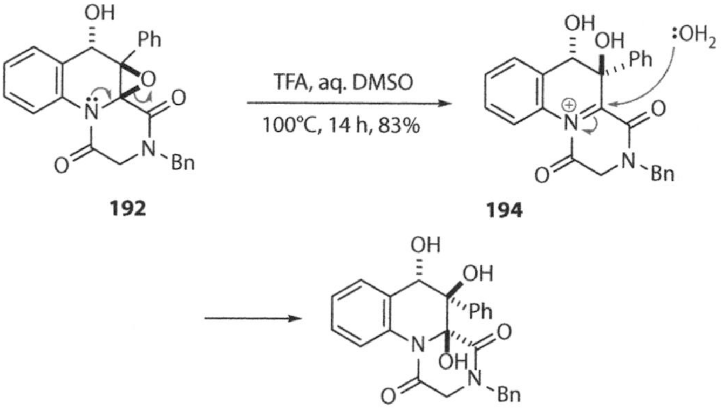

192 194

TFA, aq. DMSO
100°C, 14 h, 83%

193

Kutateladze et al. assembled oxirane **192** as a reactive versatile intermediate.[61] While in the presence of strong nucleophiles, S_N2 reactions prevailed, under the S_N1 conditions, that is, in the absence of strong nucleophiles in wet DMSO, *trans, cis*-triol **193** was generated through the intermediacy of iminium **194**.

Hydrolysis of epoxides with water generally takes place via the S_N1 mechanism because water is a weak nucleophile. For hydrolysis of (–)-α-pinene oxide (**195**), a stark contrast was observed when the reaction was carried out at different temperatures.[62] As shown below, when hydrolysis of **195** took place at room temperature, *trans*-sobrerol (**196**) was the sole product with its chiral integrity completely preserved. However, when the solvolysis was carried out in boiling water, *trans*-sobrerol (**196**) was determined to be 73% yield with complete scramble of the chirality (50:50). Meanwhile, *cis*-(±)-sobrerol (**197**) was produced in 27% yield.

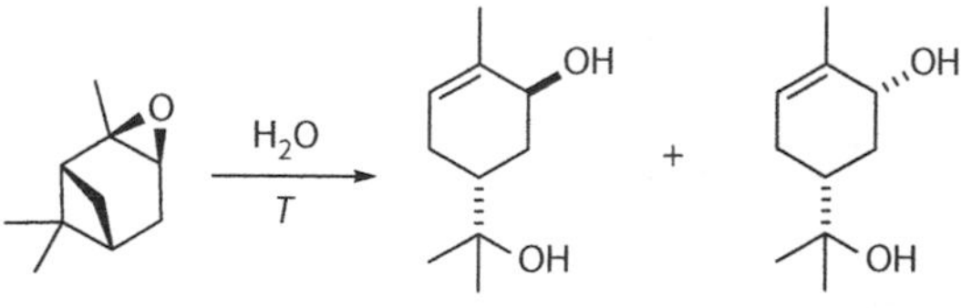

195	196	197
T = 22°C,	100	0
T = 100°C,	73	27

The aforementioned observation may be explained using the following mechanism. Epoxide ring opening of **195** is promoted by water. The resulting tertiary cation

198 rearranges to tertiary cation **199**. Presumably, the process is favored because the rearrangement released the four-membered ring strain associated with cation **198**, which traps water to afford *trans*-sobrerol (**196**). However, in hot water, *trans*-sobrerol (**196**) dehydrates easily, giving rise to allylic cation **200**, which traps water to give a mixture of diastereomers **197** and **196**.

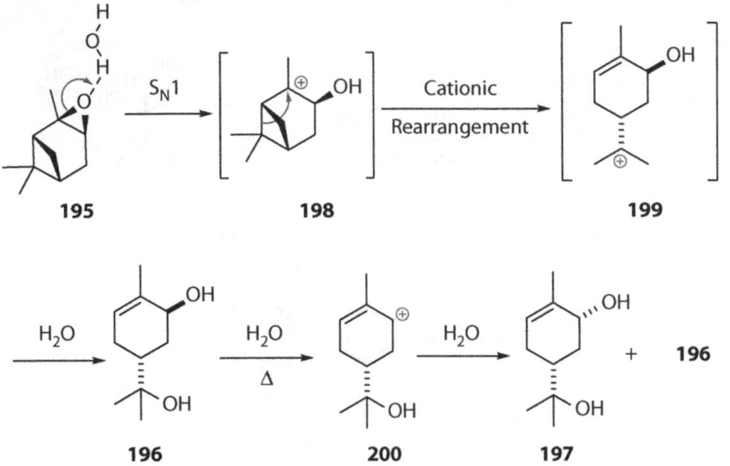

Pinacolatoboron fluoride (pinBF, **201**) was found to be an efficient fluoride transfer agent for diastereoselective synthesis of benzylic fluorides.[63] For instance, epoxide **202** chelated to the boron Lewis acid to give complex **203**, which invoked the epoxide cleavage in an S_N1 fashion to expose the benzylic cation **204**. An intramolecular fluoride transfer then delivered flurohydrin **205** in 55% yield and >95% dr.

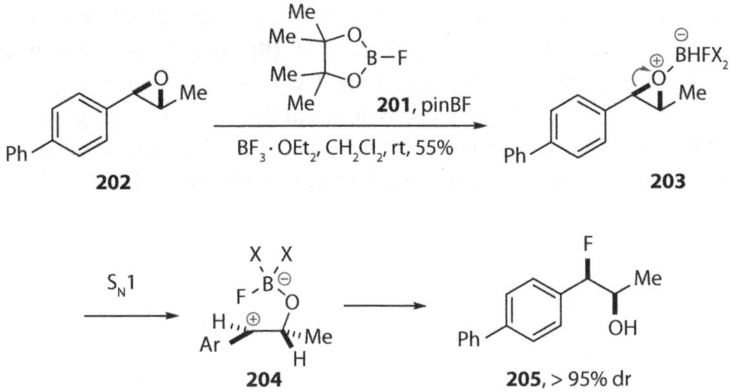

A Ritter reaction between α-methylene-β-hydroxy ester **206**, readily assembled from the Morita–Baylis–Hillman reaction, was carried out in acetonitrile and was mediated by sulfuric acid to assemble allylic acetamide **207**.[64] Mechanistically, the Ritter reaction in acidic medium proceeded via the intermediacy of an allylic cation. At first, allylic alcohol **206** is protonated by the acid to give **208**, which dehydrates to produce the key intermediate, allylic cation **209**. Experimental evidence and computational calculation indicated that carbenium ion **209′** prefers the $S_N1′$ pathway rather an S_N1 mechanism (after all, **207** is the $S_N1′$ product!) because $S_N1′$ path is *more* exergonic than the S_N1 path. The resulting $S_N1′$ adduct, *nitrilium ion* **210**, then captures water and delivers allylic acetamide **207**.

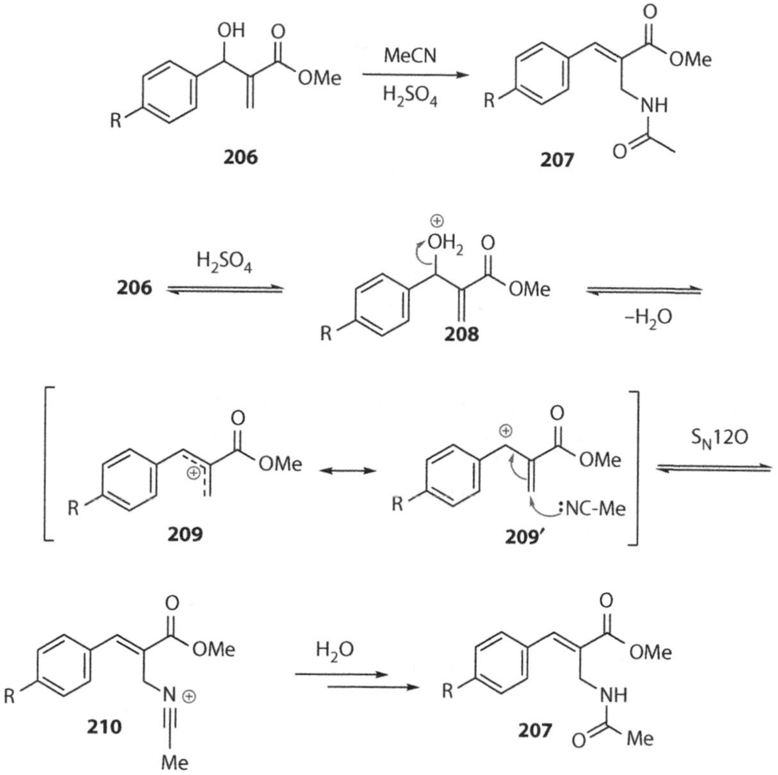

In the presence of two strong acids, H_2SO_4 and TFA, carbinol **211** was converted to 10-arylpyrrolo[1,2-b]isoquinoline **212** in a diastereo-controlled fashion with >95% de.[65] Mechanistically, benzyl alcohol **211** was at first protonated to give **213**, which loses a molecule of water to offer carbenium ion **214**. An intramolecular S_N1 reaction then afforded **212**.

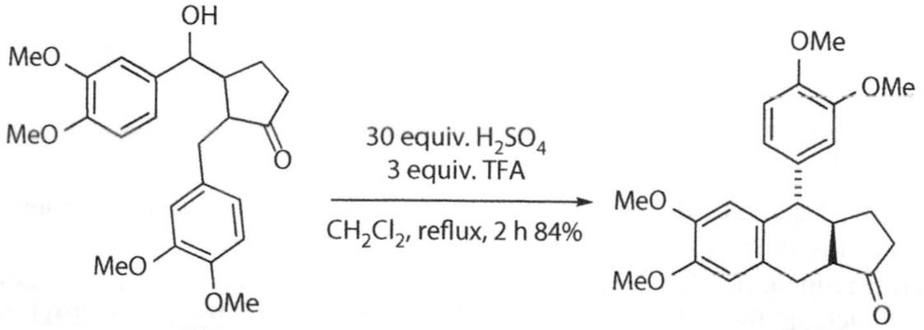

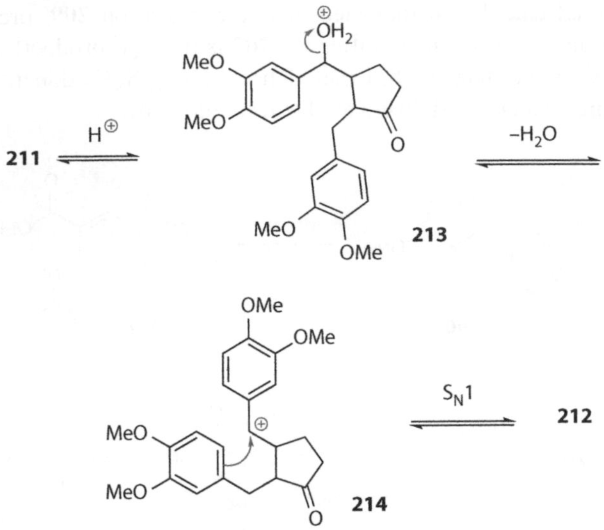

Woerpel et al. have developed an S_N1 reaction to construct the 1,2-dioxepane of 10,12-peroxycalamenene.[66] The treatment of catalytic amount of DDQ to silyl peroxide ether **215**, it led to desilylation followed by intramolecular S_N1-type seven-membered ring formation to yield **216** in a 60% overall yield. In such transformation, the mechanistic details of this reaction were unknown, but it was postulated that the DDQ played the roles as both a single-electron acceptor and a weak acid.

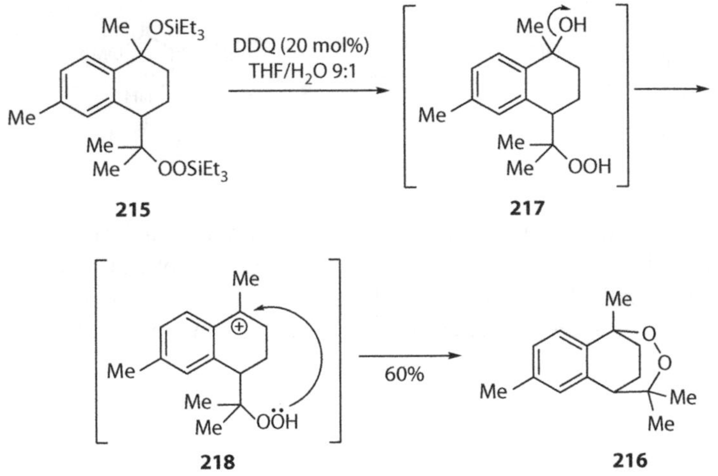

REFERENCES

1. Baeza, A.; Najera, C. Recent advances in the direct nucleophilic substitution of allylic alcohols through S_N1-type reactions. *Synthesis* **2014**, *46*, 25–34.
2. Emer, E.; Sinisi, R.; Capdevila, M. G.; Petruzziello, D.; De Vincentiis, F.; Cozzi, P. G. Direct nucleophilic S_N1-type reactions of alcohols. *Eur. J. Org. Chem.* **2011**, 647–666.

3. Alba, A.-N. R.; Calbet, T.; Font-Bardia, M.; Moyano, A.; Rios, R. Alkylation of oxazolones and related heterocycles through an S_N1 reaction. *Eur. J. Org. Chem.* **2011**, 2053–2056.

4. Gualandi, A.; Cozzi, P. G. Stereoselective organocatalytic alkylations with carbenium ions. *Synlett* **2013**, *24*, 281–296.

5. Trifonidou, M.; Kokotos, C. G. Enantioselective organocatalytic α-alkylation of ketones by S_N1-type reaction of alcohols. *Eur. J. Org. Chem.* **2012**, 1563–1568.

6. Petruzziello, D.; Gualandi, A.; Grilli, S.; Cozzi, P. G. Direct and stereoselective alkylation of nitro derivatives with activated alcohols in trifluoroethanol. *Eur. J. Org. Chem.* **2012**, 6697–6701.

7. Oukessou, M.; Genisson, Y.; El Arfaoui, D.; Ben-Tama, A.; El Hadrami, E. M.; Chauvin, R. The triazolyl analogue of the trityl cation. *Tetrahedron Lett.* **2013**, *54*, 4362–4364.

8. Zhang, T.; Qiao, Z.; Wang, Y.; Zhong, N.; Liu, L.; Wang, D.; Chen, Y.-J. Asymmetric direct [small alpha]-alkylation of 2-oxindoles with Michler's hydrol catalyzed by bis-cinchona alkaloid-Bronsted acid via an S_N1-type pathway. *Chem. Commun.* **2013**, *49*, 1636–1638.

9. Barbero, M.; Cadamuro, S.; Dughera, S.; Rucci, M.; Spano, G.; Venturello, P. Solvent-free Brønsted acid catalysed alkylation of arenes and heteroarenes with benzylic alcohols. *Tetrahedron* **2014**, *70*, 1818–1826.

10. Fernández-Valparís, J.; Romo, J.M.; Romea, P.; Urpí, F.; Kowalski, H.; Font-Bardia, M. Stereoselective alkylation of (*S*)-*N*-acyl-4-isopropyl-1,3-thiazolidine-2-thiones catalyzed by $(Me_3P)_2NiCl_2$. *Org. Lett.* **2015**, *17*, 3540–3543.

11. Bunrit, A.; Dahlstrand, C.; Olsson, S. K; Srifa, P.; Huang, G.; Orthaber, A.; Sjöberg, P. J. R.; Biswas, S.; Himo, F.; Samec, J. S. M. Brønsted acid-catalyzed intramolecular nucleophilic substitution of the hydroxyl group in stereogenic alcohols with chirality transfer. *J. Am. Chem. Soc.* **2015**, *137*, 4646 – 4649.

12. Mühlthau, F.; Schuster, O.; Bach, T. High facial diastereoselectivity in intra- and inter-molecular reactions of chiral benzylic cations. *J. Am. Chem. Soc.* **2005**, *127*, 9348–9349.

13. Mühlthau, F.; Stadler, D.; Goeppert, A.; Olah, G. A.; Prakash, G.-K. S.; Bach, T. Chiral α-branched benzylic carbocations: Diastereoselective intermolecular reactions with arene nucleophiles and NMR spectroscopic studies. *J. Am. Chem. Soc.* **2006**, *128*, 9668–9675.

14. Das, B. G.; Nallagonda, R.; Ghorai, P. Direct substitution of hydroxy group of π-activated alcohols with electron-deficient amines using Re_2O_7 catalyst. *J. Org. Chem.* **2012**, *77*, 5577–5583.

15. Nitsch, D.; Huber, S. M.; Poethig, A.; Narayanan, A.; Olah, G. A.; Prakash, G. K. S.; Bach, T. Chiral propargylic cations as intermediates in S_N1-type reactions: Substitution pattern, nuclear magnetic resonance studies, and origin of the diastereoselectivity. *J. Am. Chem. Soc.* **2014**, *136*, 2851–2857.

16. Hellal, M.; Falk, F. C.; Wolf, E.; Dryzhakov, M.; Moran, J. Breaking the dichotomy of reactivity vs. chemoselectivity in catalytic S_N1 reactions of alcohols. *Org. Biomol. Chem.* **2014**, *12*, 5990–5994.

17. Mo, X.; Yakiwchuk, J.; Dansereau, J.; McCubbin, J. A.; Hall, D. G. Unsymmetrical diarylmethanes by ferroceniumboronic acid catalyzed direct Friedel–Crafts reactions with deactivated benzylic alcohols: Enhanced reactivity due to ion-pairing effects. *J. Am. Chem. Soc.* **2015**, *137*, 9694–9703.

18. Ho, X.-H.; Mho, S.-I.; Kang, H.; Jang, H.-Y. Electro-organocatalysis: Enantioselective α-alkylation of aldehydes. *Eur. J. Org. Chem.* **2010**, 4436–4441.

19. Zhang, B.; Xiang, S.-K.; Zhang, L.-H.; Cui, Y.; Jiao, N. Organocatalytic asymmetric intermolecular dehydrogenative α-alkylation of aldehydes using molecular oxygen as oxidant. *Org. Lett.* **2011**, *13*, 5212–5215.

20. Xiao, J.; Zhao, K.; Loh, T.-P. Highly enantioselective intermolecular alkylation of aldehydes with alcohols by cooperative catalysis of diarylprolinol silyl ether with Brønsted acid. *Chem. Asian J.* **2011**, *6*, 2890–2894.
21. Gualandi, A.; Mengozzi, L.; Wilson, C. M.; Cozzi, P. G. Synergistic stereoselective organocatalysis with indium(III) salts. *Synthesis* **2014**, *46*, 1321–1328.
22. Gualandi, A.; Mengozzi, L.; Giacobni, J.; Saulnier, S.; Ciadi, M.; Cozzi, P. G. A practical and stereoselective organocatalytic alkylation of aldehydes with benzodithiolylium tetrafluoroborate. *Chirality* **2014**, *26*, 607–613.
23. Cozzi, P. G.; Benfatti, F.; Zoli, L. Organocatalytic asymmetric alkylation of aldehydes by S_N1-type reaction of alcohols. *Angew. Chem. Int. Ed.* **2009**, *48*, 1313–1316.
24. Benfatti, F.; Benedetto, E.; Cozzi, P. G. Organocatalytic stereoselective α-alkylation of aldehydes with stable carbocations. *Chem. Asian J.* **2010**, *5*, 2047–2052.
25. Brown, A.R.; Kuo, W.-H.; Jacobsen, E.N. Enantioselective catalytic α-alkylation of aldehydes via an S_N1 pathway. *J. Am. Chem. Soc.* **2010**, *132*, 9286–9288.
26. Xing, C.; Sun, H.; Zhang, J.; Li, G.; Chi, Y. R. Brønsted acid catalyzed α-alkylation of aldehydes with diaryl methyl alcohols. *Chem. Eur. J.* **2011**, *17*, 12272–12275.
27. Weng, Z.-T.; Li, Y.; Tian, S.-K. Catalytic asymmetric α-alkylation of ketones and aldehydes with *N*-benzylic sulfonamides through carbon–nitrogen bond cleavage. *J. Org. Chem. Soc.* **2011**, *76*, 8095–8099.
28. Stiller, J.; Vorholt, A. J.; Ostrowski, K. A.; Behr, A.; Christmann, M. Enantioselective tandem reactions at elevated temperatures: One-pot hydroformylation/S_N1 alkylation. *Chem. Eur. J.* **2012**, *18*, 9496–9499.
29. Song, J.; Guo, C.; Adele, A.; Yin, H.; Gong, L.-Z. Enantioselective organocatalytic construction of hexahydropyrroloindole by means of α-alkylation of aldehydes leading to the total synthesis of (+)-gliocladin C. *Chem. Eur. J.* **2013**, *19*, 3319–3323.
30. Sinisi, R.; Vita, M. V.; Gualandi, A.; Emer, E.; Cozzi, P. G. S_N1-type reactions in the presence of water: Indium(III)-promoted highly enantioselective organocatalytic propargylation of aldehydes. *Chem. Eur. J.* **2011**, *17*, 7404–7408.
31. Li, F.; Ma, J.; Wang, N. α-Alkylation of ketones with primary alcohols catalyzed by a Cp*Ir complex bearing a functional bipyridonate ligand. *J. Org. Chem.* **2014**, *79*, 10447–10455.
32. Trillo, P.; Baeza, A.; Nájera, C. Copper-catalyzed asymmetric alkylation of β-keto esters with xanthydrols. *Adv. Synth. Catal.* **2013**, *355*, 2815–2821.
33. Song, L.; Guo, Q.-X.; Li, X.-C.; Tian, J.; Peng, Y.-G. The direct asymmetric α alkylation of ketones by Brønsted acid catalysis. *Angew. Chem. Int. Ed.* **2012**, *51*, 1899–1902.
34. Bohé, L.; Crich, D. A propos of glycosyl cations and the mechanism of chemical glycosylation; the current state of the art. *Carbohydr. Res.* **2015**, *403*, 48–59.
35. Nigudkar, S. S.; Demchenko, A. V. Stereocontrolled 1,2-*cis* glycosylation as the driving force of progress in synthetic carbohydrate chemistry. *Chem. Sci.* **2015**, *6*, 2687–2704.
36. Tian, Q.; Dong, L.; Ma, X.; Xu, L.; Hu, C.; Zou, W.; Shao, H. Stereoselective synthesis of 2-*C*-branched (acetylmethyl) oligosaccharides and glycoconjugates: Lewis acid-catalyzed glycosylation from 1,2-cyclopropaneacetylated sugars. *J. Org. Chem.* **2011**, *76*, 1045–1053.
37. Huang, M.; Retailleau, P.; Bohé, L.; Crich, D. Cation clock permits distinction between the mechanisms of α- and β-O- and β-C-glycosylation in the mannopyranose series: Evidence for the existence of a mannopyranosyl oxocarbenium ion. *J. Am. Chem. Soc.* **2012**, *134*, 14746–14749.
38. Kendale, J. C.; Valentin, E. M.; Woerpel, K. A. Solvent effects in the nucleophilic substitutions of tetrahydropyran acetals promoted by trimethylsilyl trifluoromethanesulfonate: Trichloroethylene as solvent for stereoselective *C*- and *O*-glycosylations. *Org. Lett.* **2014,** *16*, 3684–3687.

39. Feng, Y.; Jiang, X., De Brabander, J. K. Studies toward the unique pederin family member psymberin: Full structure elucidation, two alternative total syntheses, and analogs. *J. Am. Chem. Soc.* **2012**, *134*, 17083–17093.

40. Tai, C.-A.; Kulkarni, S. S.; Hung, S.-C. Facile Cu(OTf)$_2$-catalyzed preparation of per-*O*-acetylated hexopyranoses with stoichiometric acetic anhydride and sequential one-pot anomeric substitution to thioglycosides under solvent-free conditions. *J. Org. Chem.* **2003**, *68*, 8719–8722.

41. Du, W.; Kulkarni, S. S.; Gervay-Hague, J. Efficient, one-pot syntheses of biologically active α-linked glycolipids. *Chem. Commun.* **2007**, 2336–2338.

42. (a) Walvoort, M. T. C.; Lodder, G.; Mazurek, J.; Overkleeft, H. S.; Codée, J. D. C.; van der Marel, G. A. Equatorial anomeric triflates from mannuronic acid esters. *J. Am. Chem. Soc.* **2009**, *131*, 12080–12081; (b) Walvoort, M. T. C.; Lodder, G.; Overkleeft, H. S.; Codée J. D. C.; van der Marel, G. A. Mannosazide methyl uronate donors. Glycosylating properties and use in the construction of β-ManNAcA-containing oligosaccharides. *J. Org. Chem.* **2010**, *75*, 7990–8002.

43. Kale, R. R.; McGannon, C. M.; Fuller-Schaefer, C.; Hatch, D. M.; Flagler, M. J.; Gamage, S. D.; Weiss, A. A.; Iyer, S. S. Differentiation between structurally homologous Shiga 1 and Shiga 2 toxins by using synthetic glycoconjugates. *Angew. Chem. Int. Ed.* **2008**, *47*, 1265–1268.

44. Matsuzawa, H.; Miyake, Y.; Nishibayashi, Y. Ruthenium-catalyzed enantioselective propargylation of aromatic compounds with propargylic alcohols via allenylidene intermediates. *Angew. Chem. Int. Ed.* **2007**, *46*, 6488–6491.

45. Watanabe, N.; Matsugi, A.; Nakano, K.; Ichikawa, Y.; Kotsuki, H. Organocatalytic Friedel–Crafts benzylation of heteroaromatic and aromatic compounds via an S$_N$1 pathway. *Synlett* **2014**, *25*, 438–442.

46. Wen, H.; Wang, L.; Xu, L.; Hao, Z.; Shao, C.-L.; Wang, C.-Y.; Xiao, J. Fluorinated alcohol-mediated S$_N$1-type reaction of indolyl alcohols with diverse nucleophiles. *Adv. Synth. Catal.* **2015**, *357*, 4023–4030.

47. Desroches, J.; Champagne, P. A.; Benhassine, Y.; Paquin, J.-F. *In situ* activation of benzyl alcohols with XtalFluor-E: Formation of 1,1-diarylmethanes and 1,1,1-triarylmethanes through Friedel–Crafts benzylation. *Org. Biomol. Chem.* **2015**, *13*, 2243–2246.

48. Sun, H.-B.; Li, B.; Hua, R.; Yin, Y. An efficient and selective hydroarylation of styrenes with electron-rich arenes, catalyzed by bismuth(III) chloride and affording Markovnikov adducts. *Eur. J. Org. Chem.* **2006**, 4231–4236.

49. Shirakawa, S.; Kobayashi, S. Surfactant-type Brønsted acid catalyzed dehydrative nucleophilic substitutions of alcohols in water. *Org. Lett.* **2007**, *9*, 311–314.

50. Kennedy-Smith, J. J.; Young, L. A.; Toste, F. D. Rhenium-catalyzed aromatic propargylation. *Org. Lett.* **2004**, *6*, 1325–1327.

51. Theresa, L.; Neumann, C. N.; Ritter, T. Introduction of fluorine and fluorine-containing functional groups. *Angew. Chem. Int. Ed.* **2013**, *52*, 8214–8264.

52. Slebocka-Tilk, H.; Ball, R. G.; Brown, R. S. The question of reversible formation of bromonium ions during the course of electrophilic bromination of olefins. 2. The crystal and molecular structure of the bromonium ion of adamantylideneadamantane. *J. Am. Chem. Soc.* **1985**, *107*, 4504–4508.

53. Stoyanov, E. S.; Stoyanova, I. V.; Tham, F. S.; Reed, C. A. Dialkyl chloronium ions. *J. Am. Chem. Soc.* **2010**, *132*, 4062–4063.

54. Hennecke, U. Revealing the positive side of fluorine. *Science* **2013**, *340*, 41–42.

55. Struble, M. D.; Scerba, M. T.; Siegler, M.; Lectka, T. Evidence for a symmetrical fluoronium ion in solution. *Science* **2013**, *340*, 57–60.

56. Struble, M. D.; Holl, M. G.; Scerba, M. T.; Siegler, M.; Lectka, T. Search for a symmetrical C-F-C fluoronium ion in solution: Kinetic isotope effects, synthetic labeling, and computational, solvent, and rate studies. *J. Am. Chem. Soc.* **2015**, *137*, 11476–11490.

57. Liu, L.-Y.; Wang, B.; Yang, H.-M.; Chang, W.-X.; Li, J. The direct substitutions of 9*H*-xanthen-9-ol with indoles in a room temperature ionic liquid medium BmimBF$_4$. *Tetrahedron Lett.* **2011,** *52*, 5636–5639.
58. Godina, T. A.; Lubell, W. D. Mimics of peptide turn backbone and side-chain geometry by a general approach for modifying azabicyclo[5.3.0]alkanone amino acids. *J. Org. Chem.* **2011,** *76*, 5846–5849.
59. Rajeshwaran, G. G.; Nandakumar, M.; Sureshbabu, R.; Mohanakrishnan, A. K. Lewis Acid-mediated Michaelis–Arbuzov reaction at room temperature: A facile preparation of arylmethyl/heteroarylmethyl phosphonates. *Org. Lett.* **2011,** *13*, 1270–1273.
60. Maity, A. K.; Roy, S. A. Multimetallic piano-stool Ir–Sn$_3$ catalyst for nucleophilic substitution reaction of γ-hydroxy lactams through *N*-acyliminium ions. *J. Org. Chem.* **2012,** *77*, 2935–2941.
61. Kumar, N. N. B.; Kuznetsov, D. M.; Kutateladze, A. G. Intramolecular cycloadditions of photogenerated azaxylylenes with oxadiazoles provide direct access to versatile polyheterocyclic ketopiperazines containing a spiro-oxirane moiety. *Org. Lett.* **2015,** *17*, 438–441.
62. Xu, Z.-B.; Qu, J. Hot water-promoted S$_N$1 solvolysis reactions of allylic and benzylic alcohols. *Chem. Eur. J.* **2013,** *19*, 314–323.
63. Cresswell, A. J.; Davies, S. G.; Figuccia, A. L. A.; Fletcher, A. M.; Heijnen, D.; Lee, J. A.; Morris, M. J.; Kennett, A. M. R.; Roberts, P. M.; Thomson, J. E. Pinacolatoboron fluoride (pinBF) is an efficient fluoride transfer agent for synthesis of benzylic fluorides. *Tetrahedron Lett.* **2015,** *56*, 3373–3377.
64. Sá, M. M.; Ferreira, M.; Caramori, G. F.; Zaramello, L.; Bortoluzzi, A. J.; Faggion, D., Jr.; Domingos, J. B. Investigating the Ritter type reaction of α-methylene-β-hydroxy esters in acidic medium: Evidence for the intermediacy of an allylic cation. *Eur. J. Org. Chem.* **2013,** 5180–5187.
65. Couto, I.; Pardo, L. M.; Tellitu, I.; Domínguez, E. A Diastereocontrolled route to 10-arylpyrrolo[1,2-*b*]isoquinolines. *J. Org. Chem.* **2012,** *77*, 11192–11199.
66. Hilf, J. A.; Witthoft, L. W.; Woerpel, K. A. An S$_N$1-type reaction to form the 1,2-dioxepane ring: Synthesis of 10,12-peroxycalamenene. *J. Org. Chem.* **2015,** *80*, 8262–8267.

3 Nucleophilic Aliphatic Substitution

S_N2

John W. Lippert III

CONTENTS

The inversion of stereochemistry in a synthetic scheme is a powerful tool for an organic chemist. The synthetic intermediate in this mechanism lacks true carbocation character, and only a partial positive charge exists. Nonetheless, within this chapter are multiple examples from the recent literature that use synthetic modifications with the S_N2 mechanism under a variety of different conditions.

3.1 CONSTRUCTION OF QUATERNARY STEREOGENIC CENTERS

The S_N2 reaction, one of the most fundamental reactions in organic transformation, can be used for the construction of enantioenriched chiral carbon centers because the substitution proceeds with the complete inversion of stereochemistry.[1] Shibatomi et al.[1] used optically active acid chlorides for the preparation of a variety of chiral molecules having a quaternary stereogenic center. These intermediates would be useful in a synthetic scheme if S_N2 substitution were to occur at the chlorinated tertiary carbon.[1] Prior to this chemistry, the S_N2 displacement of chlorinated carbons has failed due to two major issues.[1] The first is that tertiary halides rarely undergo S_N2 substitution.[1] The second is that there are relatively few catalytic methods that afford highly enantioenriched tertiary chlorides.[1] The researchers afforded the enantioselective chlorination of a variety of β-keto esters, and subsequently used

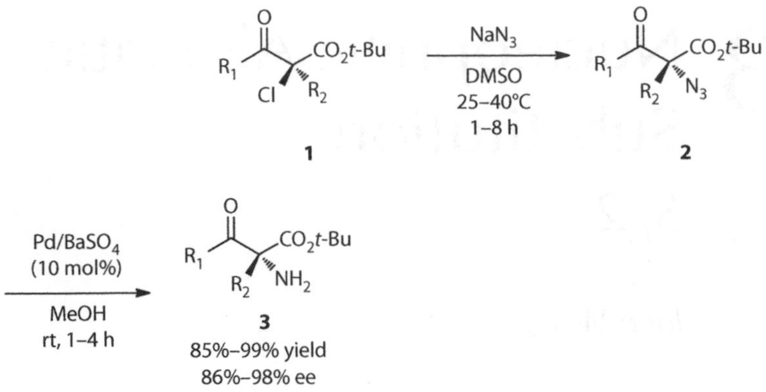

SCHEME 3.1 S_N2 substitution with sodium azide and subsequent hydrogenolysis to give α-amino acid derivatives.[1]

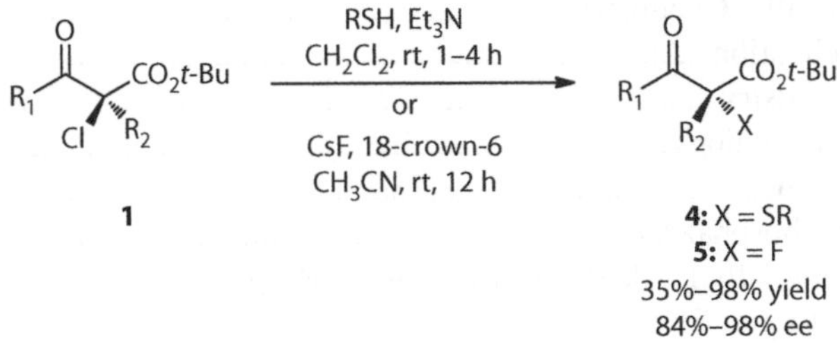

SCHEME 3.2 S_N2 reaction with alkylthiols and cesium fluoride.[1]

an unimpeded S_N2 substitution at the chlorinated tertiary carbon.[1] This chemistry is illustrated in Schemes 3.1 and 3.2.

3.2 SULFUR CHEMISTRY

Avenoza et al.[2] did extensive synthetic work on the reaction of sulfur nucleophiles with hindered sulfamidates. Cyclic sulfamidates have been used as reactive intermediates in organic synthesis because most ring-opening reactions of cyclic sulfamidates with nucleophiles proceed by the S_N2 pathway with total inversion at the stereogenic center.[2] Scheme 3.3 illustrates the synthetic utility of this methodology using a cyclic sulfamidate.

Robertson and Wu[3] prepared a variety of chiral tetrahydrothiophenes (THTs) in high enantiopurity via phosphorothioic acids and related compounds. The authors consider the methods developed to be a safer alternative to the use of H_2S, which is a highly toxic gas.[3] Each THT is derived from a common intermediate, thereby greatly enhancing the flexibility of the synthesis.[3] The key transformation is a base promoted, intramolecular, carbon–sulfur bond-forming event.[3] These reactions are highly stereospecific as they operate through a double S_N2 displacement mechanism.[3] A representation of the synthesis is shown in Scheme 3.4.

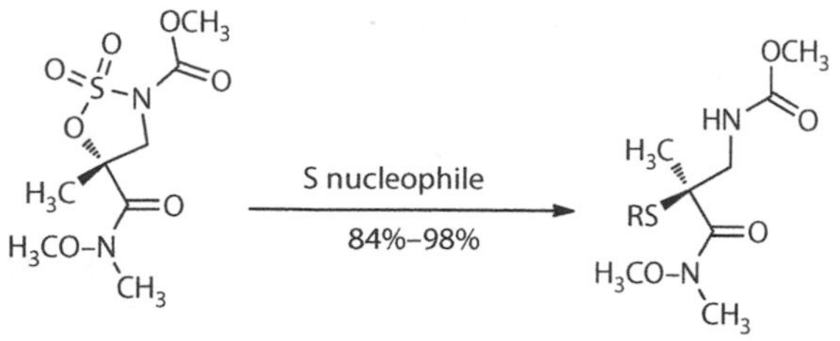

SCHEME 3.3 S_N2 reaction of cyclic sulfamidate (*R*)-**6** with different sulfur nucleophiles.[2]

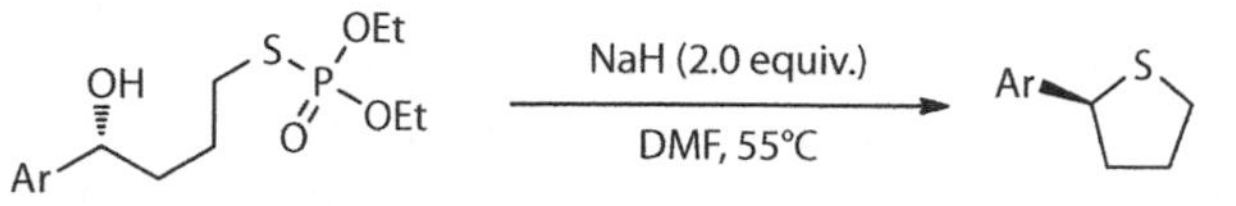

SCHEME 3.4 Scope of chiral THT formation.[3]

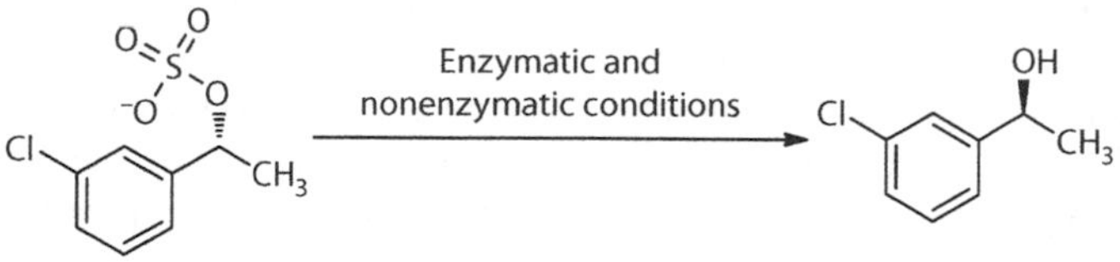

SCHEME 3.5 Hydrolysis of benzylic *sec*-sulfate esters.[4]

Toesch et al.[4] showed that the stereochemistry of enzymatic and nonenzymatic hydrolysis of benzylic *sec*-sulfate esters proceeded by S_N2 inversion at the benzylic carbon to give the major product shown (Scheme 3.5). The suppression of the other possible products was carried out with the addition of DMSO as the cosolvent.[4]

3.3 ORGANOMETALLIC CHEMISTRY

Zhu et al.[5] developed a palladium-catalyzed intramolecular aminohydroxylation of alkenes, in which H_2O_2 was applied as the sole oxidant. A variety of alkyl alcohols could be successfully obtained with good yields and excellent diastereoselectivities, which directly derived from oxidation cleavage of alkyl C–Pd bond by H_2O_2.[5] Facile transformation of these products provided a powerful tool toward the synthesis of 2-amino-1,3-diols and 3-ol amino acids.[5] Preliminary mechanistic studies revealed that major nucleophilic attack of water (S_N2 type) at high-valent Pd center contributes to the final C–O(H) bond formation.[5] Scheme 3.6 illustrates the aforementioned chemistry and the reaction scope.

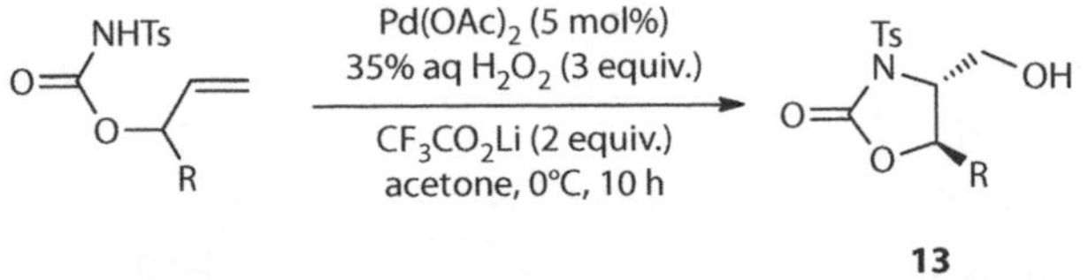

12

13
50%–95% yield
>20:1 de

SCHEME 3.6 Pd-catalyzed intramolecular aminohydroxylation of alkenes with hydrogen peroxide as oxidant and water as nucleophile.[5]

Fang et al.[6] afforded a general inversion-stereospecific, *N*-selective alkylation of substituted 2-pyridones (and analogs), amides, and carbamates using chiral α-chloro or bromocarboxylic acids in the presence of KO*t*-Bu (or KHMDS) and Mg(O*t*-Bu)$_2$. The researchers isolated the resulting α-chiral carboxylic acid products in high chemical yields and high ee (>90% ee).[6] Mechanistic evidence suggested that the reaction proceeds through 2-pyridone 0-coordinated Mg carboxylate intermediates, which afford the product through an intramolecular S_N2 alkylation.[6] Scheme 3.7 illustrates this chemistry.[6]

Katcher and Doyle[7] described the enantioselective fluorination of readily available cyclic allylic chlorides with AgF using a Pd(0) catalyst and Trost bisphosphine ligand. The reactions proceed with unprecedented ease of operation for Pd-mediated nucleophilic fluorination, allowing access to highly enantioenriched cyclic allylic fluorides that bear diverse functional groups.[7] The mechanism proceeds by an S_N2-type attack of fluoride on a Pd(II)-allyl intermediate to afford the C–F bond.[7] Scheme 3.8 summarizes this chemistry.

Ghorai et al.[8] described a highly efficient strategy for the Cu(OTf)$_2$-mediated S_N2-type nucleophilic ring-opening followed by the [4 + 2] cycloaddition reactions of a

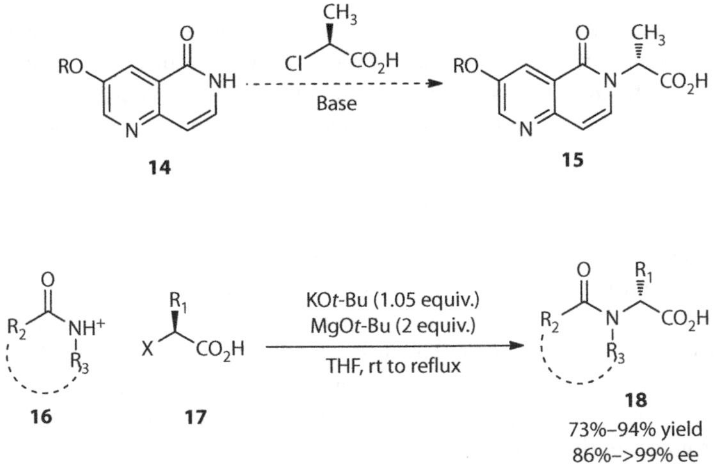

14

15

16

17

18
73%–94% yield
86%–>99% ee

SCHEME 3.7 Synthetic route and reaction scope for enantioenriched *N*-alkylated products.[6]

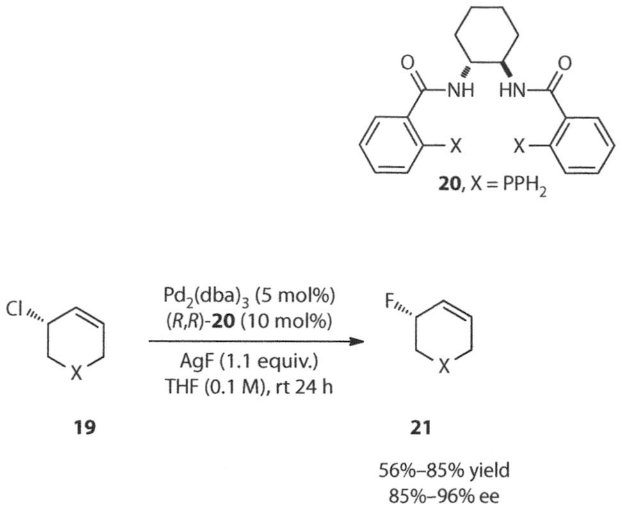

20, X = PPH$_2$

19 → 21

Pd$_2$(dba)$_3$ (5 mol%)
(R,R)-**20** (10 mol%)

AgF (1.1 equiv.)
THF (0.1 M), rt 24 h

56%–85% yield
85%–96% ee

SCHEME 3.8 Palladium-catalyzed fluorination of allylic chlorides.[7]

number of 2-aryl-N-tosylazetidines with nitriles to afford a variety of substituted tetrahydropyrimidines in excellent yields. The resulting tetrahydropyrimidines could easily be transformed into synthetically important 1,3-diamines by acid-catalyzed hydrolysis.[8] The strategy has been extended to the synthesis of enantiomerically pure tetrahydropyrimidines from enantiopure disubstituted azetidines.[8] Scheme 3.9 illustrates the cycloaddition of 2,4-disubstituted-N-tosylazetidines with acetonitrile which proceeds through a S$_N$2 mechanism.[8]

Trost et al.[9] published an article describing the palladium-catalyzed decarboxylative asymmetric allylic alkylation (DAAA) of allyl enol carbonates as highly chemo-, region-, and enantioselective process for the synthesis of ketones bearing either a quaternary or a tertiary α-stereogenic center. Their mechanistic studies revealed that, similar to the direct allylation of lithium enolates, the DAAA reaction proceeds through an "outer sphere" S$_N$2-type attack on the π-allylpalladium complex by the enolate.[9] Scheme 3.10 illustrates this chemistry.

Geier et al.[10] presented a reaction of a complex Pt organometallic species with electrophilic halogen sources in the presence of X⁻ ligands which changes the mechanism of reductive elimination from a concerted reductive coupling type to

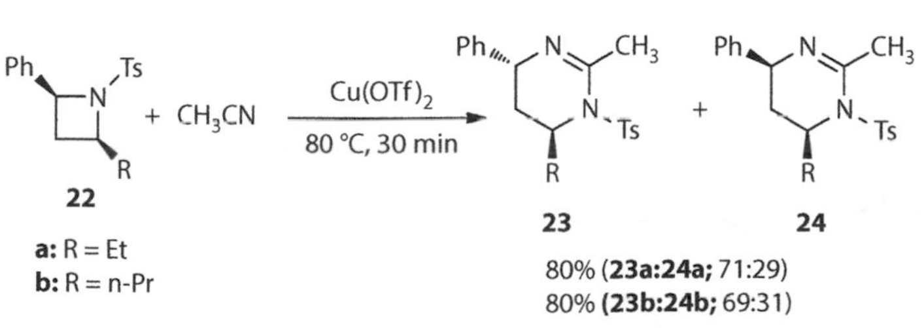

22 + CH$_3$CN $\xrightarrow{\text{Cu(OTf)}_2}$ 23 + 24
80 °C, 30 min

a: R = Et
b: R = n-Pr

80% (**23a:24a**; 71:29)
80% (**23b:24b**; 69:31)

SCHEME 3.9 Cycloaddition of 2,4-disubstituted-N-tosylazetidines with acetonitrile.[8]

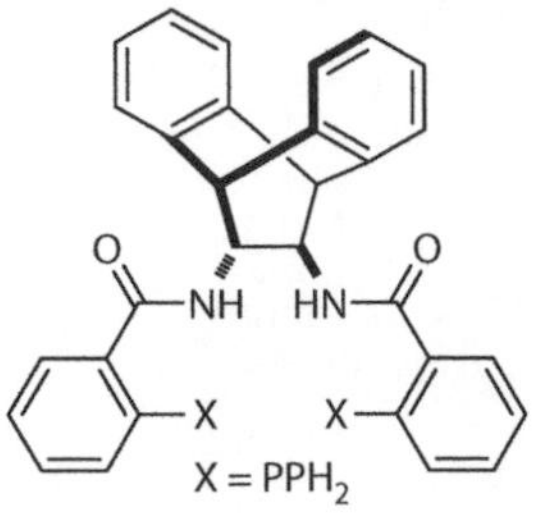
O
NH
HN
O
X
X
X = PPH$_2$
(*R,R*)-**26**

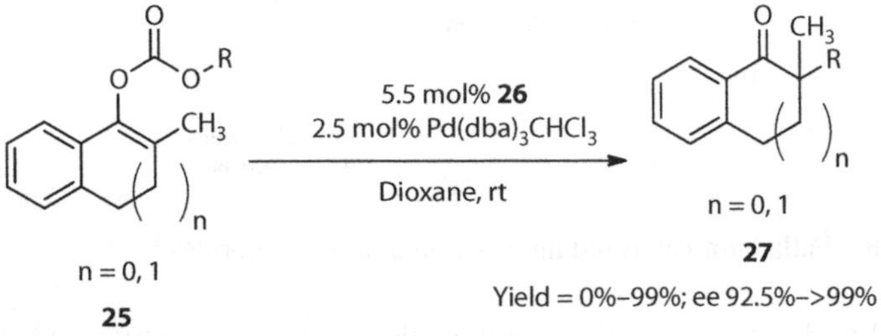
O
O
O
R
CH$_3$
n
n = 0, 1
25
5.5 mol% **26**
2.5 mol% Pd(dba)$_3$CHCl$_3$
Dioxane, rt
O
CH$_3$
R
n
n = 0, 1
27
Yield = 0%–99%; ee 92.5%–>99%

SCHEME 3.10 Palladium-catalyzed decarboxylative AAA of substituted allyl enol carbonates.[9]

an S_N2-type reductive elimination. In the absence of the X^- ligand the reductive elimination is stereoretentive; in its presence, the process is stereoinvertive.[10] This selectivity hinges on the reactivity of a key five-coordinate Pt(IV) intermediate with the X^- ligand.[10] These findings are summarized in Scheme 3.11.

Thompson and Dong[11] described the alkylation of rhodium(III) porphyrins [Rh^{III}(por)] which was achieved under relatively mild conditions in up to 98% yields,

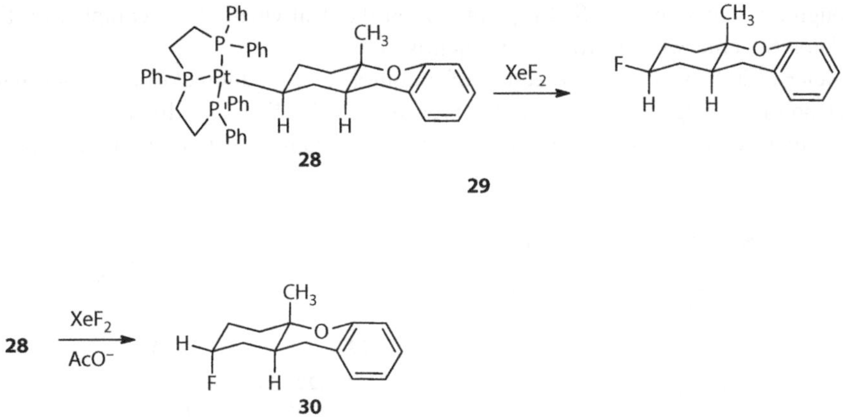
Ph
P
Ph
Ph
P
Pt
P
Ph
Ph
CH$_3$
O
H
H
28
XeF$_2$
CH$_3$
F
O
H
H
29
28
XeF$_2$
AcO$^-$
CH$_3$
H
O
F
H
30

SCHEME 3.11 Pt reductive elimination.[10]

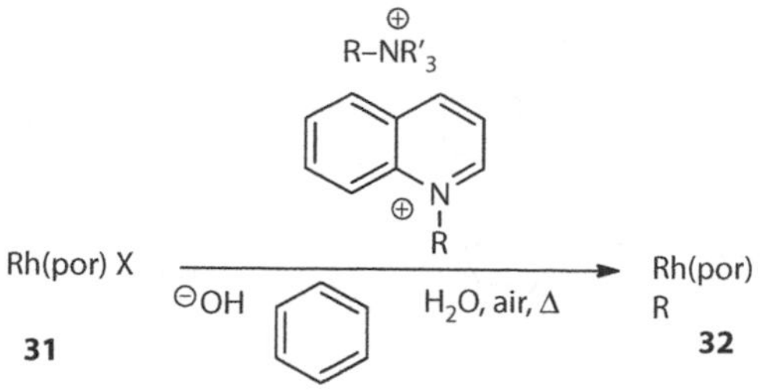

SCHEME 3.12 Alkylation of rhodium porphyrins.[11]

where readily available ammonium and quinolinium salts were utilized as the alkylating agents. This transformation was shown to tolerate air and water, thus serving as a convenient method to prepare a variety of alkyl- and benzyl-Rh^{III}(por) complexes.[11] Preliminary mechanistic studies support an S_N2-like reaction pathway involving Rh^I(por) anion intermediate.[11] Scheme 3.12 illustrates this chemistry.

Nookaraju et al.[12] presented a simple and novel synthesis of (+)-monocerin which was achieved in 15 steps and 15.5% overall yield from 3-buten-1-ol employing hydrolytic kinetic resolution, Julia olefination, intramolecular tandem Sharpless asymmetric dihydroxylation-S_N2 cyclization, and a novel copper-mediated tandem cyanation–cyclization as the key steps. Scheme 3.13 illustrates the relevant chemistry.

Shafaatian and Heidari[13] described a new organoplatinum(II) complex containing 2,2'-biquinoline ligand (biq) synthesized from the reaction of [Pt(p-MeC$_6$H$_4$)$_2$(SMe$_2$)$_2$ with 2,2'-biquinoline in a 1:1 molar ratio. In this complex the ligand was coordinated to the metal via the chelating nitrogen donor atoms.[13] Also, platinum(IV) complex was obtained by the oxidative addition reaction of methyl iodide with the platinum(II) complex in acetone.[13] The kinetic studies of the oxidative addition reaction of methyl iodide with the platinum(II) complex in different temperatures were investigated.[13] It was indicated that the reaction occurred by the S_N2 mechanism. Scheme 3.14 illustrates this oxidative addition.

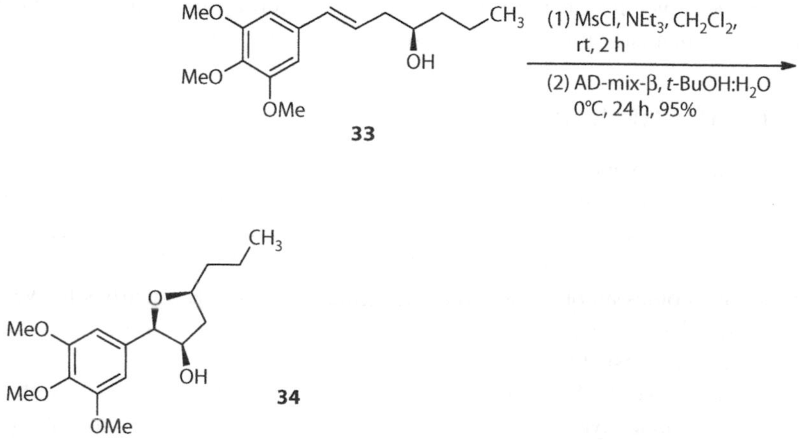

SCHEME 3.13 Tandem asymmetric dihydroxylation-S_N2 cyclization.[12]

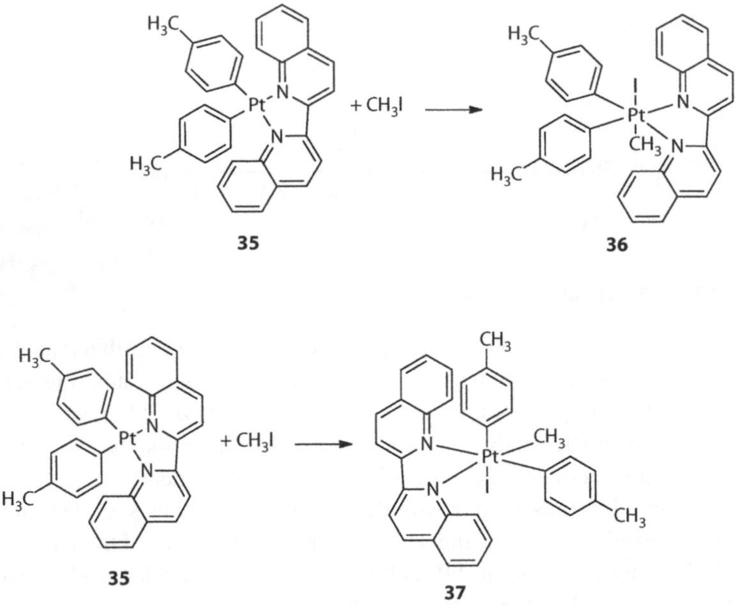

35   **36**

35   **37**

SCHEME 3.14 Oxidative addition using methyl iodide.[13]

3.4 MACROCYCLIZATION

Marti-Centelles et al.[14] synthesized a family of pseudopeptidic macrocycles containing nonnatural amino acids. The overall macrocyclization reaction was studied both
experimentally and computationally.[14] This reaction proceeded via a S_N2 reaction
mechanism.[14] The researchers studied the effects of both the amino acid side chain
and the bromide anion as it pertained to the overall macrocyclization.[14] The main
conclusion found that use of the cyclohexylalanine side chain within the amino acid
portion of the molecule gave the highest yield of macrocyclization.[14] Scheme 3.15
illustrates the chemistry used to synthesize a variety of macrocyclic products.

3.5 SUGAR CHEMISTRY

Issa and Bennett[15] worked on a tosylate S_N2-glyosylation for the direct synthesis
of β-linked 2-deoxy-sugars. Starting with 2-deoxy-sugars the researchers used
p-toluenesulfonic anhydride to afford an α-glycosyl tosylate which undergoes nucleophilic attack to afford the β-anomer exclusively (Scheme 3.16).[15] This powerful synthetic tool overcomes an obstacle which has puzzled synthetic chemists for years.[15]
Previously the β-linked deoxy sugars have been uncontrolled with the stereochemical information intrinsic to the glycosyl donor.[15]

Jiang et al.[16] described a highly efficient methodology for the stereo-controlled
synthesis of aminodeoxyalditols through a dimesylation/intramolecular S_N2 nucleophilic substitution ring-forming reaction sequence from glucose, mannose, and gla-
ctose derivatives. The degree of difficulty and the rate of the reaction of dimesylates

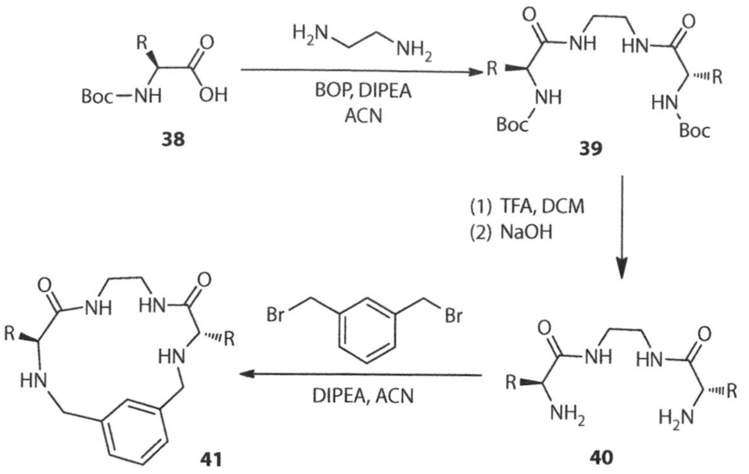

SCHEME 3.15 [1 + 1] Macrocylic structures obtained from C$_2$-symmetrical open-chain pseudopeptides.[14]

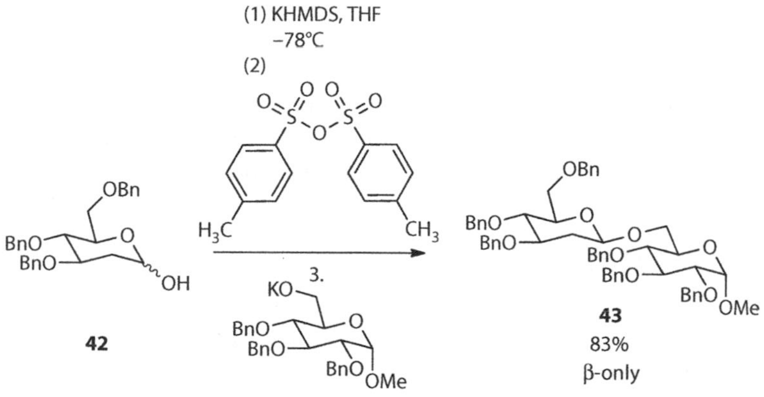

SCHEME 3.16 Reaction optimization with carbohydrate acceptors.[15]

in the cyclization process was evaluated in the text and shows the trend: galactose > glucose > mannose derivatives, which has been identified from the researchers experiments and classical theories.[16] Scheme 3.17 illustrates the synthesis using the galactose derivative.

3.6 NUCLEOSIDE ANALOGS

Prevost et al.[17] reported a highly diastereoselective route to 1′,2′-*cis*-nucleoside analogs in the D-ribo, D-lyxo, D-xylo, and D-arabinoside series. Five-membered ring lactols undergo highly selective *N*-glycosidation reactions in the presence of dimethylboron bromide with different silylated nucleobases.[17] Stereoelectronic control plays

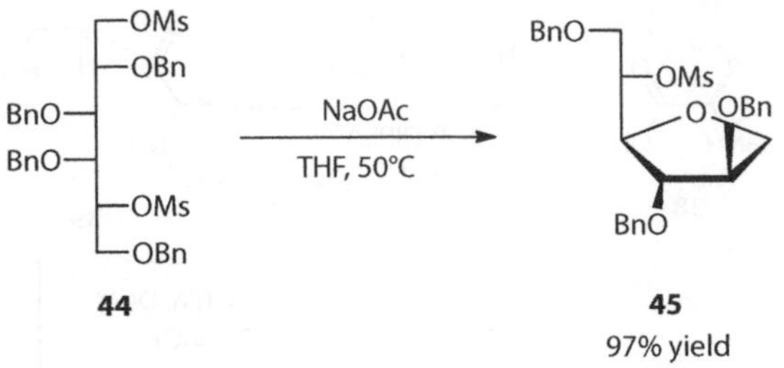

SCHEME 3.17 Formation of a galactose dervative.[16]

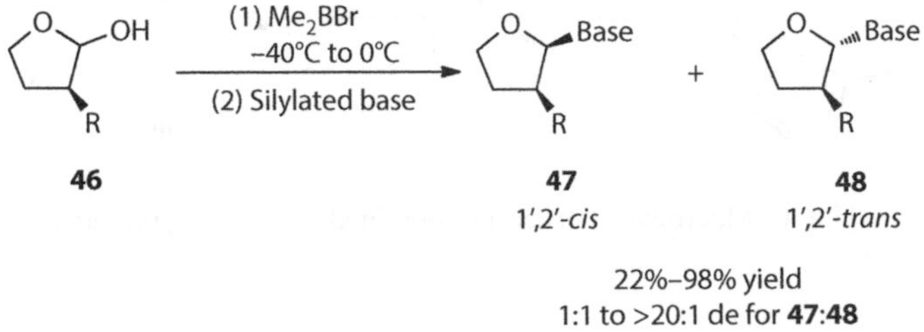

22%–98% yield
1:1 to >20:1 de for **47:48**

SCHEME 3.18 *N*-glycosidation of C-2-substituted lactols.[17]

a crucial role for the observed induction, and the products are proposed to be formed through S$_N$2 "exploded" transition states.[17] Scheme 3.18 illustrates this chemistry as it applies to the *N*-glycosidation of C-2-substituted lactols.

3.7 *N*-ALKYLATION

Zeinyeh et al.[18] showed that regioselectivities were determined for *N*-alkylations of imidazo[4,5-b]pyridine-4-oxide and 2-methyl-imidazo[4,5-b]pyridine-4-oxide with benzyl bromide or benzyl iodide at RT using K$_2$CO$_3$ in DMF as a base. Experimental attempts have shown that N-1/N-3 ratios slightly varied according to the substitution on C-2 position.[18] This was confirmed by DFT calculations in solvent phase.[18] This computational study has shown first that this *N*-benzylation reaction passed through a S$_N$2 mechanism.[18] Moreover, regioselectivity of *N*-benzylation has appeared essentially governed by "steric approach control."[18] It explained that opposite site N-1/N-3 ratios were obtained with imidazo[4,5-b]pyridine-4-oxide and its 2-methyl-substituted analog.[17] Finally, regioselectivities varied slightly with the nature of benzyl halide.[18] Scheme 3.19 illustrates the *N*-benzylation of imidazo[4,5-b]pyridine-4-oxide and its 2-methyl derivative.

3.8 CYCLOTETRAPHOSPHAZENES

Ciftci et al.[19] presented a study where 2-naphthylamine substituted cyclotetraphosphazenes were synthesized and characterized for the first time. The reaction of

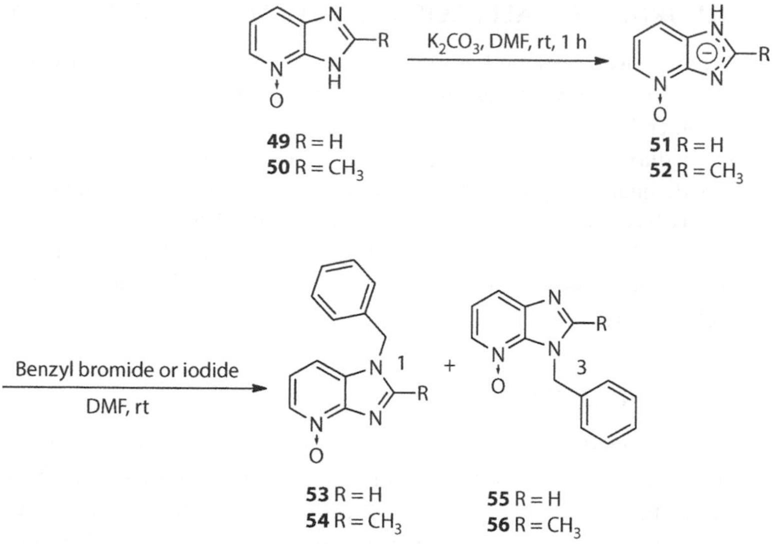
K2CO3, DMF, rt, 1 h
49 R = H
50 R = CH3
51 R = H
52 R = CH3
Benzyl bromide or iodide
DMF, rt
53 R = H
54 R = CH3
55 R = H
56 R = CH3

SCHEME 3.19 *N*-benzylation of 3*H*-imidazo[4,5-*b*]pyridine-4-oxide derivatives 53–56.[18]

octachlorocyclotetraphosphazene (**57**) with 2-naphthylamine (**58**) was performed in a THF (tetrahydrofuran) solution and gave eight products.[19] All of the 2-naphthylamine substituted cyclotetraphosphazene compounds were fully characterized.[19] A number of these products were formed by both S_N1 and S_N2 reaction mechanism.[19] The mechanisms were supported by ^{31}P NMR (nuclear magnetic resonance) and x-ray crystallography results. A representative reaction for the conversion using an S_N2 mechanism is shown in Scheme 3.20.

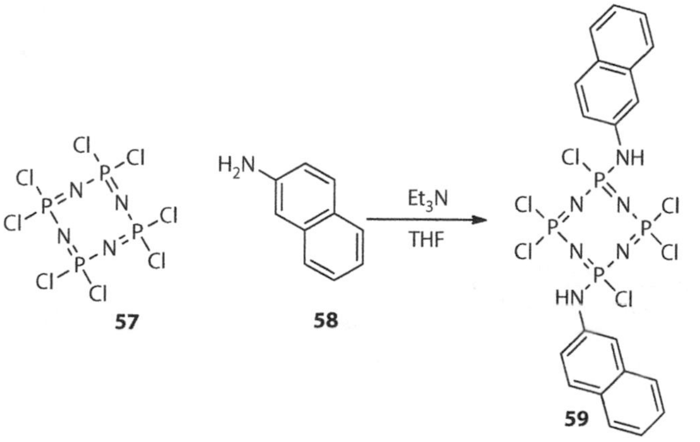
Et3N
THF
57
58
59

SCHEME 3.20 Representative formation of a 2-naphthylamine substituted cyclotetraphosphazene using S_N2 chemistry.[19]

3.9 CONFORMATIONALLY LOCKED TETRAHYDROPYRAN RING

Karban et al.[20] have presented a paper where 1,6-anhydro-2,3,4-trideoxy-2,3-(tosylepimino)-β-D-hexopyranoses underwent aziridine ring-opening reactions with halides and other heteroatom-centered nucleophiles. *Ribo*-epimine provided (**60**) provided *trans*-diaxial and *cis* products.[20] The *lyxo*-epimine (**61**) gave *trans*-diaxial and *trans*-diequatorial products, depending upon the reaction conditions (acid cleavage versus base cleavage) and the nucleophile (hard nucleophiles versus soft ones).[20] These results have been rationalized by assuming that both S_N2 and borderline S_N2 cleavage mechanisms are operative.[20] Scheme 3.21 illustrates a representation of this chemistry.

3.10 THE IONIC LIQUID EFFECT

Hallet et al.[21] presented their findings in the field of ionic liquids. The application of liquids that are salts at room temperature to chemical synthesis has become a hugely exciting field of study. The greatest promise that these ionic liquids hold is that they might offer process advantages, even novel behaviors that cannot be achieved in molecular solvents.[21] The S_N2 reaction of the trifluoromethanesulfonate and bis(trifluoromethanesulfonyl)imide salts of dimethyl-4-nitrophenylsulfonium ($[p\text{-}NO_2PhS(CH_3)_2]^+[X]^-$:$[X]^- = [CF_3SO_3]^-$, $[N(CF_3SO_2)_2]^-$) with chloride ion follow a fundamentally different pathway to when the same salts react in molecular solvents.[21] Scheme 3.22 illustrates the S_N2 mechanism in molecular solvents.

3.11 SILVER CHEMISTRY

Llaveria et al.[22] describes their synthetic work with silver complexes bearing trispyrazolylborate ligands (Tp^x) which catalyze the aziridination of 2,4-diene-1-ols in a

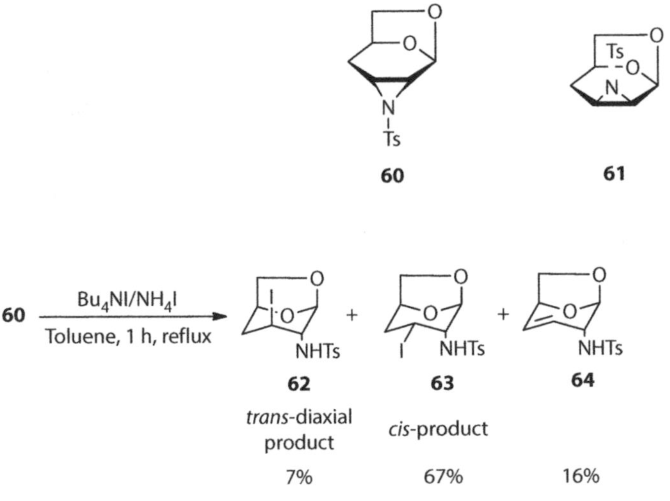

SCHEME 3.21   Ring opening of aziridines.[20]

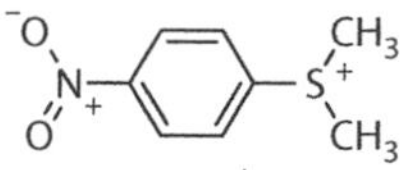

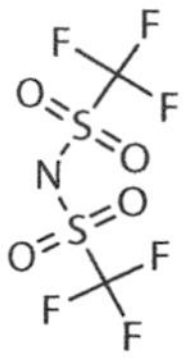

In the presence of a large excess of chloride ion exchange takes place rapidly before the reaction begins such that the initial species is the chloride salt and not the bis-trifluoromethylsulfonyl amide ion.

65

Kinetic scheme

(1) Formation of ion pair:

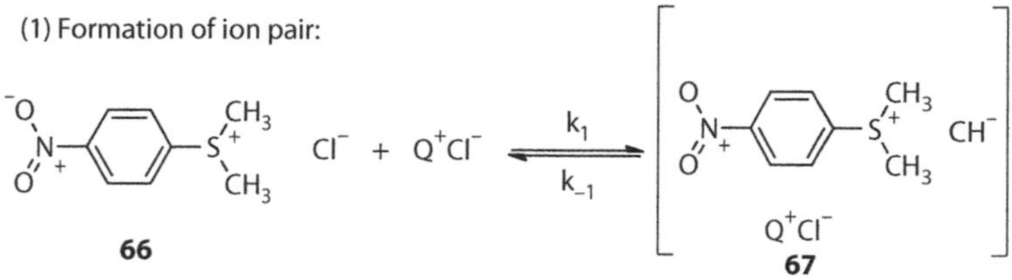

66

$$Cl^- + Q^+Cl^- \underset{k_{-1}}{\overset{k_1}{\rightleftharpoons}}$$

67

(2) Formation of products from ion pair:

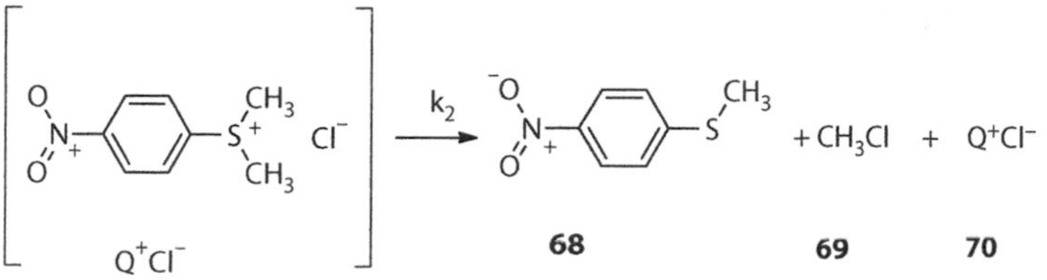

67

$$\xrightarrow{k_2}$$

$$+ CH_3Cl + Q^+Cl^-$$

68 **69** **70**

(3) Redistribution of ion pair:

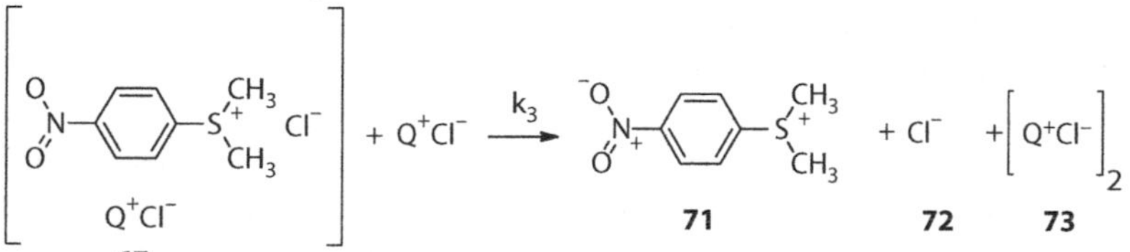

67

$$+ Q^+Cl^- \xrightarrow{k_3}$$

$$+ Cl^- + \left[Q^+Cl^- \right]_2$$

71 **72** **73**

(4) Monomer–dimer equilibrium:

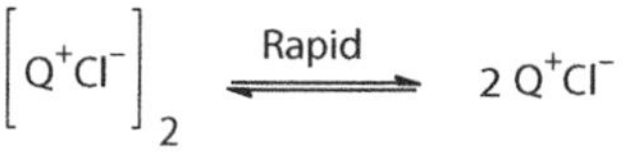

$$\left[Q^+Cl^- \right]_2 \overset{Rapid}{\rightleftharpoons} 2\,Q^+Cl^-$$

73 **74**

Final rate law: $\dfrac{d\ Products}{dt} = \dfrac{k_1 k_2 (E^+Cl^-)(Q^+Cl^-)}{k_1 + k_2 + k_3\,(Q^+Cl^-)}$

$k\ (second\text{-}order) = \dfrac{k_1 k_2}{k_1 + k_2 + k_3 (Q^+Cl^-)}$

SCHEME 3.22 S_N2 mechanism in molecular solvents.[21] *Note:* Q⁺ refers to the (quaternary) cation originally associated with the chloride ion.

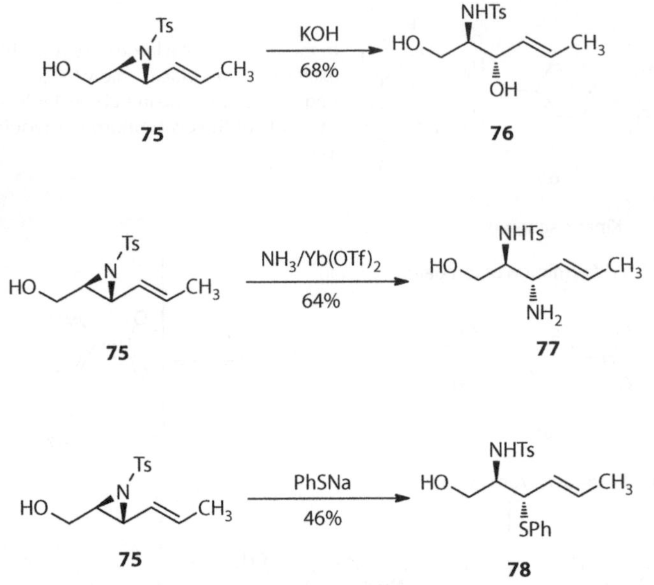

SCHEME 3.23   Ring-opening of vinylaziridine **75** by S_N2 reactions.[22]

chemo-, regio-, and stereoselective manner to give vinylaziridines in high yields by means of the metal-mediated transfer of nitrogen tosylates (NTs) (Ts = p-toluensulfonyl) units from PhI = NTs. The ring opening of vinylaziridines can be carried out under S_N2 and S_N2' conditions based on the reagents.[22] Scheme 3.23 illustrates the use of three different conditions that react with the vinylaziridine to form products via the S_N2 mechanism.[22]

Rijs et al.[23] describes their findings in researching the allylic substitution reactions with silver and gold catalysts. Copper-mediated allylic substitution reactions are widely used in organic synthesis whereas the analogous reactions for silver and gold are essentially unknown.[23] To unravel why this is the case, the gas-phase

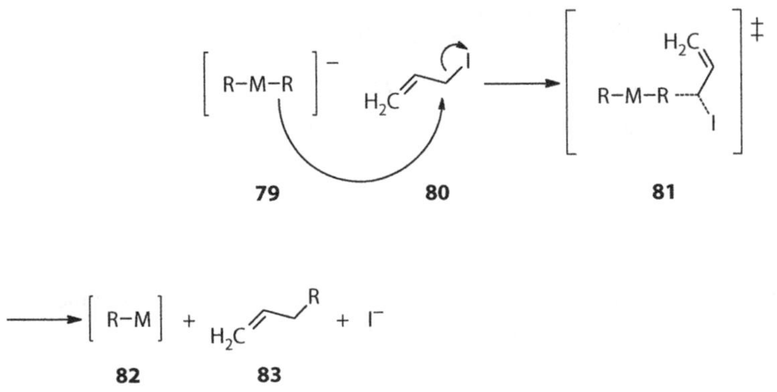

SCHEME 3.24   Possible mechanistic pathway for the C–C silver coupling reaction.[23]

reactions of allyl iodide with the coinage metal dimethylmetallates were examined under near thermal conditions of an ion trap mass spectrometer and via electronic structure calculations.[23] Based on the results described in the article, the authors find that the kinetically most probable pathway for silver is through the S_N2 reaction.[23] Scheme 3.24 illustrates this plausible mechanism based on their results.

REFERENCES

1. Shibatomi, K., Soga, Y., Narayama, A., Fujisawa, I., Iwasa, S. *J. Am. Chem. Soc.* 2012, 134, 9836–9839.
2. Avenoza, A., Busto, J. H., Jimenez-Oses G., Peregrina, J. M. *J. Org. Chem.* 2006, 71, 1692–1695.
3. Robertson, F. J., Wu, J. *J. Am. Chem. Soc.* 2012, 134, 2775–2780.
4. Toesch, M., Schober, M., Breinbauer, R., Faber, K. *Eur. J. Org. Chem.* 2014, 3930–3934.
5. Zhu, H., Chen, P., Liu, G. *J. Am. Chem. Soc.* 2014, 136, 1766–1769.
6. Fang, Y.-Q., Bio, M. B., Hansen, K. B., Potter, M. S., Clausen, A. *J. Am. Chem. Soc.* 2010, 132, 15525–15527.
7. Katcher, M. H., Doyle, A. G. *J. Am. Chem. Soc.* 2010, 132, 17402–17404.
8. Ghorai, M. K., Das, K., Kumar, A. *Tetrahedron Lett.* 2009, 50, 1105–1109.
9. Trost, B. M., Xu, J., Schmidt, T. *J. Am. Chem. Soc.* 2009, 131, 18343–18357.
10. Geier, M. J., Aseman, M. D., Gagne, M. R. *Organometallics* 2014, 33, 4353–4356.
11. Thompson, S. J., Dong, G. *Organometallics* 2014, 33, 3757–3767.
12. Nookaraju, U., Begari, E., Kumar, P. *Org. Biomol. Chem.* 2014, 12, 5973–5980.
13. Shafaatian, B., Heidari, B. *J. Organomet. Chem.* 2015, 780, 34–42.
14. Marti-Centelles, V., Burguete, M. I., Cativiela, C., Luis, S. V. *J. Org. Chem.* 2014, 79, 559–570.
15. Issa, J. P., Bennett, C. S. *J. Am. Chem. Soc.* 2014, 136, 5740–5744.
16. Jaing, Y., Fang, Z., Zheng, Q., Jia, H., Cheng, J., Zheng, B. *Synthesis* 2009, 16, 2756–2760.
17. Prevost, M., St-Jean, O., Guidon, Y. *J. Am. Chem. Soc.* 2010, 132, 12433–12439.
18. Zeinyeh, W., Pilme, J., Radix, S., Walchshofer, N. *Tetrahedron Lett.* 2009, 50, 1828–1833.
19. Ciftci, G. Y., Senkuytu, E., Yuksel, F., Kilic, A. *Polyhedron* 2014, 77, 1–9.
20. Karban, J., Kroutil, J., Budesinsky, M., Sykora, J., Cisarova, I. *Eur. J. Org. Chem.* 2009, 6399–6406.
21. Hallett, J. P., Liotta, C. L., Ranieri, G., Welton, T. *J. Org. Chem.* 2009, 74, 1864–1868.
22. Llaveria, J., Beltran, A., Sameera, W. M. C., Locati, A., Diaz-Requejo, M. M., Matheu, M. I., Castillon, S., Maseras, F., Perez, P. J. *J. Am. Chem. Soc.* 2014, 136, 5342–5350.
23. Rijs, N. J., Yoshikai, N., Nakamura, E., O'Hair, R. A. J. *J. Am. Chem. Soc.* 2012, 134, 2569–2580.

4 Electrophilic Addition to Alkenes

Adam M. Azman

CONTENTS

4.1   INTRODUCTION

Electrophilic additions to alkenes are ubiquitous and at the same time invisible in much of natural product synthesis. With the exception of a few named reactions, electrophilic additions to alkenes are often utilized as a necessary step in functional group transformation to set up the key reaction—not necessarily acting as the key reaction in a total synthesis. As such, classical electrophilic additions to alkenes rarely receive attention in article abstracts or titles. This chapter deals with electrophilic additions to alkenes as used in the synthesis of natural or biologically active targets and, wherever possible, highlight their use as the key step in that synthesis.

4.2   CYCLOPROPANATION

Carbenes can be electrophiles for electrophilic addition to alkenes.[1–6] The carbene lone pair of electrons then captures the forming carbocation to install a cyclopropane ring. The carbene can be generated through a variety of methods and utilized in a variety of reactions to generate cyclopropane products. Cossy and coworkers opened cyclopropenes with a rhodium catalyst to form a transient carbene which cyclized onto an intramolecular alkene to form tricyclic cyclopropane-containing products. Yields and diastereoselectivities are typically high, and the authors used this methodology in the synthesis of benzazocanes and benzoxocanes.[7]

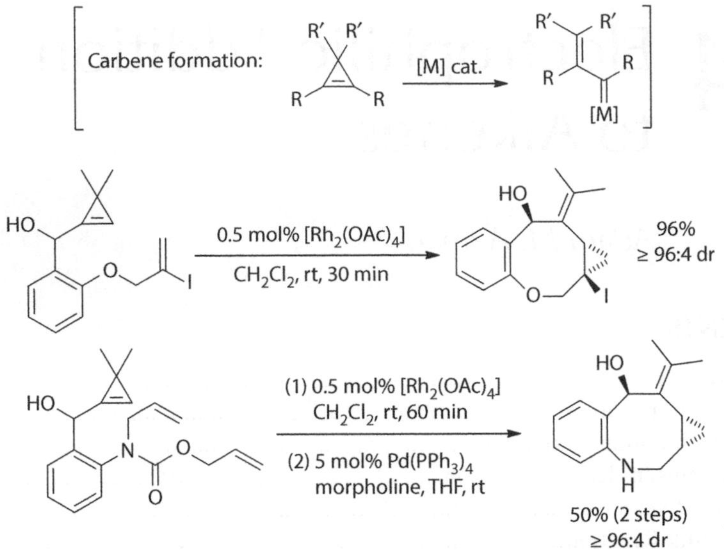

The cyclopropane generated in these reactions also holds interest as reactive functional groups. (+)-Barekoxide and (+)-barekol have been synthesized by Davies and coworkers through tandem carbene cyclopropanation/Cope rearrangement.[8] This tandem reaction (a formal [4 + 3]-cycloaddition) is often highly diastereoselective. The tricyclic fused cycloheptane framework generated is present in several terpene natural products. Depending on the setup of the precursors, three new stereocenters can be set in the course of this reaction. For this particular reaction, the stereochemistry of the diene and the stereochemistry of the catalyst create a matched system in model studies, whereby the cyclopropanation occurs from one face selectively. In the actual system, the methodology allows access to a congested, all-carbon quaternary center with good diastereocontrol.

Model system

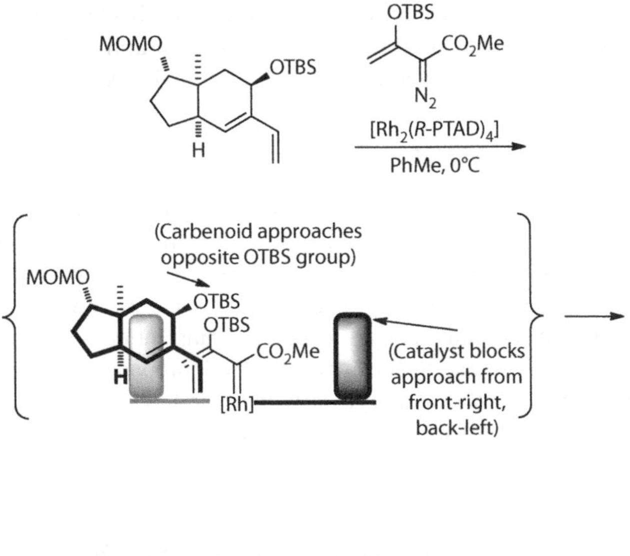

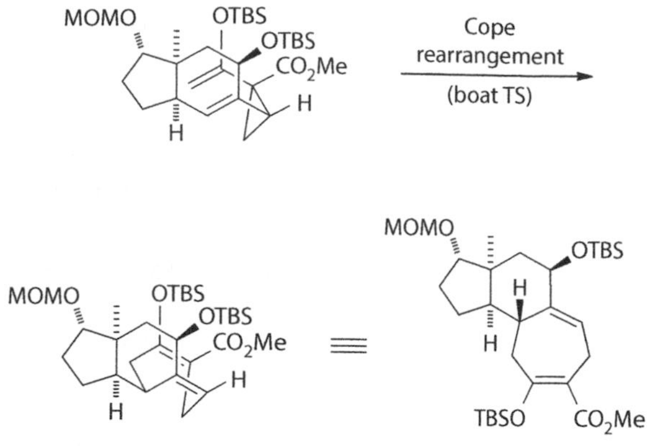

Total synthesis

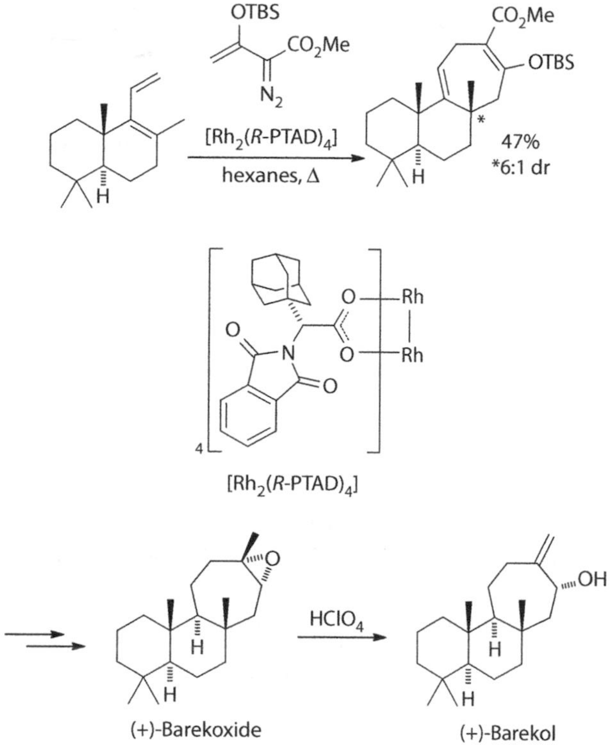

$[Rh_2(R\text{-PTAD})_4]$

(+)-Barekoxide

(+)-Barekol

Nakada and coworkers successfully employed intramolecular copper-catalyzed cyclopropanation of α-diazocarbonyls to synthesize the structurally related natural products nemorosone,[9] hyperforin,[10] and garsubellin A.[11] An enantioselective intramolecular cyclopropanation by the same group was a key step in the synthesis of (+)-colletoic acid.[12]

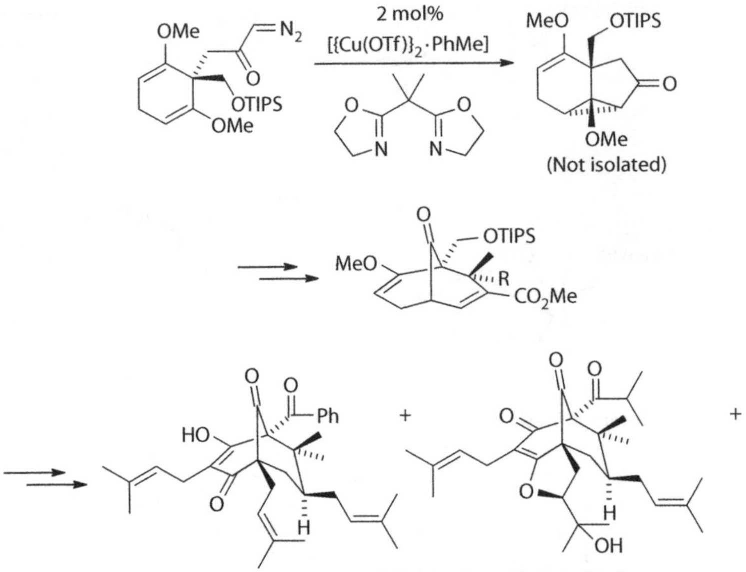

Nemorosone

Garsubellin A

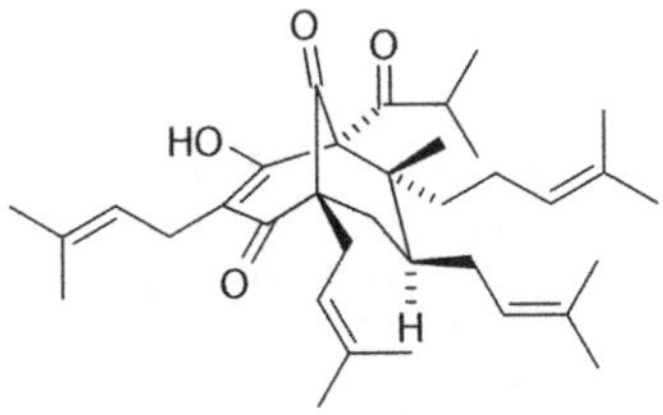

Hyperforin

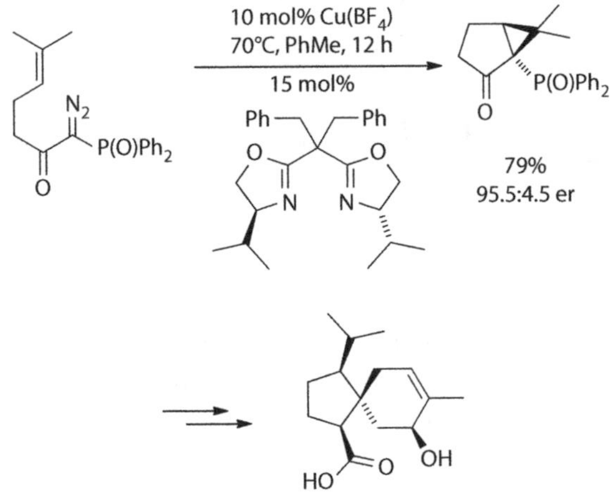

79%
95.5:4.5 er

(+)-Colletoic acid

Rhodium-catalyzed cyclopropanation of α-diazocarbonyls is also synthetically useful, and Taber et al.[13] prepared (+)-coronafacic acid in such a manner. The stereoselectivity is controlled by a menthyl ester chiral auxiliary. (+)-Coronafacic acid induces tubers and cell expansion while promoting senescence and inhibiting cell division in plants.

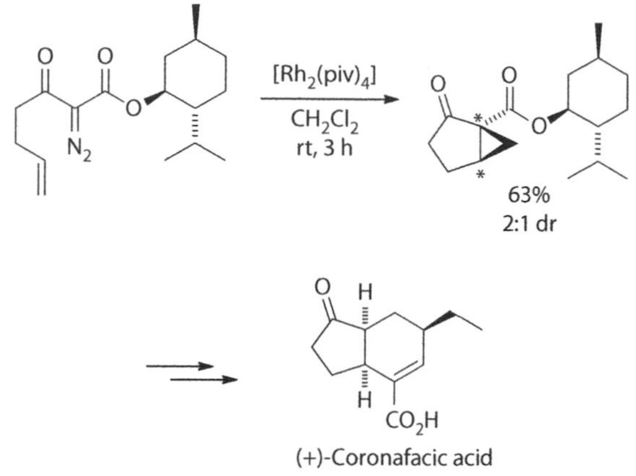

63%
2:1 dr

(+)-Coronafacic acid

(−)-Ardeemin was isolated from a strain of *Aspergillus fischeri*[14] and has been shown to reverse multidrug resistance in some tumor cell lines. It has increased the cytotoxicity of vinblastine 1000-fold against the KBV-1 tumor cell line. Qin and coworkers used copper salts to facilitate carbene formation and effect cyclopropanation. The conditions for the cyclopropanation event were exploited to induce a cascade reaction consisting of cyclopropanation, ring opening, and iminium ion cyclization. The product was carried on to complete the synthesis of (−)-ardeemin.[15] The same group utilized this cascade reaction on different indole scaffolds to synthesize (±)-vincorine,[16] (±)-minfiensine,[17] and (±)-communesin F.[18]

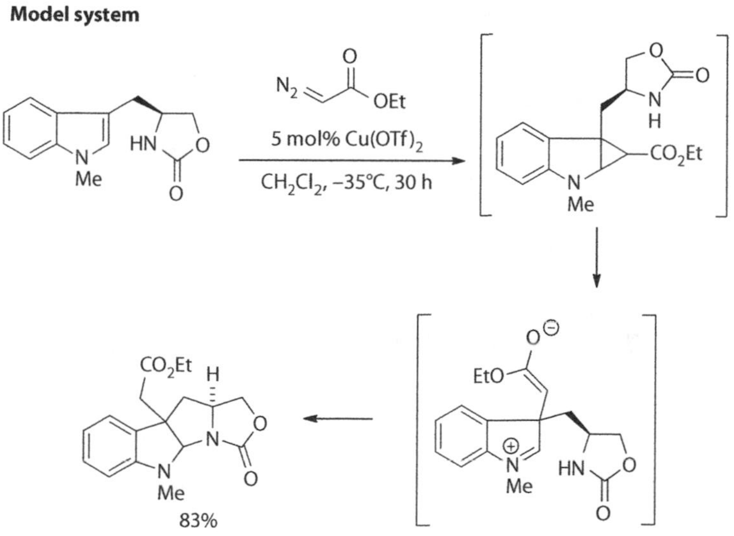

Model system

83%

Total synthesis

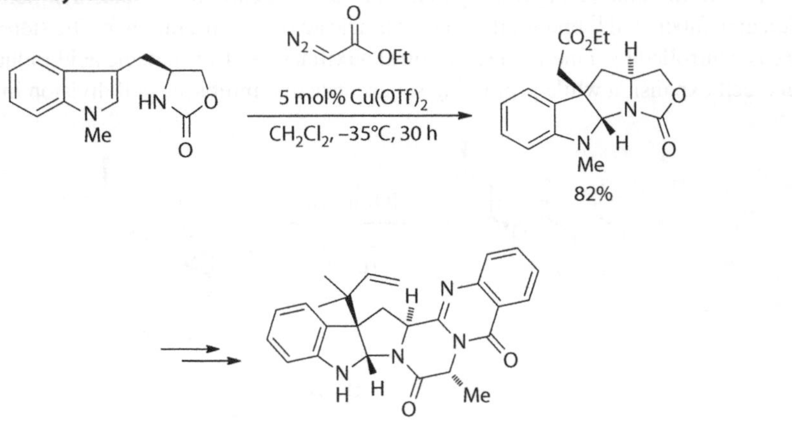

(−)-Ardeemin

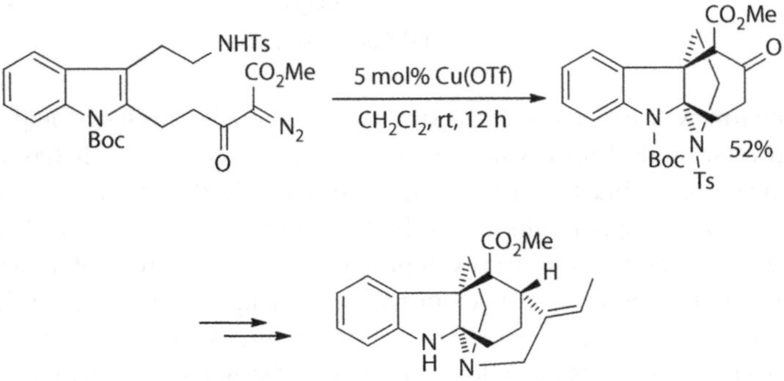

(±)-Vincorine

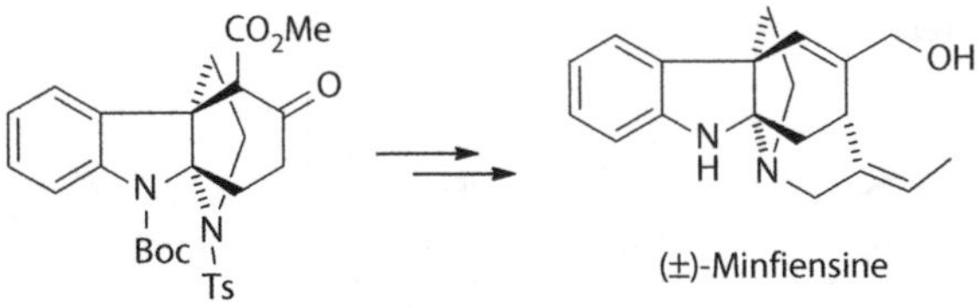

(±)-Minfiensine

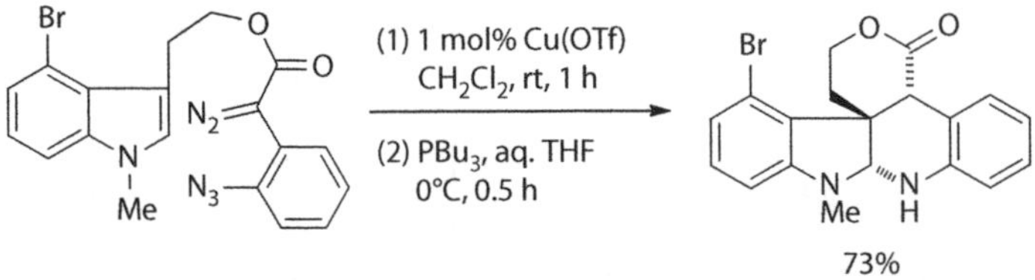

73%
(Two steps)

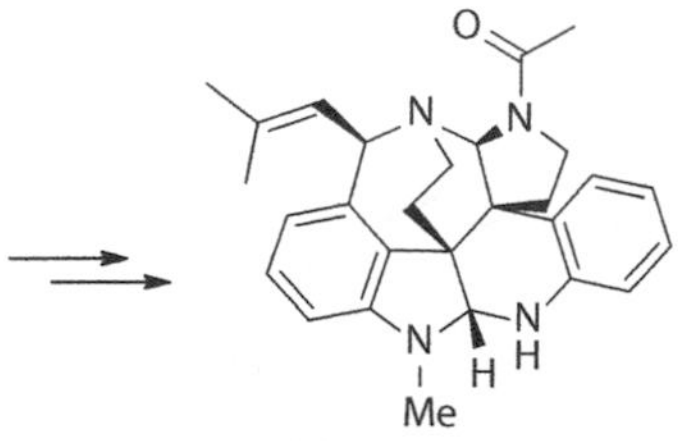

(±)-Communesin F

4.3 HYDROBORATION–OXIDATION

The hydroboration–oxidation of terminal olefins in organic synthesis is often used to set the stereochemistry of the internal carbon atom more so than necessarily functionalizing the terminal position. Often, the newly installed hydroxy group is a placeholder until further functional group transformations convert it into the desired functionality of the ultimate target. Like other electrophilic additions, hydroboration–oxidation events often receive less focus in articles than the key bond-forming events.[19,20]

The fungus *Claviceps paspali* produces the indole diterpene alkaloid paspaline. The Johnson lab has disclosed the synthesis of this alkaloid utilizing hydroboration–oxidation.[21] The hydroboration–oxidation step of an exocyclic alkene set the C4b (paspaline numbering) stereocenter. The resulting alcohol was oxidized and used as an electrophile for an aldol condensation. Paspaline was completed in 28 steps and 0.4% yield.

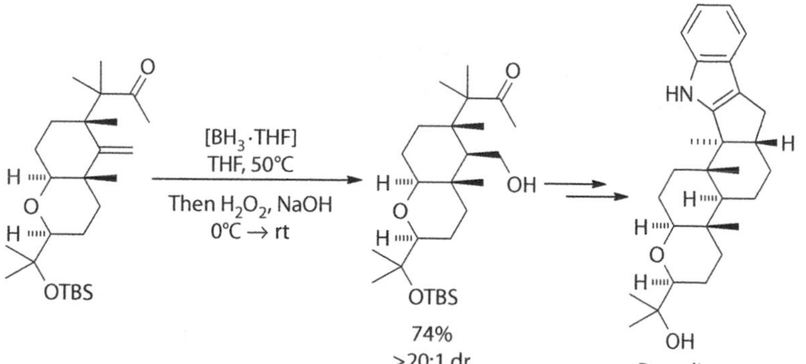

In much the same way, the Carreira lab set a key stereocenter through hydroboration–oxidation and oxidized the alcohol to use as an electrophile for an aldol condensation. With this sequence, the researchers completed the synthesis of gelsemoxonine, an alkaloid isolated from *Gelsemium elegans* Bentham.[22]

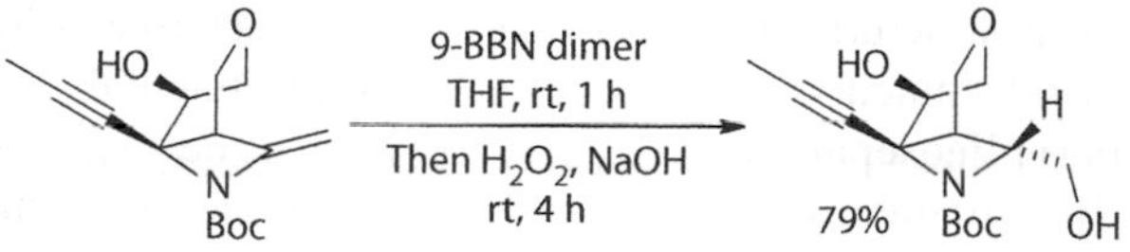

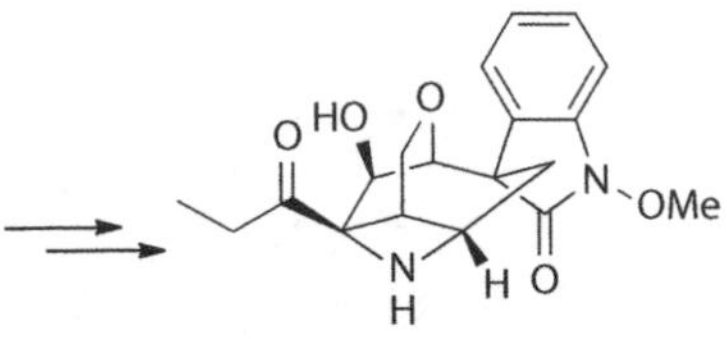

Gelsemoxonine

Two hydroboration–oxidations were utilized in the synthesis of (+)-daphmanidin E by the Carreira group.[23] (+)-Daphmanidin E has been shown to be a vasorelaxant in studies of rat aorta. A high concentration (>1 M) of BH$_3$ in THF was required to complete the slow hydroboration of a protected cyclohexenone, and two contiguous stereocenters were set with high diastereoselectivity. Later in the synthesis, the researchers carried out a second hydroboration of a monosubstituted olefin in the presence of an exocyclic 1,1-disubstituted olefin. The completion of the synthesis marked the first synthesis of any member of the daphmanidin alkaloids.

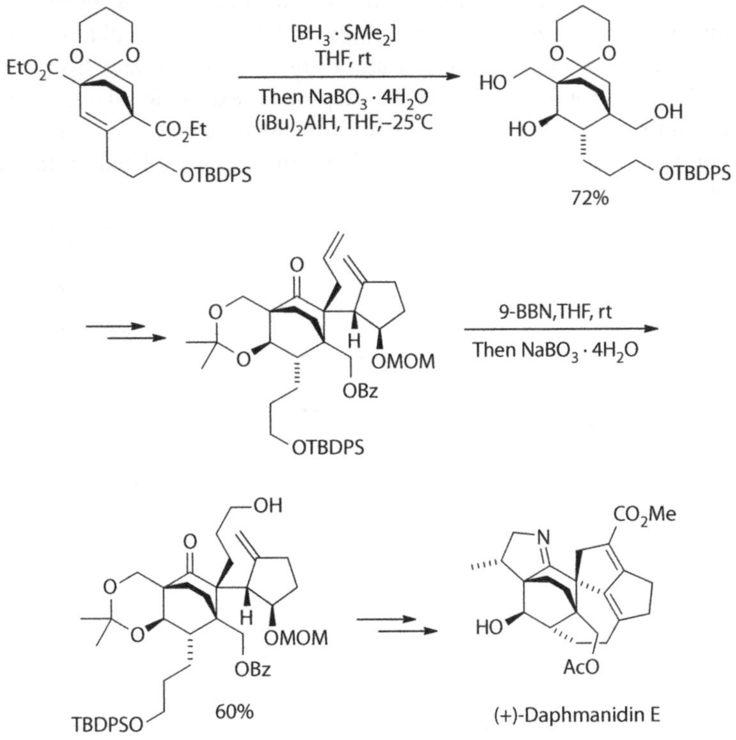

Stereoselective hydroboration–oxidation of a 1,1-disubstituted olefin is difficult without any inherent structural bias. Tietze et al. instead performed a hydroboration–oxidation on a trisubstituted vinyl silane as a 1,1-disubstituted alkene surrogate in their synthesis of (+)-linoxepin.[24] (–)-(ipc)BH$_2$ provided the best yield, and protodesilylation with TBAF (tetrabutylammonium fluoride) revealed the primary alcohol.

Enantiopure (+)-linoxepin was synthesized in 11 steps and 27% yield through this method.

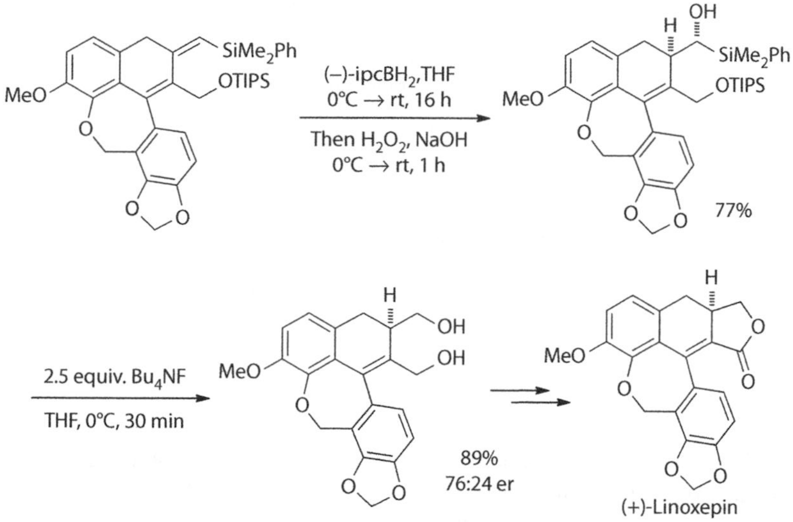

(+)-Linoxepin

One of the strategies for sequential formation of the ring ethers of marine ladder toxins involves hydroboration of a cyclic alkene and ring closure through an intermediate acetal. Recently, the ladder toxin gambieric acid A was produced in this manner by Fuwa et al.[25] The C-ring was formed by this sequence. In addition, a temporary F-ring was functionalized through hydroboration–oxidation to introduce the oxygen atom which would eventually take part in closing the E-ring. The free hydroxy group on the B-ring was also introduced earlier in the synthesis through hydroboration–oxidation.

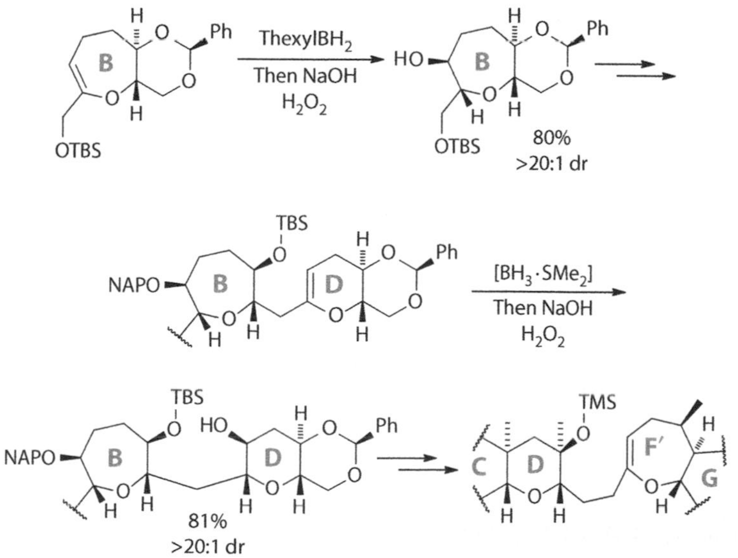

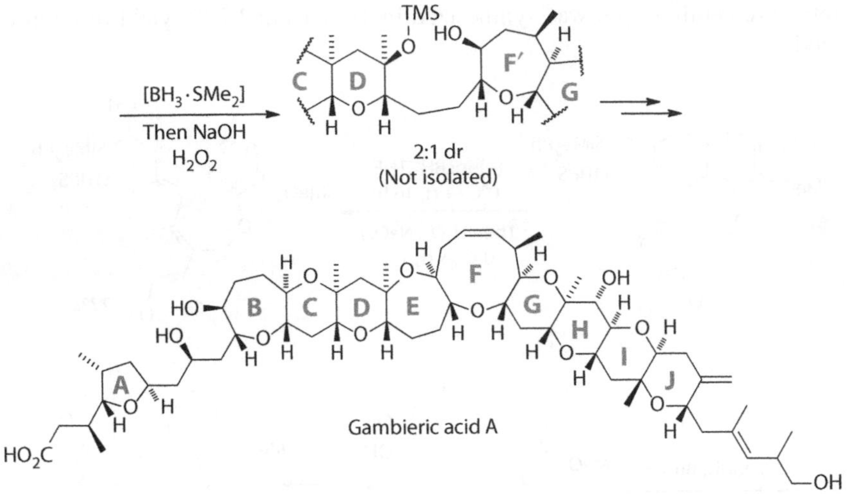

2:1 dr
(Not isolated)

[BH₃·SMe₂]
Then NaOH
H₂O₂

Gambieric acid A

Hydroboration–oxidation quietly set the C6 stereocenter of (–)-callipeltoside throughout several iterations of the synthesis by Hoye et al. While late-stage issues with cyclization of the cytotoxic macrolide called for rethinking the synthetic strategy, the hydroboration–oxidation furnished the C6 stereocenter each time. (–)-Callipeltoside was finally achieved in 21 linear steps and 0.7% overall yield.[26]

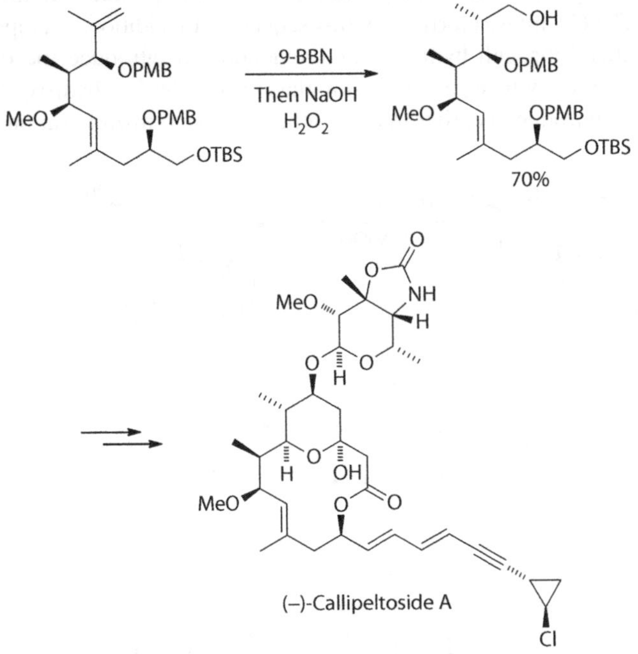

9-BBN
Then NaOH
H₂O₂

70%

(–)-Callipeltoside A

For Cho et al.,[27] a Diels–Alder reaction was a key step in their synthesis of (±)-pancratistatin, however, the dienophile would have needed to be an unstable enol.

To work around this issue, a terminal alkyne was treated with pinacolborane. The resulting vinylborane was utilized as the dienophile to form an alkylborane, which was subsequently oxidized to the (*S*)-alcohol. Unfortunately pancratistatin contains the (*R*)-alcohol. This could be remedied by an oxidation–reduction strategy at one of (+/–)-lycorine, which does contain the (*S*)-alcohol. Cho and his group again utilized this strategy in the more recent synthesis of (±)-lycorine.[28]

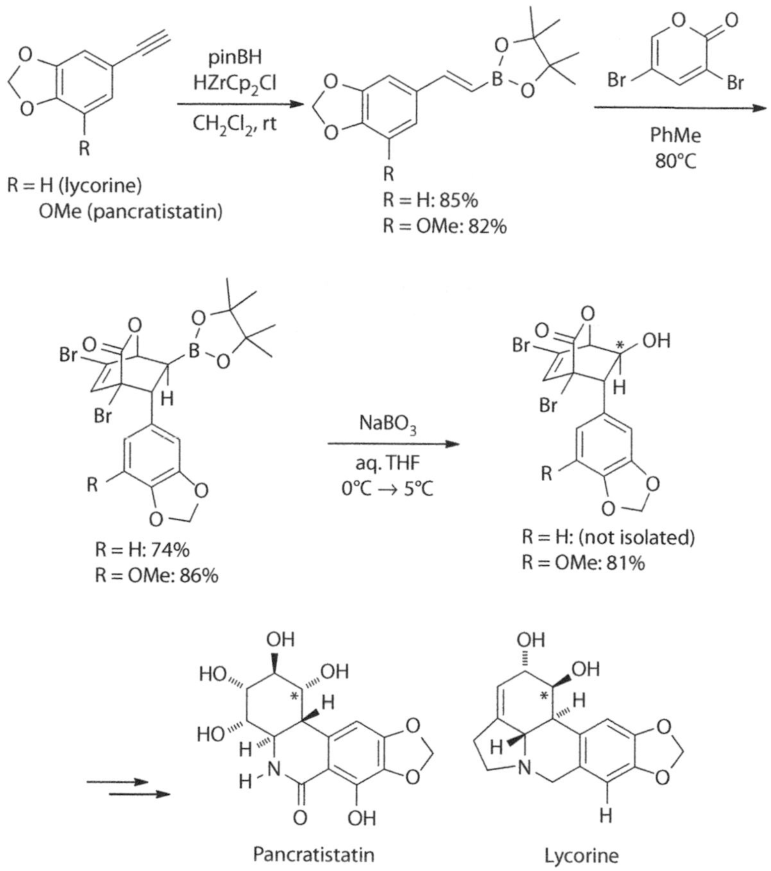
pinBH
$HZrCp_2Cl$
CH_2Cl_2, rt
R = H (lycorine)
OMe (pancratistatin)
R = H: 85%
R = OMe: 82%
PhMe
80°C
R = H: 74%
R = OMe: 86%
$NaBO_3$
aq. THF
0°C → 5°C
R = H: (not isolated)
R = OMe: 81%
Pancratistatin
Lycorine

4.4 THE PAUSON–KHAND REACTION

The Pauson–Khand reaction is well known for its ability to form cyclopentenones from enynes and a metal carbonyl. The metal center coordinates with the alkyne for use as the electrophile in the electrophilic addition to an alkene. The Pauson–Khand reaction has been reviewed several times recently,[29–32] and examples of its use in total synthesis will be highlighted here.

Chung and his team prepared a library of steroids with modified D-rings.[33] The D-rings were prepared through the Pauson–Khand reaction. The ring formation proceeded with comparable yields utilizing either cobalt or rhodium carbonyl catalysts. Steroid derivatives were more recently prepared by Helaja and team with an appended E-ring.[34,35] A D-ring alkene served as the nucleophile for an intermolecular Pauson–Khand reaction.

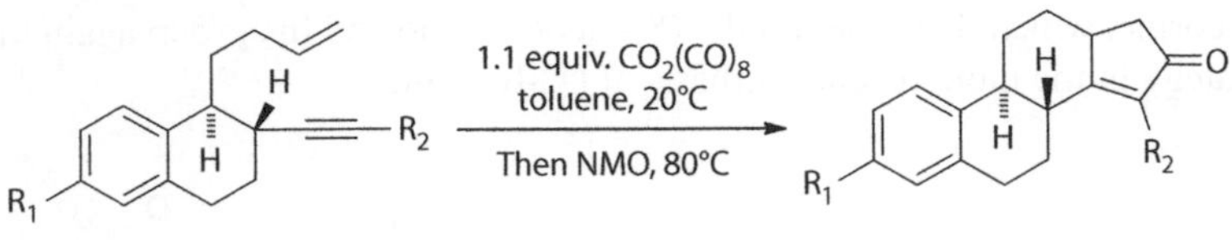

$R_1 = H$, $R_2 = Ph$,   91%
$R_1 = H$, $R_2 = Tolyl$,   77%
$R_1 = H$, $R_2 = nBu$,   44%
$R_1 = H$, $R_2 = TMS$,   19%
$R_1 = H$, $R_2 = H$,   23%
$R_1 = OMe$, $R_2 = Ph$,   81%

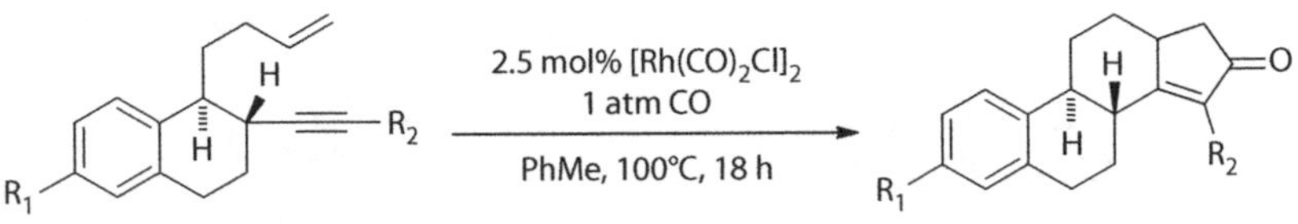

$R_1 = H$, $R_2 = Ph$,   63%
$R_1 = H$, $R_2 = Tolyl$,   82%
$R_1 = H$, $R_2 = nBu$,   46%
$R_1 = H$, $R_2 = TMS$,   29%
$R_1 = OMe$, $R_2 = Ph$,   80%

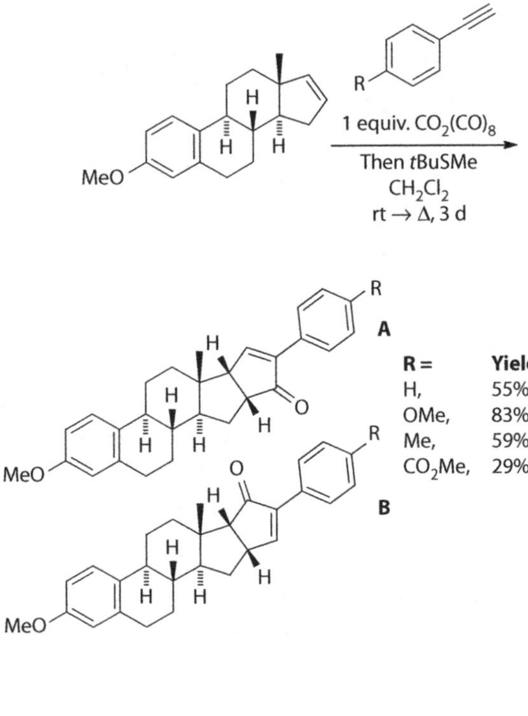

R =	Yield	A:B
H,	55%,	1:1.6
OMe,	83%,	1.4:1
Me,	59%,	1.5:1
CO$_2$Me,	29%,	1.3:1

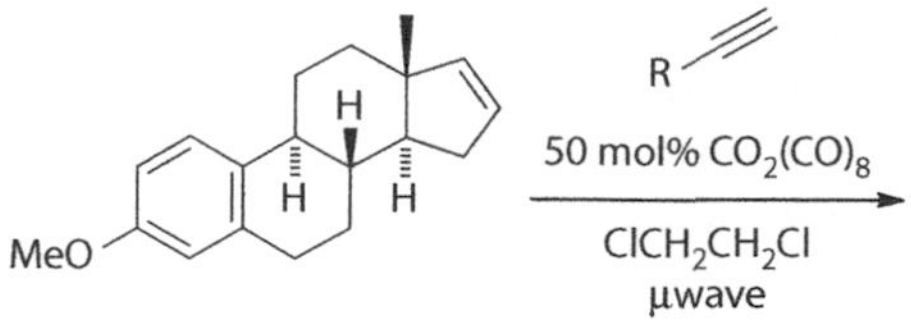

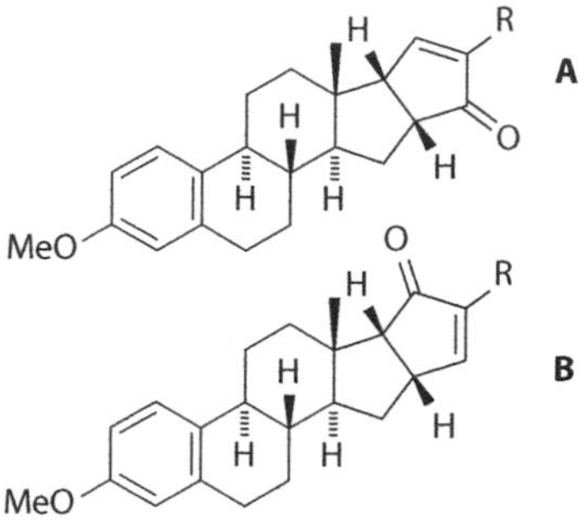

R =	Yield	A:B
CH$_2$OMe,	60%	1.3:1
CH$_2$OH,	44%	2:1
nPr,	33%	1.7:1
Ph,	57%	1.3:1
TMS,	6%	1:0

Two similar monoterpene piperidine alkaloids were prepared by Honda and Kaneda[36] utilizing an intramolecular Pauson–Khand reaction. The two rings of (−)-incarvilline were closed diastereoselectively in the Pauson–Khand reaction, with the MOM (methoxymethyl ether) group sterically blocking one of the diastereomers from forming to an appreciable extent. Similarly, a diastereoselective Pauson–Khand reaction closed the two rings of (+)-α-skytanthine.[37]

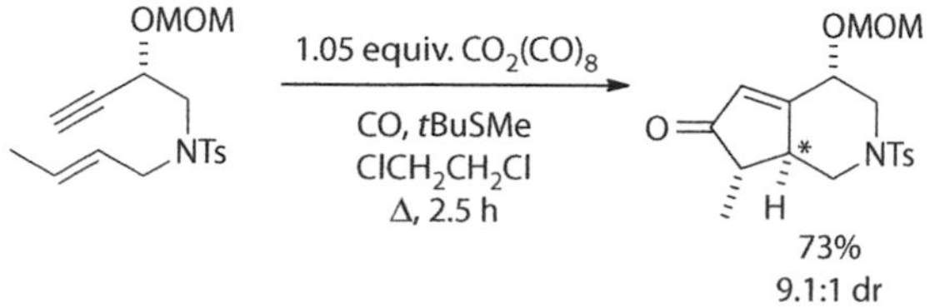

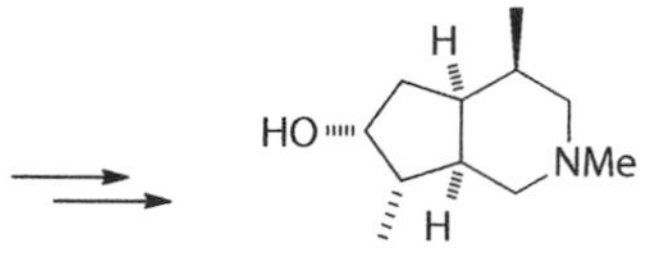

(−)-Incarvilline

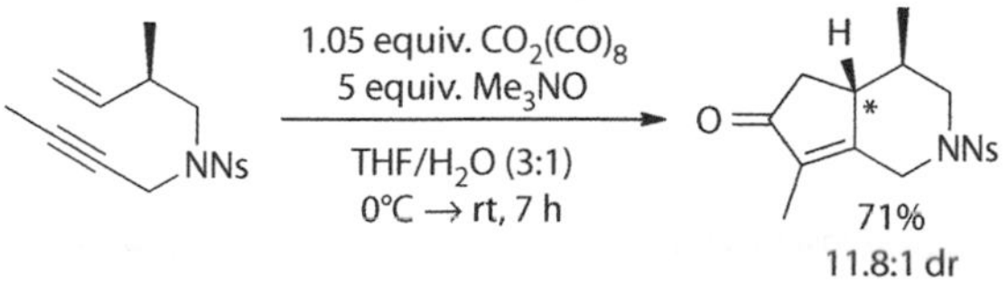

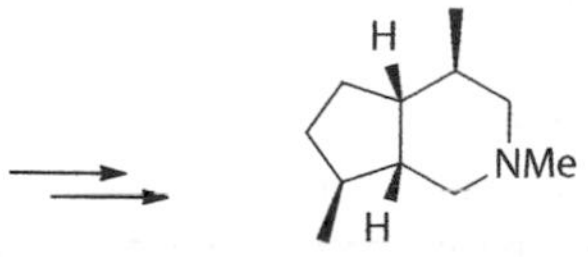

(+)-α-Skytanthine

The carbanucleosides (–)-abacavir and (–)-carbovir,[38] as well as prostaglandin B_1 and a botanical analog phytoprostane B_1[39] were prepared enantioselectively by Verdaguer, Rieira, and coworkers. Abacavir has been FDA (Food and Drug Administration) approved for the treatment of HIV. All of the syntheses rely on an intermolecular Pauson–Khand reaction to create the cyclopentene ring necessary for the synthesis.

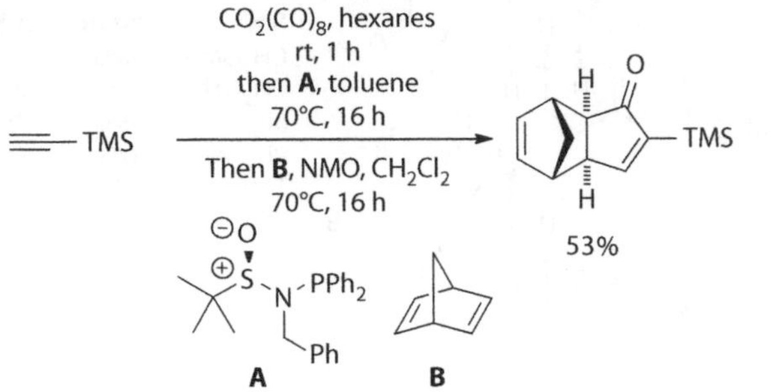

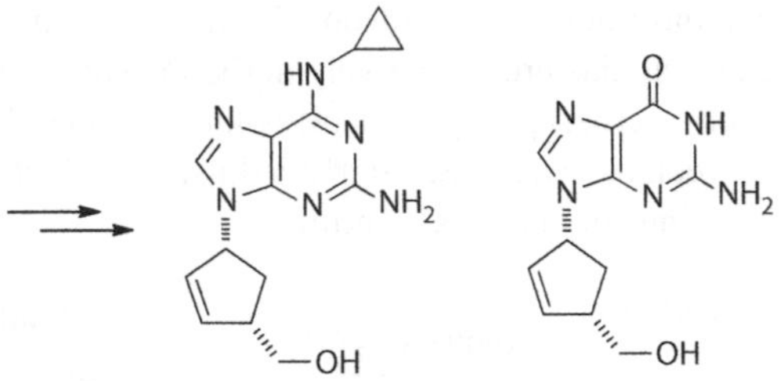

(–)-Abacavir   (–)-Carbovir

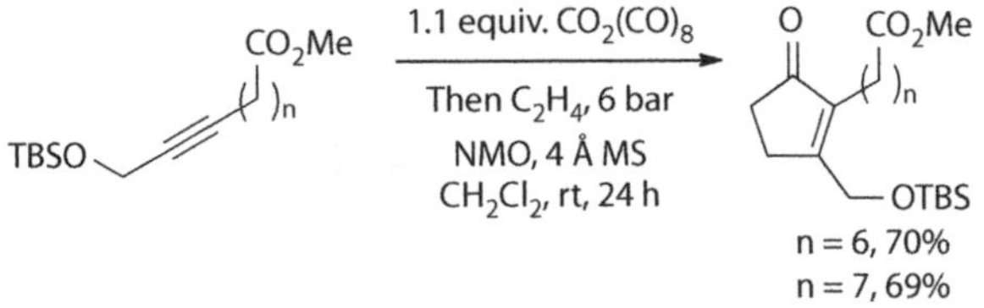

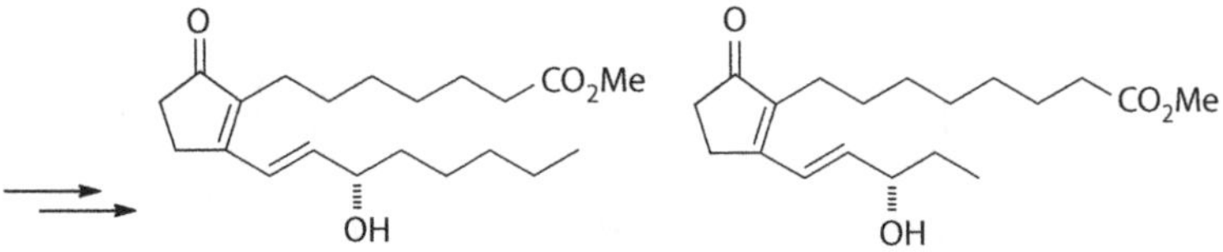

n = 6, 70%
n = 7, 69%

Prostaglandin B_1
methyl ester

Phytoprostane B_1
methyl ester

The structurally related (+)-achalensolide[40] and (+)-indicanone[41] were prepared through a Pauson–Khand reaction of a terminal allene by Mukai and coworkers. The terminal π bond of the allene is the reactive π bond, while the internal π bond remains intact as part of the newly formed cycloheptene.

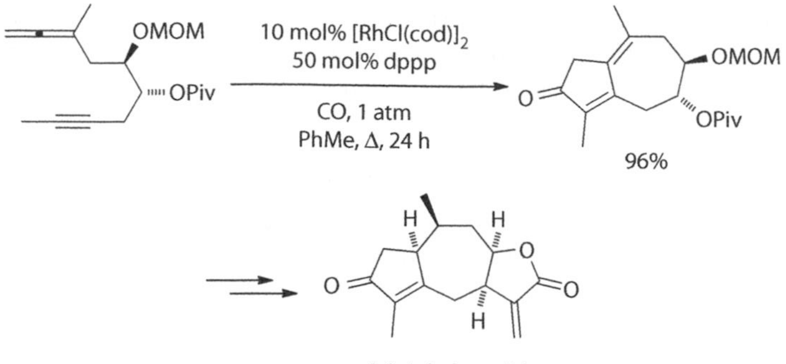

96%

(+)-Achalensolide

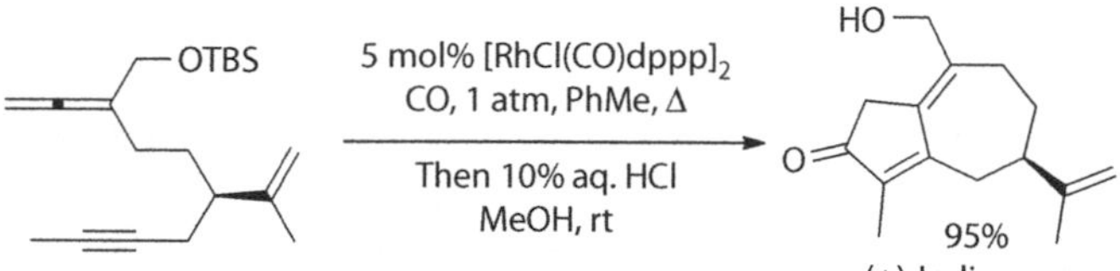

95%
(+)-Indicanone

Baran and coworkers recently reported the gram-scale synthesis of (+)-ingenol utilizing an intramolecular Pauson–Khand reaction of an allene.[42] Again, the distal π bond of the allene ends up in the newly formed cyclopentenone. A degassed, anhydrous solvent, *bis*-protection of the two hydroxy groups, and high dilution (0.005 M) were required for the Pauson–Khand reaction to proceed in appreciable yield.

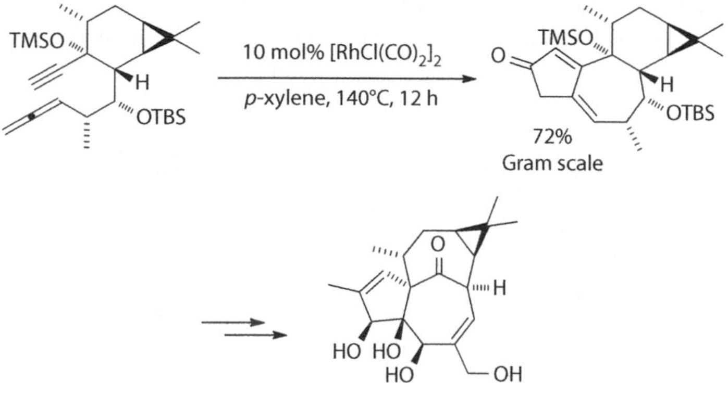

72%
Gram scale

(+)-Ingenol

4.5 PRINS REACTION

The Prins reaction is well known for constructing dihydropyran rings,[43–45] especially in the formation of macrocycles. The Prins reaction is an established methodology as the ring-closing step to make many of these macrocycles. The 14-membered macrolide (+)-neopeltolide was synthesized in 2010 by Yadav and Narayana Kumar.[46] (+)-Neopeltolide was isolated from a deepwater Caribbean sponge and showed cytotoxicity against several cancer cell lines by inhibiting mitochondrial ATP (adenosine triphosphate) synthesis. An initial intermolecular Prins reaction formed a temporary tetrahydropyran ring. Further elaboration to a straight-chain ester set the stage for a second Prins macrocyclization as the final step to complete the synthesis.

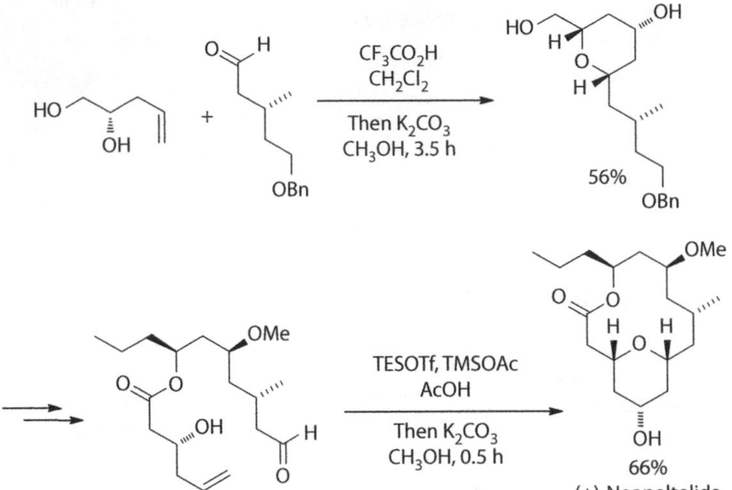

(+)-Neopeltolide

Polycavernoside A was isolated from the edible red alga *Polycavernosa tsudai* and confirmed to be the cause of a fatal food poisoning event in Guam several decades ago. The marine toxin has been the subject of several syntheses, and recently Lee and team utilized the Prins reaction in their synthesis.[47] The Prins reaction of a late-stage diethyl acetal closed the macrolide and formed the first tetrahydropyran ring with 5.5:1 dr. The tetrahydrofuran ring was closed through an alkyne hydration strategy (also an electrophilic addition, see Section 4.10).

(–)-Polycavernoside A

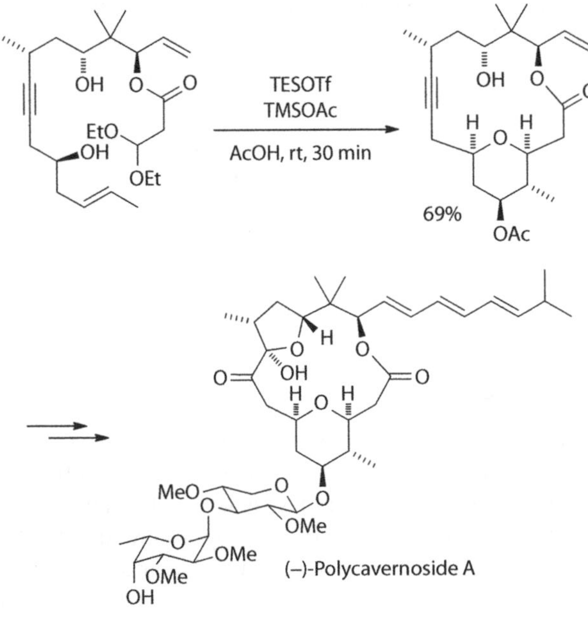

Iterative Prins cyclizations allowed Scheidt and coworkers to assemble the macrocyclic ester (–)-exiguolide,[48] isolated in 2006 from the marine sponge *Geodia exigua* Thiele. The biological profile of the natural product demonstrates potentially useful anticancer activity. Extensive optimization was required to complete the first Prins cyclization. Robust protecting groups, the Lewis acid borontrifluoride and calcium sulfate as a dehydrating reagent were required, and the reaction proceeded in 63% overall yield on gram scale. The 16-member macrolactone was closed through the Prins reaction using scandium triflate as the Lewis acid. With this strategy, (–)-exiguolide and a number of analogs were prepared for biological testing.

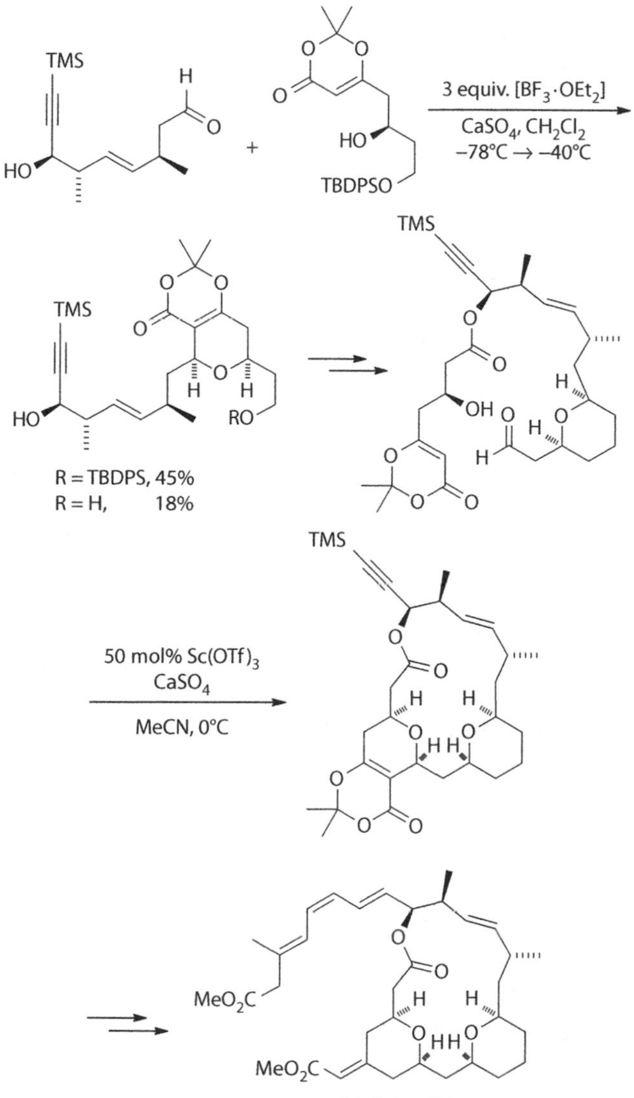

(–)-Exiguolide

4.6   SCHMIDT REACTION

In one of the variations of the Schmidt reaction, an alkene undergoes electrophilic addition with a protic acid, and an azide traps the resulting carbocation. Alkyl migration and evolution of nitrogen gas completes the Schmidt reaction. Pearson and Fang[49] utilized this variation of the Schmidt reaction in their formal synthesis of gephyrotoxin. Treatment of an azide-tethered indane with triflic acid afforded isomeric iminium ions. Without isolation of the iminium ion, the intermediate was elaborated to an azotricycle, intercepting an intermediate in Ito's total synthesis.[50]

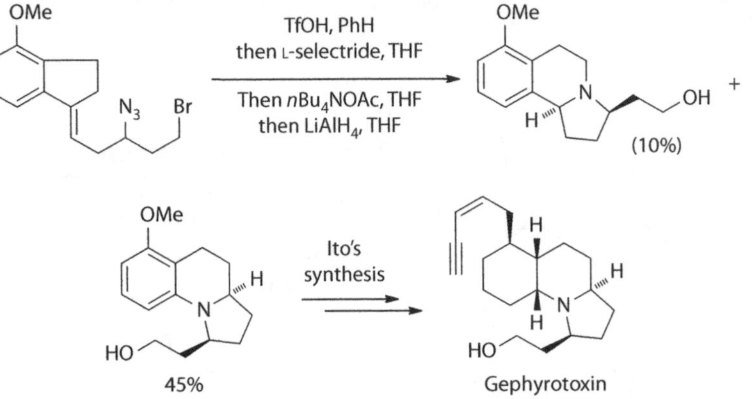

4.7   HALOGENATION

Adding molecular halogens in total synthesis often involves opening the intermediate halonium ion with an oxygen- or nitrogen-based nucleophile, rather than with the matching halide ion typically generated during formation of the halonium ion. Jamison and group treated an alkene with the tetrafluoroborate salt of *bis*(trimethylpyridine) bromine as the electrophilic source of bromine. The bromonium ion was opened with a tethered oxygen atom to form the oxepane ring of armatol A.[51] The synthesis of several diastereomers utilizing this methodology allowed for the confirmation of the absolute configuration of armatol A. In the same paper, they detail a bromonium-initiated, epoxide-opening cascade reaction utilizing *N*-bromosuccinimide to construct the three, fused, cyclic ethers of *ent*-dioxepandehydrothyrsiferol.

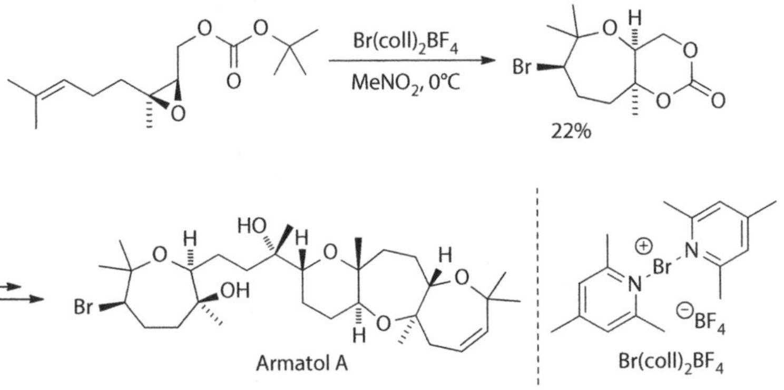

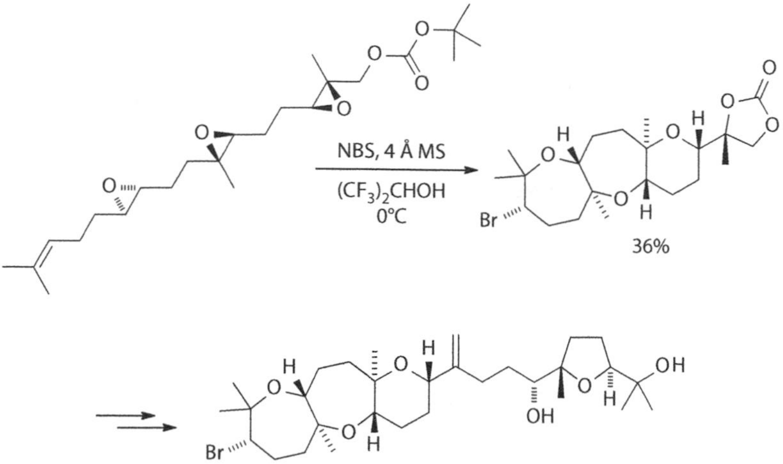

ent-dioxepandehydrothyrsiferol

Podlech and coworkers required a vinyl iodide for a key cross-coupling reaction in the synthesis of altenuic acid III.[52] They prepared this vinyl iodide from an allene. Treatment with elemental iodine and potassium carbonate allowed the intermediate iodonium ion to be trapped by a tethered carboxylic acid to create an iodobutenolide. The synthesis of altenuic acid III allowed for confirmation of the structure elucidated via NMR (nuclear magnetic resonance) studies by the same group.

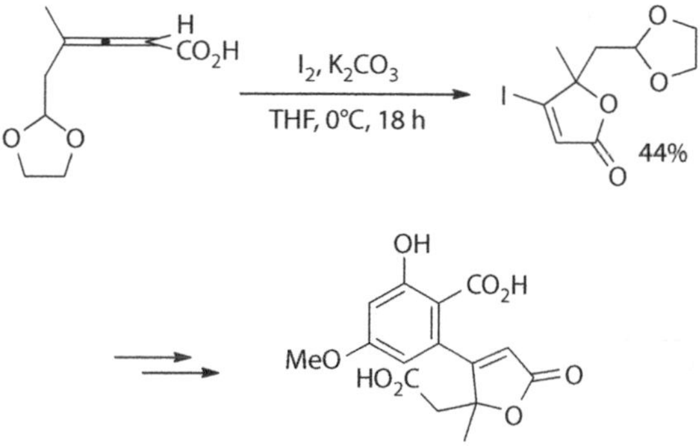

Altenuic acid III

Barrett and coworkers formed an iodoether as a protecting group for an alkene in their synthesis of mycophenolic acid.[53] The synthesis required an aryl bromide for a cross-coupling reaction, yet attempts at direct halogenation led to intractable mixtures due to reaction at the tethered alkene. Treatment with elemental iodine and *t*-butylamine afforded the iodo ether, which could be brominated cleanly at the desired position.

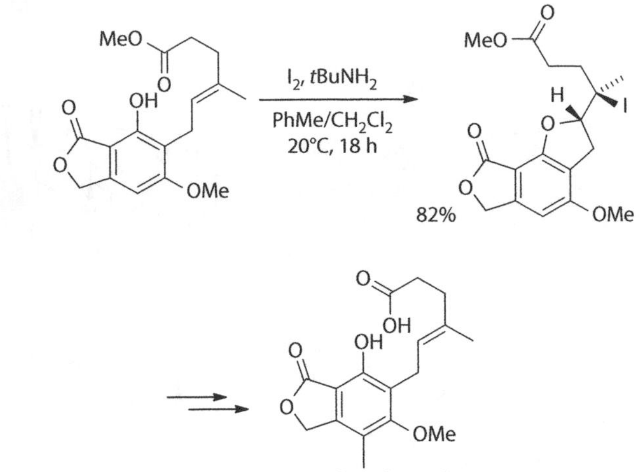

Mycophenolic acid

In the synthesis of dendrodolide K by Mohapatra et al.,[54] a 1,3,5-triol substructure was required. The group took advantage of successive iodocarbonate cyclization reactions to set the alcohols with the desired regio- and stereochemistry. Treatment of a Boc-protected homoallylic alcohol with *N*-iodosuccinimide and elemental iodine led to the decomposition of the Boc group and concomitant intramolecular opening of the iodonium ion intermediate. The newly installed alcohol was again turned into a Boc-protected alcohol for a second iodocyclization.

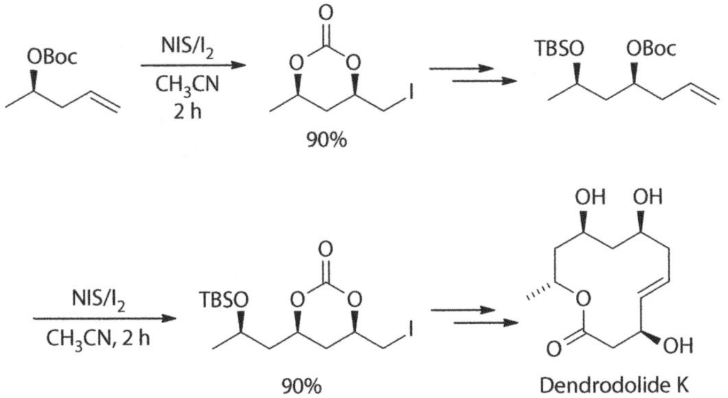

4.8   OXYMERCURATION–REDUCTION

While mostly abandoned due to toxicity issues, oxymercuration–reduction is still the desired alkene hydration method for certain syntheses. Carboxymercuration of a conjugated carboxylic acid formed the second ring of a fused bicyclic lactone in Hayashi and coworkers synthesis of epi-*ent*-EI-1941-2.[55] Carboxypalladation formed the same fused ring system with the natural configuration of the epimeric side chain stereocenter, though in a lower yield than the 99% achieved through carboxymercuration.

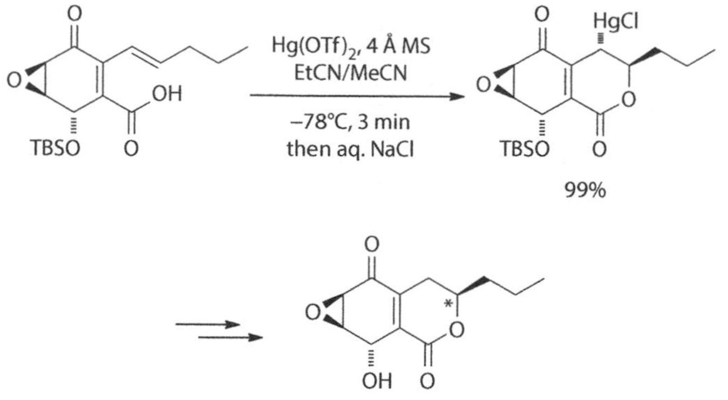

epi-*ent*-EI-1941-2

Late-stage hydration of a dihydropyran ring in (−)-apicularen A by Uenishi and coworkers required regio- and stereocontrol at the resulting carbinol position.[56] Hydroboration–oxidation and epoxidation–hydride reduction both produced an alcohol with the undesired regio- and stereochemistry. However, oxymercuration–reduction of the same alkene did provide the desired alcohol at the desired position with the desired stereochemistry.

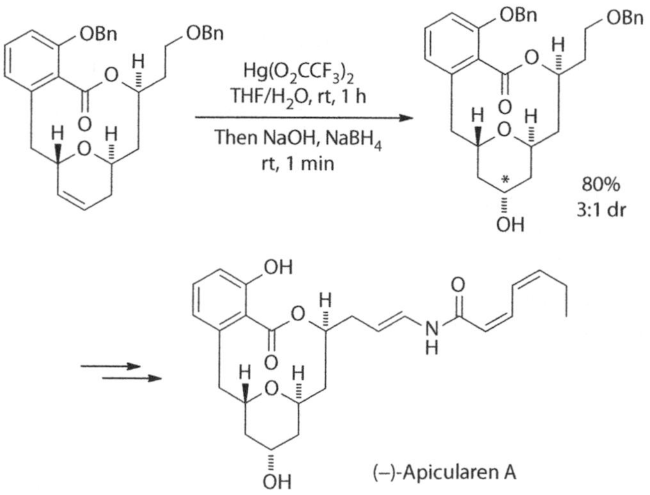

(−)-Apicularen A

4.9 EPOXIDATION

Many natural products contain epoxides, and epoxides are also useful electrophiles in synthesis. Strategies for the stereoselective installation of epoxides are well studied and are used widely in synthesis. Bittman's group used the Sharpless asymmetric epoxidation protocol to install an epoxide with >20:1 diastereomeric ratio. The epoxide was opened regioselectively with sodium azide to ultimately provide a hydroxy amide in the preparation of α-1*C*-galactosylceramide.[57]

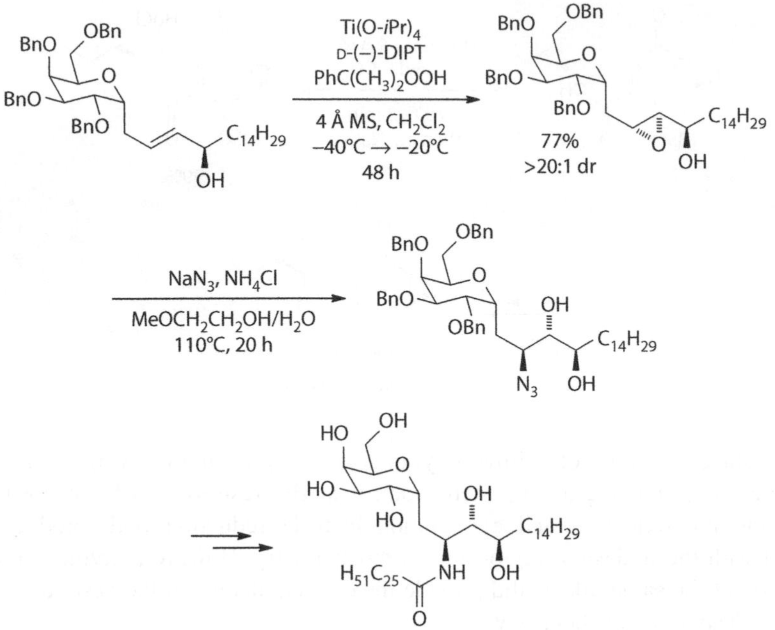

α-1C-Galactosylceramide

Illicidione A and illihendione A are related natural products derived from the stem bark of *Illicium oligandrum* and *Illicium henryi*, respectively. They were prepared by Yu and coworkers[58] through the Diels–Alder reaction of a dihydropyran-fused *o*-quinol with either itself or with a structurally similar dihydrofuran-fused *o*-quinol. Both of the Diels–Alder precursors could be prepared through epoxide opening of a tethered phenol. Acid-catalyzed epoxide opening provided the chromane, and the base-promoted epoxide opening yielded the dihydrobenzofuran. Oxidation of a mixture of the Diels–Alder precursors with a stabilized formulation of IBX safe for scale up (SIBX)[59] formed the *o*-quinols. Conveniently, the dihydrofuran-fused *o*-quinol served only as a dienophile, leading only to the two desired Diels–Alder adducts instead of the four possible adducts.

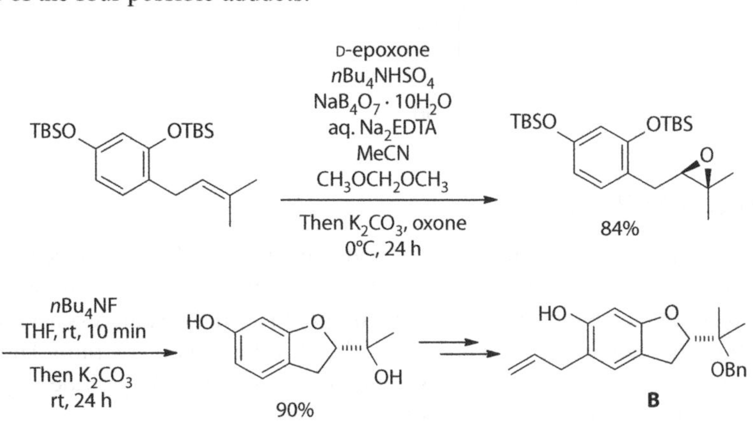

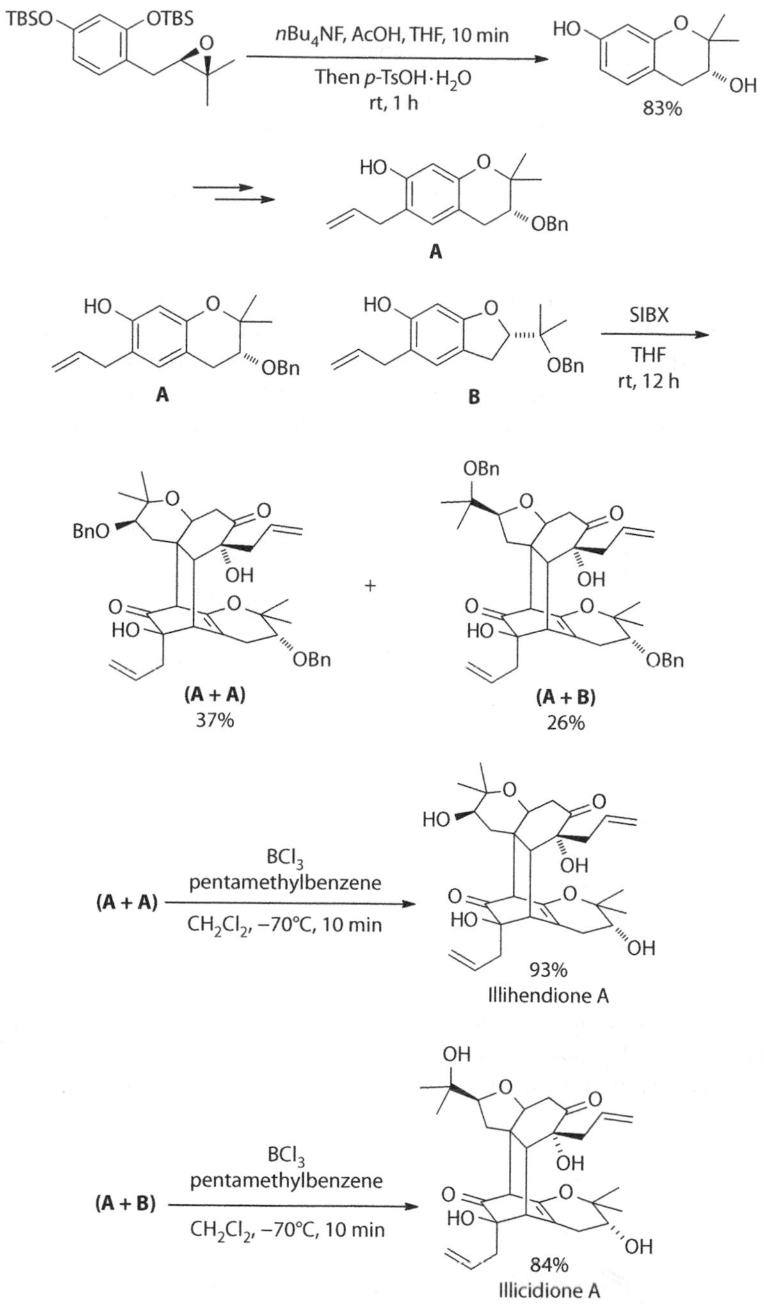

(−)-Amphidinolide K is an epoxide-containing natural product. The epoxide was installed in the final step of the recent synthesis by Vilarrasa and coworkers.[60] After macrolactonization, a Sharpless asymmetric epoxidation with (+)-diethyl tartrate provided the natural product.

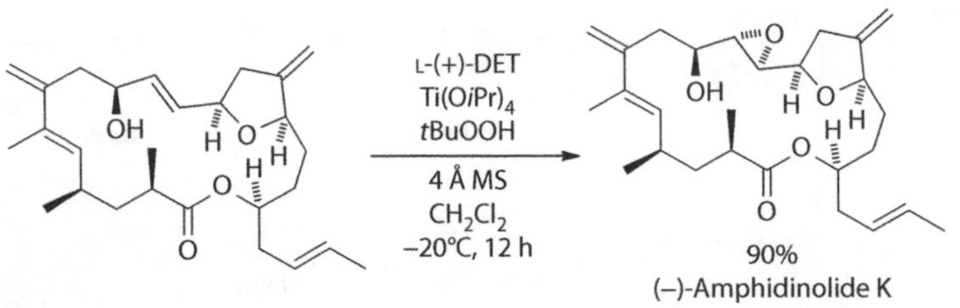

90%
(−)-Amphidinolide K

An *N*-acyl-*N*-alkoxyaziridinium ion provided the necessary site of reactivity to form the lactam and lactone necessary for the formation of (−)-swainsonine by Wardrop and Bowen.[61] Oxidation of an *N*-alkoxy enamide with *bis*(trifluoroacetoxy) iodobenzene induced intramolecular aziridinium ion formation. The aziridinium ion was opened with a tethered methyl ester to form both a lactone and a lactam in the same step.

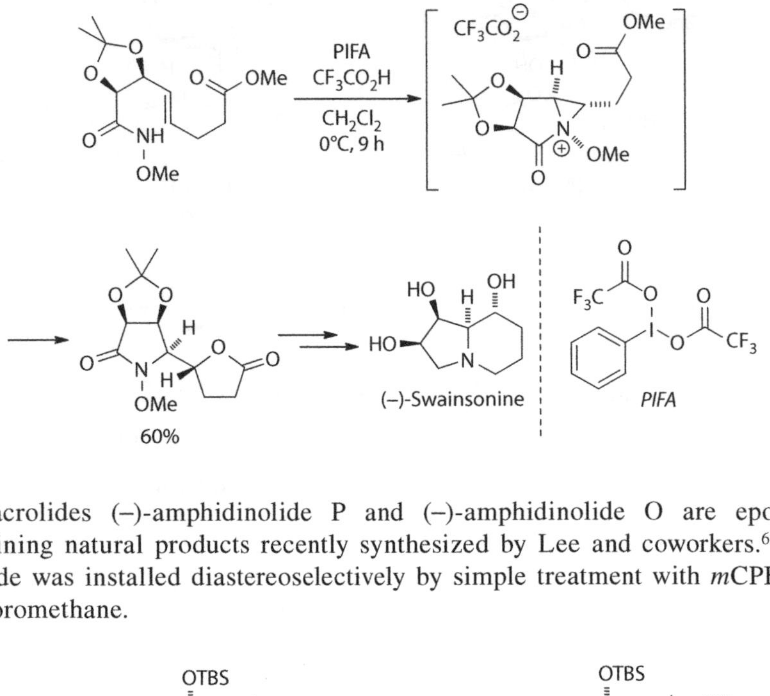

(−)-Swainsonine   PIFA

60%

Macrolides (−)-amphidinolide P and (−)-amphidinolide O are epoxide-containing natural products recently synthesized by Lee and coworkers.[62] The epoxide was installed diastereoselectively by simple treatment with *m*CPBA in dichloromethane.

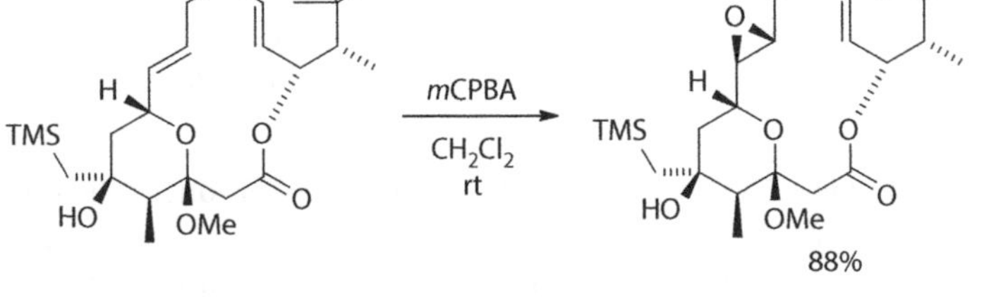

88%

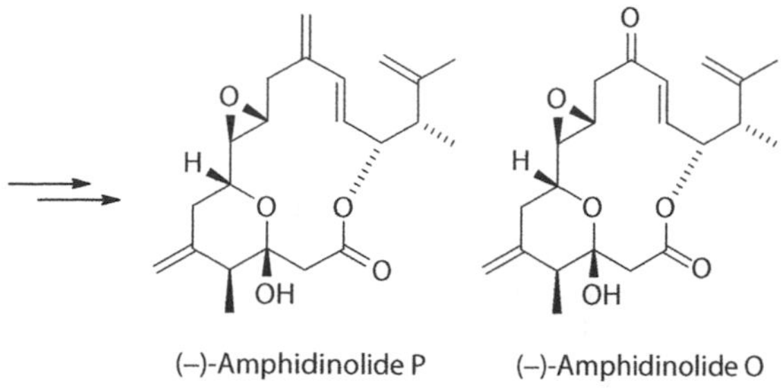

(−)-Amphidinolide P   (−)-Amphidinolide O

4.10   GOLD-CATALYZED ALKYNE HYDRATION

Gold-catalyzed alkyne hydration has proven to be a successful strategy for the formation of cyclic ethers and ketones. A tandem gold-catalyzed electrophilic addition of both an alkyne and a tethered alkene was accomplished by Echavarren and coworkers in their synthesis of (−)-nardoaristolone B.[63] The alkyne underwent electrophilic addition to the gold center, followed by electrophilic addition of the alkene onto the alkyne to form a fused tricycle.

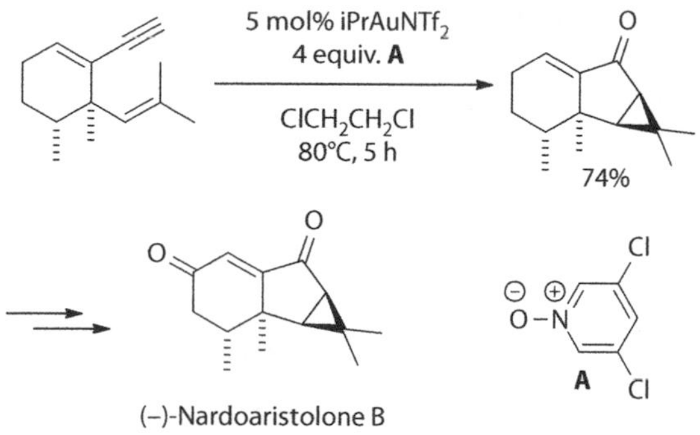

(−)-Nardoaristolone B

In Shair and coworkers' synthesis of (+)-cephalostatin 1, the gold–alkyne complex was attacked by a pendant homopropargylic alcohol to form a cyclic enol ether.[64] The reactive site was further cyclized to ultimately form a spiroketal.

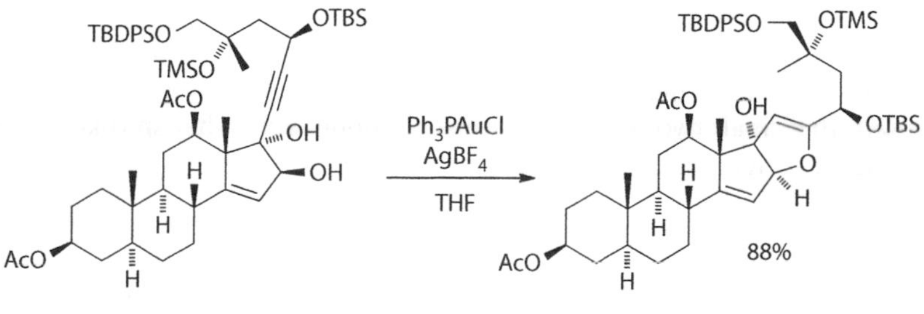

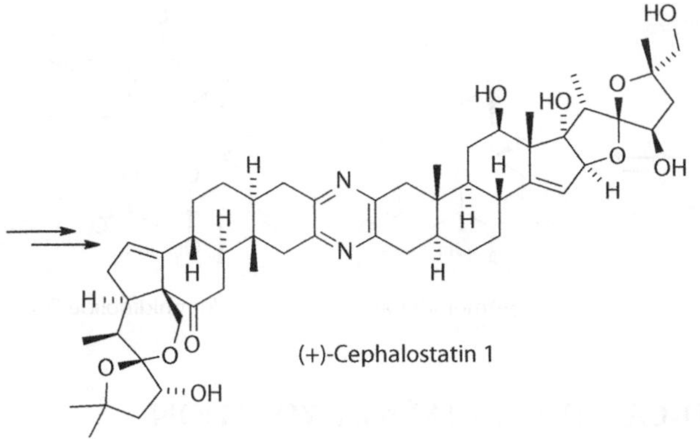

(+)-Cephalostatin 1

Forsyth and coworkers utilized the gold-catalyzed alkyne hydration twice in their formal synthesis of okadaic acid.[65] A trihydroxyalkyne was subjected to 10 mol% gold(I) chloride to close both rings of two separate spiroketals in one step.

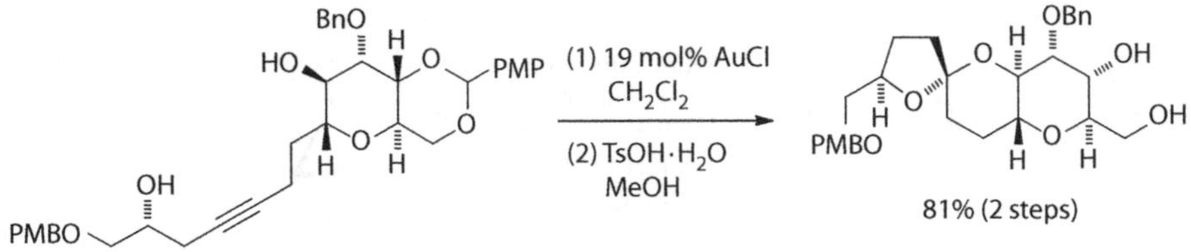

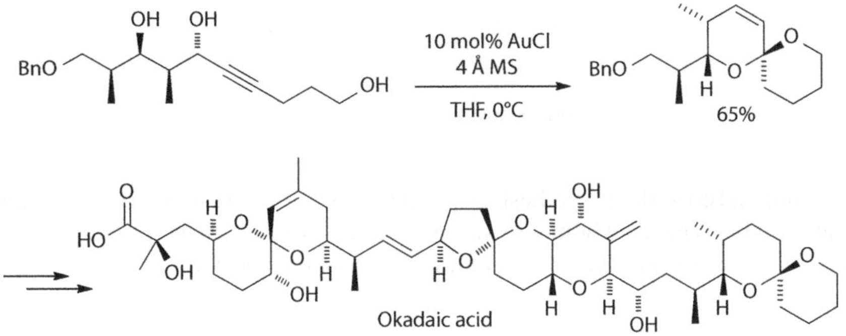

An impressive late-stage spiroketalization reaction by Fürstner and coworkers closed the final two rings of spirastrellolide F.[66] The spiroketalization reaction occurs after macrocyclization and proceeds through an intermediate enol ether.

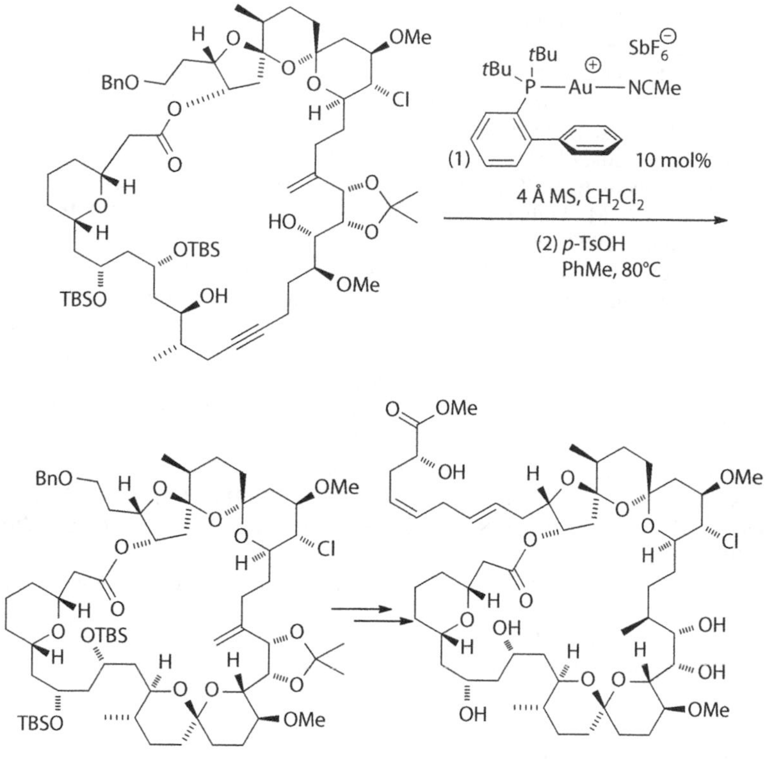

Spirastrellolide F
methyl ester

4.11   CONCLUSION

The number of examples of electrophilic additions to alkenes is quite large. Considering that many electrophilic additions are not necessarily the key steps of a synthesis, these transformations are often not given as much attention in abstracts or the text of articles. There is no doubt that many more examples could have been included in this chapter. Alkenes are versatile functional groups, and their transformations are often critical in setting up the key steps of many synthetic efforts.

REFERENCES

1. Kaschel, J.; Schneider, T. F.; Werz, D. B. *Angew. Chem. Int. Ed.*, **2012**, 51, 7085–7086.
2. Archambeau, A.; Miege, F.; Meyer, C.; Cossy, J. *Acc. Chem. Res.*, **2015**, 48, 1021–1031.
3. Ford, A.; Miel, H.; Ring, A.; Slattery, C. N.; Maguire, A. R.; McKervey, M. A. *Chem. Rev.*, **2015**, 115, 9981–10080.
4. Zhang, D.; Song, H.; Qin, Y. *Acc. Chem. Res.*, **2011**, 44, 447–458.
5. Concellón, J. M.; Rodríguez-Solla, H.; Concellón, C.; del Amo, V. *Chem. Soc. Rev.*, **2010**, 39, 4103–4113.
6. Qian, D.; Zhang, J. *Chem. Soc. Rev.*, **2015**, 44, 677–698.
7. Miege, F.; Meyer, C.; Cossy, J. *Angew. Chem. Int. Ed.*, **2011**, 50, 5932–5937.

8. Lian, Y.; Miller, L. C.; Born, S.; Sarpong, R.; Davies, H. M. L. *J. Am. Chem. Soc.*, **2010**, 132, 12422–12425.
9. Uwamori, M.; Saito, A.; Nakada, M. *J. Org. Chem.*, **2012**, 77, 5098–5107.
10. Uwamori, M.; Nakada, M. *Tetrahedron Lett.*, **2013**, 54, 2022–2025.
11. Uwamori, M.; Nakada, M. *J. Antibiot.*, **2013**, 66, 141–145.
12. Sawada, T.; Nakada, M. *Org. Lett.*, **2013**, 15, 1004–1007.
13. Taber, D. F.; Sheth, R. B.; Tian, W. *J. Org. Chem.*, **2009**, 74, 2433–2437.
14. Hochlowski, J. E. M.; Mullally, M.; Spanton, S. G.; Whittern, D. N.; Hill, P.; McAlpine, J. B. *J. Antibiot.*, **1993**, 46, 380–386.
15. He, B.; Song, H.; Du, Y.; Qin, Y. *J. Org. Chem.*, **2009**, 74, 298–304.
16. Zhang, M.; Huang, X.; Shen, L.; Qin, Y. *J. Am. Chem. Soc.*, **2009**, 131, 6013–6020.
17. Shen, L.; Zhang, M.; Wu, Y.; Qin, Y. *Angew. Chem. Int. Ed.*, **2008**, 47, 3618–3621.
18. Yang, J.; Wu, H.; Shen, L.; Qin, Y. *J. Am. Chem. Soc.*, **2007**, 129, 13794–13795.
19. McNeill, E.; Ritter, T. *Acc. Chem. Res.*, **2015**, 48, 2330–2343.
20. Thomas, S. P.; Aggarwal, V. K. *Angew. Chem. Int. Ed.*, **2009**, 48, 1896–1898.
21. Sharpe, R. J.; Johnson, J. S. *J. Org. Chem.*, **2015**, 80, 9740–9766.
22. Diethelm, S.; Carreira, E. M. *J. Am. Chem. Soc.*, **2015**, 137, 6084–6096.
23. Weiss, M. E.; Carreira, E. M. *Angew. Chem. Int. Ed.*, **2011**, 50, 11501–11505.
24. Tietze, L. F.; Clerc, J.; Biller, S.; Duefert, S.-C.; Bischoff, M. *Chem. Eur. J.*, **2014**, 20, 17119–17124.
25. Fuwa, H.; Ishigai, K.; Hashizume, K.; Sasaki, M. *J. Am. Chem. Soc.*, **2012**, 134, 11984–11987.
26. Hoye, T. R.; Danielson, M. E.; May, A. E.; Zhao, H. *J. Org. Chem.*, **2010**, 75, 7052–7060.
27. Cho, H.-K.; Lim, H.-Y.; Cho, C.-G. *Org. Lett.*, **2013**, 15, 5806–5809.
28. Shin, H.-S.; Jung, Y.-G.; Cho, H.-K.; Park, Y.-G.; Cho, C.-G. *Org. Lett.*, **2014**, 16, 5718–5720.
29. Kotora, M.; Hessler, F.; Eignerová, B. *Eur. J. Org. Chem.*, **2012**, 2012, 29–42.
30. Lee. H.-W.; Kwong, F.-Y. *Eur. J. Org. Chem.*, **2010**, 2010, 789–811.
31. Kitagaki, S.; Inagaki, F.; Mukai, C. *Chem. Soc. Rev.*, 2014, 43, 2956–2978.
32. Reddy, L. V. R.; Kumar, V.; Sagar, R.; Shaw, A. K. *Chem. Rev.*, **2013**, 113, 3605–3631.
33. Kim, D. H.; Kim K.; Chung, Y. K. *J. Org. Chem.*, **2006**, 71, 8264–8267.
34. Kaasalaine, E.; Tois, J.; Russo, L.; Rissanen, K.; Helaja, J. *Tetrahedron Lett.*, **2006**, 47, 5669–5672.
35. Fager-Jokela, E.; Kaasalainen, E.; Leppänen, K.; Tois, J.; Helaja, J. *Tetrahedron*, **2008**, 64, 10381–10387.
36. Honda, T.; Kaneda, K. *J. Org. Chem.*, **2007**, 72, 6541–6547.
37. Kaneda, K.; Honda, T. *Tetrahedron*, **2008**, 64, 11589–11593.
38. Vázquez-Romero, A.; Rodriguez, J.; Lledó, A.; Verdaguer, X.; Riera, A. *Org. Lett.*, **2008**, 10, 4509–4512.
39. Vázquez-Romero, A.; Cárdenas, L.; Blasi, E.; Verdaguer, X.; Riera, A. *Org. Lett.*, **2009**, 11, 3104–3107.
40. Hirose, T.; Miyakoshi, N.; Mukai, C. *J. Org. Chem.*, **2008**, 73, 1061–1066.
41. Hayashi, Y.; Ogawa, K.; Inagaki, F.; Mukai, C. *Org. Biomol. Chem.*, **2012**, 10, 4747–4751.
42. Jørgensen, L.; McKerrall, S. J.; Kuttruff, C. A.; Ungeheuer, F.; Felding, J.; Baran, P. S. *Science*, **2013**, 341, 878–882. Corrections: Jørgensen, L.; McKerrall, S. J.; Kuttruff, C. A.; Ungeheuer, F.; Felding, J.; Baran, P. S. *Science*, **2015**, 350.
43. McDonald, B. R.; Scheidt, K. A. *Acc. Chem. Res.*, **2015**, 48, 1172–1183.
44. Ellis, J. M.; Crimmins, M. T. *Chem. Rev.*, **2008**, 108, 5278–5298.
45. Crane, E. A.; Scheidt. K. A. *Angew. Chem. Int. Ed.*, **2010**, 49, 8316–8326.
46. Yadav, J. S.; Narayana Kumar, G. G. K. S. *Tetrahedron*, **2010**, 66, 480–487.
47. Woo, S. K.; Lee, E. *J. Am. Chem. Soc.*, **2010**, 132, 4564–4565.

48. Crane, E. A.; Zabawa, T. P.; Farmer, R. L.; Scheidt, K. A. *Angew. Chem. Int. Ed.*, **2011**, 50, 9112–9115.

49. Pearson, W. H.; Fang, W. *J. Org. Chem.*, **2000**, 65, 7158–7174. Corrections: Pearson, W. H.; Fang, W. *J. Org. Chem.*, **2001**, 66, 6838.

50. Ito, Y.; Nakajo, E.; Nakatsuka, M.; Saegusa, T. *Tetrahedron Lett.*, **1983**, 24, 2881–2884.

51. Underwood, B. S.; Tanuwidjaja, J.; Ng, S.-S.; Jamison, T. F. *Tetrahedron*, **2013**, 69, 5205–5220.

52. Nemecek, G.; Thomas, R.; Goesmann, H.; Feldmann, C.; Podlech, J. *Eur. J. Org. Chem.*, **2013**, 2013, 6420–6432.

53. Brookes, P. A.; Cordes, J.; White, A. J. P.; Barrett, A. G. M. *Eur. J. Org. Chem.*, **2013**, 2013, 7313–7319.

54. Mohapatra, D. K.; Pu lluri, K.; Gajula, S.; Yadav, J. S. *Tetrahedron Lett.*, **2015**, 56, 6377–6380.

55. Shoji, M.; Uno, T.; Hayashi, Y. *Org. Lett.*, **2004**, 6, 4535–4538.

56. Palimkar, S. S.; Uenishi, J.; Hiromi, I. *J. Org. Chem.*, **2012**, 77, 388–399.

57. Liu, Z.; Byun, H.-S.; Bittman, R. *J. Org. Chem.*, **2011**, 76, 8588–8598.

58. Ren, X.-D.; Zhao, N.; Ux, S.; Lü, H.-N.; Ma, S.-G.; Liu, Y.-B.; Li, Y.; Qu, J.; Yu, S.-S. *Tetrahedron*, **2015**, 71, 4821–4829.

59. Ozanne, A.; Pouységu, L.; Depernet, D.; François, B.; Quideau, S. *Org. Lett.*, **2003**, 5, 2903–2906.

60. Sánchez, D.; Andreou, T.; Costa, A. M.; Meyer, K. G.; Williams, D. R.; Barasoain, I.; Díaz, J. F.; Lucena-Agell, D.; Vilarrasa, J. *J. Org. Chem.*, **2015**, 80, 8511–8519.

61. Wardrop, D. J.; Bowen, E. G. *Org. Lett.*, **2011**, 13, 2376–2379.

62. Hwang, M.-H.; Han, S.-J.; Lee, D.-H. *Org. Lett.*, **2013**, 15, 3318–3321.

63. Homs, A.; Muratore, M. E.; Echavarren, A. M. *Org. Lett.*, **2015**, 17, 461–463.

64. Fortner, K. C.; Kato, D.; Tanaka, Y.; Shair, M. D. *J. Am. Chem. Soc.*, **2010**, 132, 275–280.

65. Fang, C.; Pang, Y.; Forsyth, C. J. *Org. Lett.*, **2010**, 12, 4528–4531.

66. Benson, S.; Collin, M.-P.; Arlt, A.; Gabor, B.; Goddard, R.; Fürstner, A. *Angew. Chem. Int. Ed.*, **2011**, 50, 8739–8744.

5 Electrophilic Aromatic Substitution

Jie Jack Li

CONTENTS

5.1 INTRODUCTION

Electrophilic aromatic substitution is also known as EAS or S_EAr for short. On this topic, in 1990, Taylor published a book titled *Electrophilic Aromatic Substitution.*[1] In this chapter, key advancements in the field of EAS are summarized.

The most fundamental mechanism for EAS reaction is the Ingold–Hughes mechanism shown below.[2] More sophisticated mechanisms have been put forward,[3] with the most popular one invoking π complex **4** and σ complex **2**.

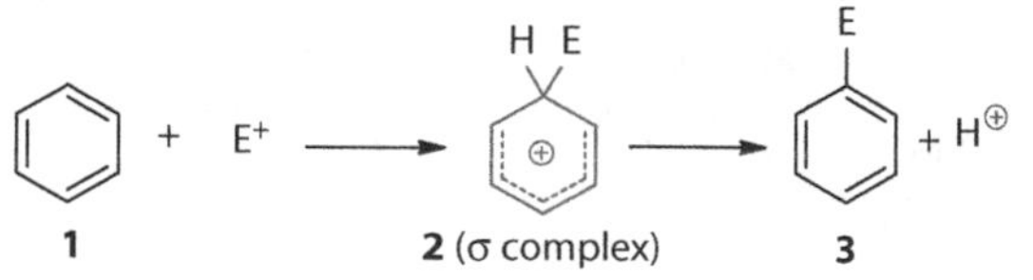

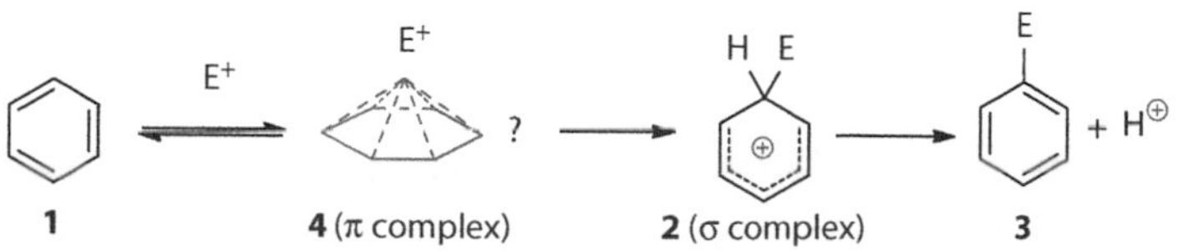

In 2003, Lenoir highlighted the electrophilic substitution of arenes.[4] The textbook perception of the mechanism of such reactions may be represented as **1 → 3** with the key intermediate as π complex **4**. The nature of the σ complex is relatively straightforward, the nature of the π complex **4** is less clear although Benesi and Hildebrand first established the existence of π or charge-transfer (CT) complexes in 1949 from the characteristic CT bands in the UV/Vis spectrum. In 1991, Kochi et al. derived a symmetrical *over-ring structure* similar to **4** for the complex from mesitylene and NO. In early 2000s, Kochi et al. demonstrated the existence of CT complexes as a key intermediate in the bromination of benzene and toluene by isolating different arene–bromine CT complexes and determined the x-ray crystal structure at –150°C. Another mechanism involves the intermediacy of radical pair **5** while a third mechanism involves a combination of π complex **4** and radical pair **5**.

1 **5** (radical pair) **2** (σ complex)

1 **4** (π complex) **5** (radical pair) **2** (σ complex)

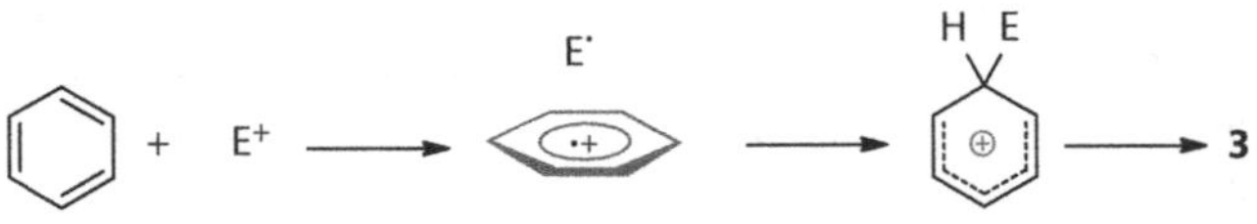

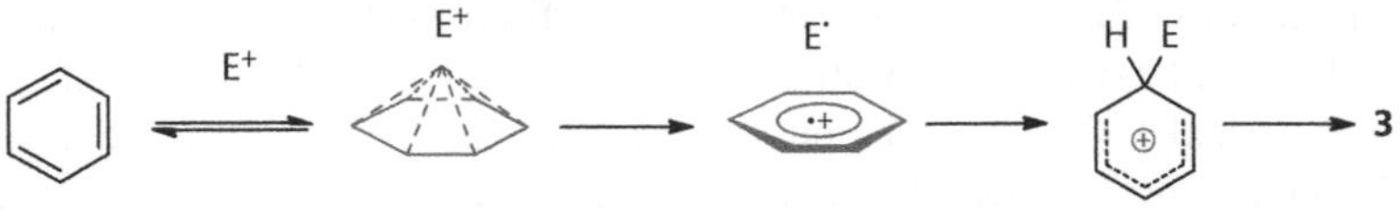

One of the unconventional outcomes of EAS is the formation of dienones and quinones.[5] When 4-bromo-1,2-dimethylbenzene (**6**) was treated with nitric acid in acetic anhydride, 3,4-dimethyl-4-nitrocyclohexa-2,5-dien-1-one (**9**) was produced in addition to conventional nitration product via intermediates **7** and **8**. As a matter of fact, in the case of 3,4-dimethylanisole, its adduct **11** has been isolated and shown to decompose readily to nitrocyclohexadienone **9**.

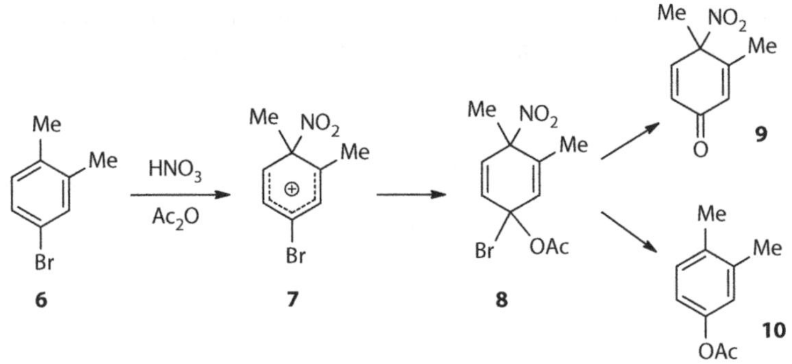

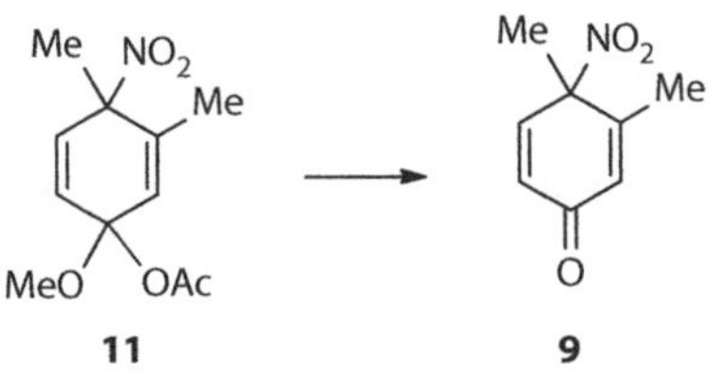

Another unconventional outcome of EAS features the *t*-butyl group as the leaving group.[6] Bromination of 1,3,5-tri-*t*-butylbenzene (**12**) in acetic acid gives rise to four products **15–18**, presumably through intermediates **13** and **14**. Acetoxylation product **17** is the product from intermediate **13** and bromination product **18** is the product from intermediate **14**, respectively. Both cases have the *t*-butyl cation as the leaving group!

A review published in 2014 highlighted recent advances in mechanistic studies of the aromatic C–H bond substitution and related chemistry.[7]

5.2 NITRATION

EAS is one of the most important types of reactions in undergraduate education. Nitration using sulfonitric mixture as an experiment for undergraduates is far from "green." Attempts have been made to make this more environmentally friendly by using tyrosine (**19**) as the substrate and water as the solvent.[8]

Tyrosine (**19**)   **20**

Nitration of phenylacetic acid (**21**) affords an opportunity for undergraduates to solve a puzzle of the products.[9] Only one major product **22** was obtained when phenylacetic acid (**21**) was heated together with 90% HNO_3. The melting point and NMR (nuclear magnetic resonance) were consistent with those of 2,4-dinitro phenylacetic acid. The nitration product was more puzzling when phenylacetic acid (**21**) was heated together with 70% HNO_3. Since the resulting product had a melting point lower than all three possible mono-nitrated products, it was decided that the product was actually a mixture of 2-nitro-phenylacetic acid (**23**) and 4-nitro-phenylacetic acid (**24**).

Phenylacetic acid (21)   22

Phenylacetic acid (21)   23   24

25   26, 10%   27, 90%

Nitration of acetanilide (**25**) gives different regiochemical outcomes depending on the reaction conditions. When **25** was nitrated with conventional sulfonitric mixture, the *ortho*-nitration product **26** was the minor regioisomer (10%) and the *para*-nitration product **27** was the predominant regioisomer (90%).[10] This result may be explained using the electric hindrance argument because the amide adopts a resonance structure where the C=N has a positive charge on the nitrogen atom. This positive charge would repel the positive charge on the nitronium ion and drive the formation of the product the *para*-nitration. On the other hand, when **25** was nitrated with acetyl nitrate (also known as acetonitric anhydride, **28**), the *ortho*-nitration product **26** was isolated exclusively. This phenomenon may be explained by evoking the formation of the precursor complex between the amide and the nitrate reagent, bring it to the vicinity of the *ortho*-position.

25   26

2-(Arylsulfenyl)pyrrole are good substrates for nitration. After adding one or two nitro groups to the benzene ring, the resulting substrates **29** and **31** are still active enough to be nitrated at the 5-position of the pyrrole ring to produce products **30** and **32**, respectively.[11] The outcome is not completely surprising because pyrrole is one of the most electron-rich aromatic heterocycles. However, when the sulfides on **29** and **31** are oxidized to the corresponding sulfoxides, they cannot be nitrated at all because the sulfoxide group is too electron withdrawing.

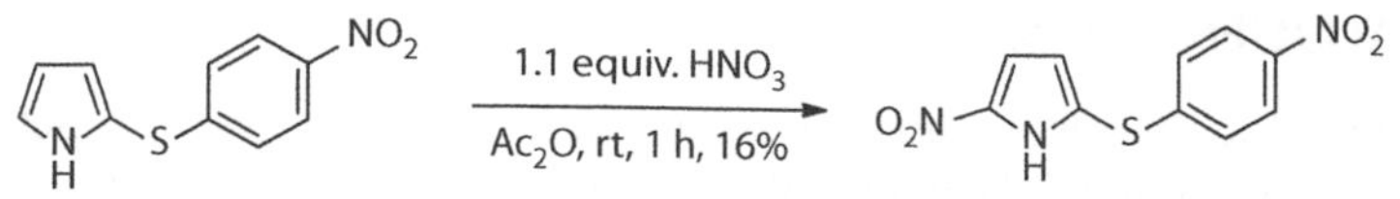

Nitration of substituted pyrido[1,2-*a*]benzimiodazoles have been reported.[12] Since the pyridyl ring is an electron-deficient heterocycle, it is not surprising that EAS reactions take place on the benzene ring. In case of substrate **33**, the nitration still took place on the benzene ring in spite of the presence of a nitro group to provide dinitro product **34**.

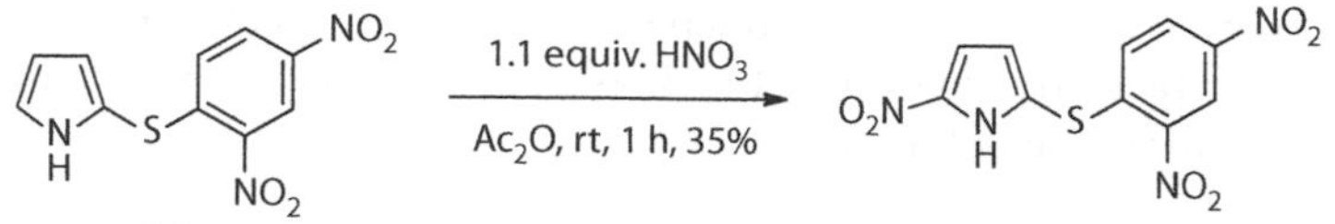

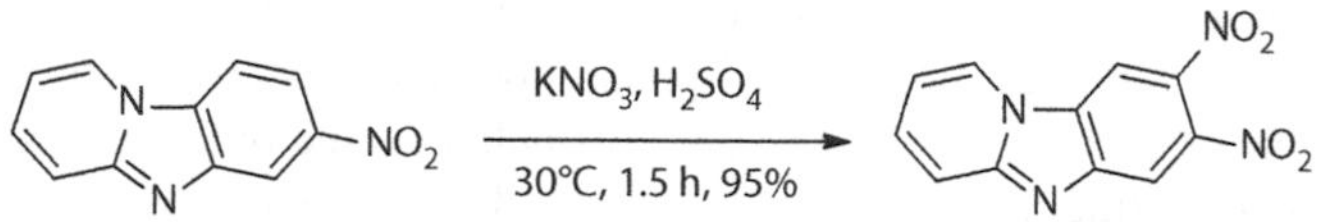

When 1,2-dialkoxybenzene (**35**, R = –Me, –Et, –*n*-Bu, –*n*-Hex) was treated with concentrated nitric acid, the 4,5-dinitro product **36** was obtained almost exclusively.[13] On the other hand, nitration of 1,4-dialkoxybenzene (**37**, R = –Me, –Et, –*n*-Bu, –*n*-Hex) with concentrated nitric acid also displayed good regioselectivity. However, instead of the expected regioisomers, dinitro products **38** and **39** were the only isomers detected in an approximately 9:1 ratio. The interesting regioselectivity was investigated using the density functional theory (DFT) analysis, which suggests that the nitration is likely to involve the single-electron transfer (SET) mechanism. For the nitration of 1,2-dialkoxy-benzene **35**, the regioselectivity is mainly determined by the symmetry of the HOMO (highest occupied molecular orbital) of the aromatic moiety that defines the structure of the SHOMO (second highest occupied molecular orbital) of the aromatic radical cation formed by the SET process. For the nitration of 1,4-dialkoxybenzene **37**, salvation effects were the determining factor to give rise to the observed regioselectivity.

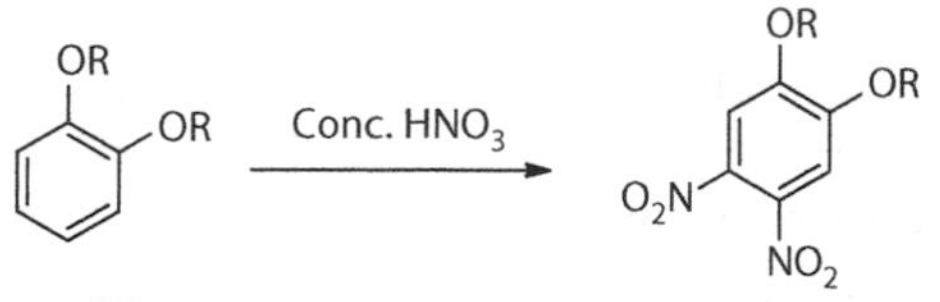

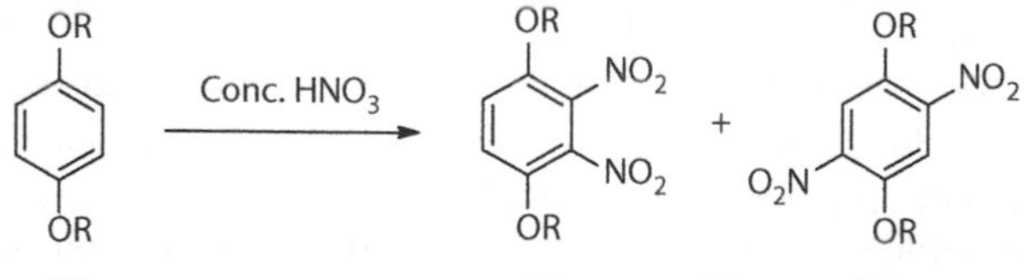

5.3 HALOGENATION

Fluorination of aromatics was known as early as 1927 when the Balz–Schiemann reaction was discovered.[14] Therefore, diazotization of aniline **40** gives rise to diazonium tetrafluoroborate **41**, which readily decomposes to fluorobenzene **42** upon heating. However, regardless of the minutia of the detailed mechanism, the overall outcome of the transformation from **41** to **42** is a *nucleophilic* aromatic substitution. This is outside the scope of this chapter.

Fluorination via *electrophilic* aromatic substitution was known in the early 1970s when Grakauskas reported direct liquid-phase fluorination of aromatic compounds using molecular fluorine.[15] Also in 1970, Shaw et al. reported that fluorination of aromatic compounds using xenon difluoride followed the EAS mechanism.[16] For electrophilic fluorination of aromatic compounds using F_2, Hehre and Hiberty[17] studied the intermediates and Wolfe and colleagues investigated[18] the substrate selectivity and orientation. A review on elective direct fluorination using elemental fluorine was published in 2001.[19]

The amine group on aniline is an activating and an *ortho-* and *para*-directing group (DG). However, when anilines were treated with elemental fluorine in triflic acid, *meta*-fluorination was the major product.[20] The reaction was sometimes promoted by a catalytic amount of SbF_5 and the regioselectivity was increased when an electron-donating substituent was situated at the *para*-position.

One of the first synthetically useful electrophilic fluorination of aromatic compounds was reported by Fifolt et al.[21] Fluoroxytrifluoromethane (**43**, FTM, CF_3OF) was prepared from carbon monoxide and fluorine, followed by cesium fluoride catalysis. On the other hand, bis(fluoroxy)difluoromethane (**44**, BDM) was prepared from carbon dioxide and fluorine, followed by cesium fluoride catalysis. When *N*-(4-(trifluoromethyl)phenyl)acetamide (**45**) as a substrate was treated with fluoroxytrifluoromethane (**43**) in $CHCl_3$, *ortho*-monofluorinated product **46** was obtained in greater than 57% yield because the *para*-position was blocked. In addition to electrophilic fluorination of aromatic compounds, fluoroxytrifluoromethane (**43**) was used to achieve direct trifluoromethoxylation although the reaction went through a SET mechanism.[22]

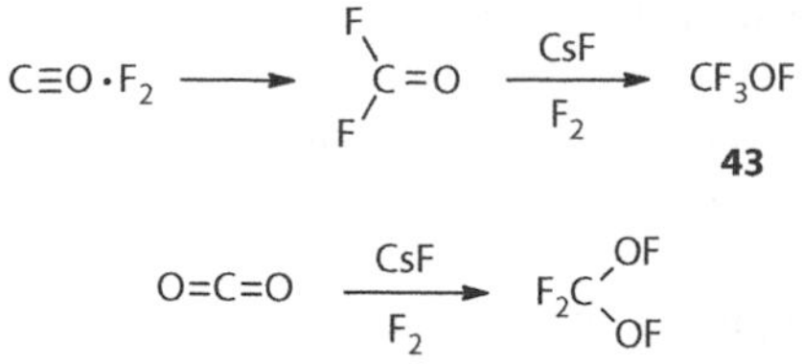

$$C \equiv O \cdot F_2 \longrightarrow \underset{F}{\overset{F}{\diagdown}} C = O \xrightarrow[F_2]{CsF} CF_3OF$$

43

$$O = C = O \xrightarrow[F_2]{CsF} F_2C \overset{OF}{\underset{OF}{\diagdown}}$$

44

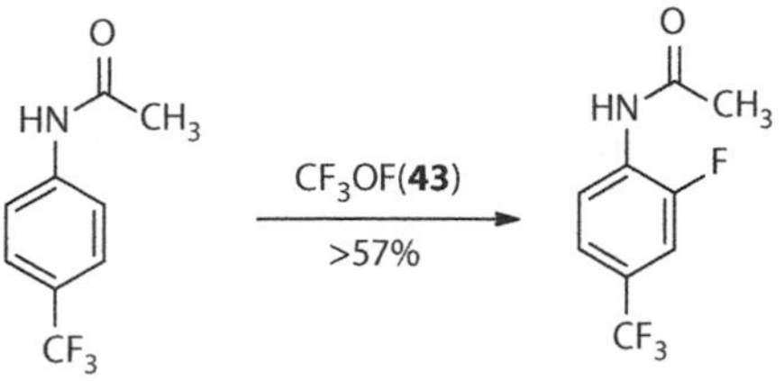

Fast forward to 2003. Klapötke and colleagues prepared a more convenient fluorinating reagent, 1-fluoro-2,4,6-trochloro-1,3,5-triazinium tetrafluroborate ([(ClCN)$_3$F]$^+$[BF$_4$]$^-$, **47**) using 2,4,6-trichlorotriazine, fluorine, and boron trifluoride.[23] Electrophilic fluorination of chlorobenzene gave the *ortho-*, *meta-*, and *para*-fluoro-chlorobenzne in a 1:0.3:2 ratio. Indeed, although chlorine atom is a deactivating functional group, it also is an *ortho-* and *para*-DG.

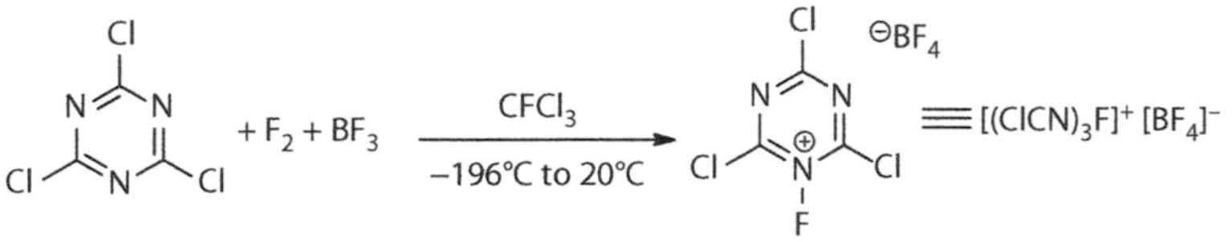

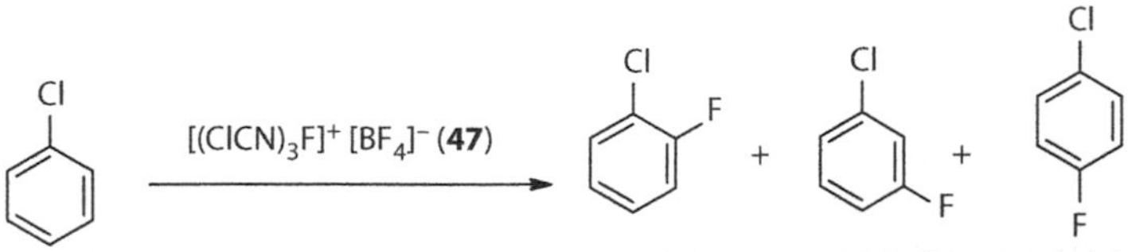

Interestingly, fluorofullerene C$_{60}$F$_{18}$ was found to be an efficient fluorinating agent via the EAS mechanism.[24] Even more interestingly, the reaction between 4-substituted trimethyltin aromatics with perchloryl fluoride (FClO$_3$) did not yield the desired fluorinated aromatic compounds,[25] yet it was accepted for publication. It is refreshing that negative results were published. Some may argue that more negative results should be published but they rarely passed muster with referees and associate editors.

Borodkin et al. investigated the kinetic isotope effects (KIE) and mechanism for electrophilic fluorination of aromatic compounds with nitrogen–fluorine (NF)-type reagents such as 1-chloro-2,2-methyl-4-fluoro-1,4,diazoniabicyclo[2.2.2]octane bis(tetrafluoroborate) (Selectfluor, F-TMEDA-BF$_4$), 1,1′-difluro-2,2′-bispyridinium bis(tetrafluoroborate), and *N*-fluorobenzenesulfonimide (NFSI).[26] The deuterium isotope effects was small ($k_H/k_D = 0.86$–0.99) thus indicating that decomposition of a Wheland-type intermediate was not rate determining. It was concluded that the mechanism is consistent with a polar S_EAr mechanism.

Selectfluor is probably the most popular electrophilic fluorinating agent. With the aid of microwave, diphenylether **51** was fluorinated product **52** in 14% yield.[27] Although the yield was low, but it was convenient because it was the last-stage functionalization of a complex molecule.

Using ionic liquid as the solvent for electrophilic fluorination may be considered as green chemistry since the solvent can be recycled. Electrophilic fluorination of *N*-(naphthalen-2-yl)acetamide (**53**) using Selectfluor in a mixture of [bmim][PF$_6$] and EtOH as the solvent formed *N*-(1-fluoronaphthalen-2-yl)acetamide (**54**) as the predominant product.[28]

The selectivity problem in electrophilic fluorination of aromatic compounds was reviewed in 2010.[29]

Radiosynthesis of [¹⁸F]-Selectfluor bis(triflate) was accomplished and was evaluated to make the positron (β^+) emission tomography (PET) reagent.[30] Tobias and colleagues also reported a fluoride-derived electrophilic late-stage fluorination reagent for PET imaging.[31] The reagents are arylpalladium complexes where the palladium atom is chelated by three nitrogen atoms. In 2012, a review was published delineating the conversion of fluoride to fluronium ion.[32]

The methoxyl group is an *ortho*- and *para*-DG. When electron-rich substrate **55** was treated with Selectfluor in acetonitrile, the reaction offered the expected *ortho*-fluorination product **56** and an unusual *para*-fluorination product in the form of 4-methyl-4-fluoro-cyclohexa-2,5-dienone (**57**).[33] Needless to say, compound **57** was the consequence of the electron-pushing nature of the methoxyl group.

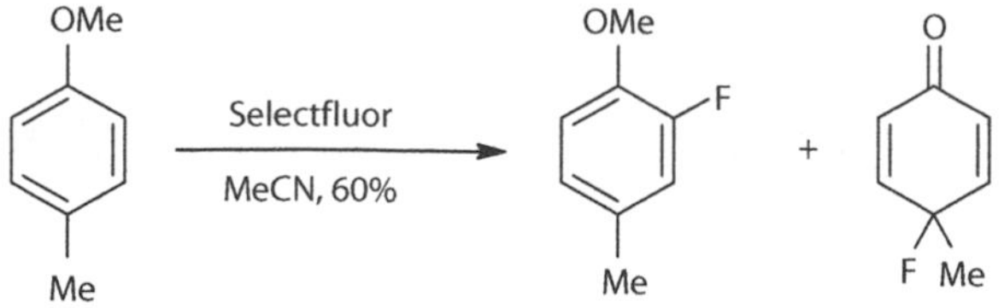

Imidazo[1,2-*a*]pyridines such as **58** are electron-rich aromatic heterocycles. Treatment of **59** with Selectfluor in aqueous condition in the presence of 4-dimethylaminopyridine (DMAP) afforded regioselective fluorination product **59** in 83% yield.[34]

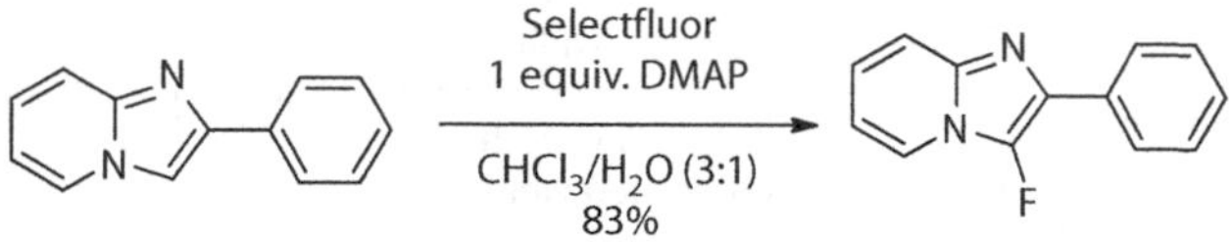

In 2009, Knochel et al. reported a convenient electrophilic fluorination of Grignard reagents.[35] Grignard reagent **60** was prepared from the corresponding aryl bromide via halogen–metal exchange with *i*-PrMgCl · LiCl. Treatment of **60** with *N*-fluorobenzenesulfonimide (NFSI) then produced aryl fluoride **61** in 91% yield. Addition of fluorinated solvent perfluorodecalin was helpful.

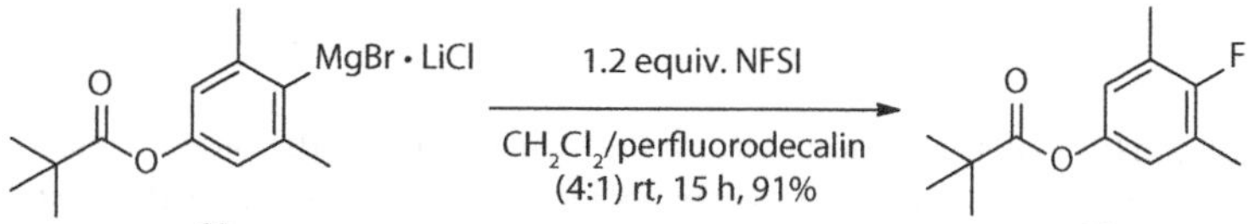

A curious side reaction was reported when NFSI was used as the electrophilic fluorinating agent. Instead of the corresponding fluorinated product, *N*-methyl-*N*-(phenylsulfonyl)benzenesulfonamide (**63**) was observed when tetramethylnaphthalene-1,8-diamine **62** was treated with NFSI.[36]

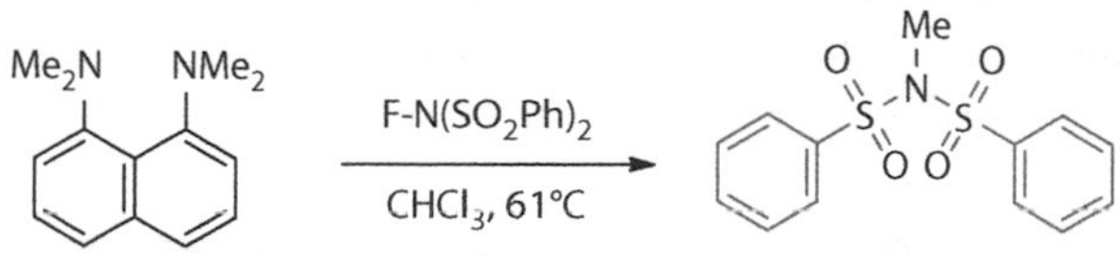

Palladium-catalyzed electrophilic fluorination of carboranes has been reported.[37] With the assistance of a Pd(II) catalyst, *o*-carborane **64** was treated with 10 equiv. of 1-fluoro-2,4,6-trimethylpyridinium triflate (**65**) to lead to tetrafluoro-*o*-carborane **66** in 78% yield.

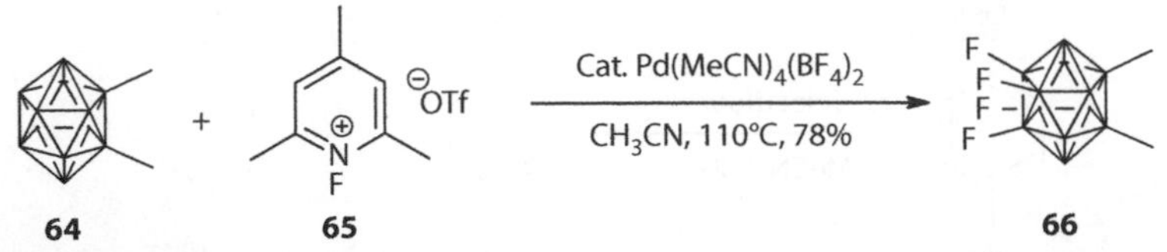

Using pyrazole as the directing group (DG), *ortho*-substituted phenyl-pyrazole **67** was fluorinated with Pd(OAc)$_2$ as the catalyst and *N*-fluorobenzenesulfonimide (NSFI) as the electrophilic fluorinating agent, that is, the fluoronium ion (F$^{\oplus}$) source, to synthesize monofluorinated product **68**.[38] When there was no *ortho*-substitution on substrate **69**, electrophilic fluorination gave rise to difluorinated product **70**.

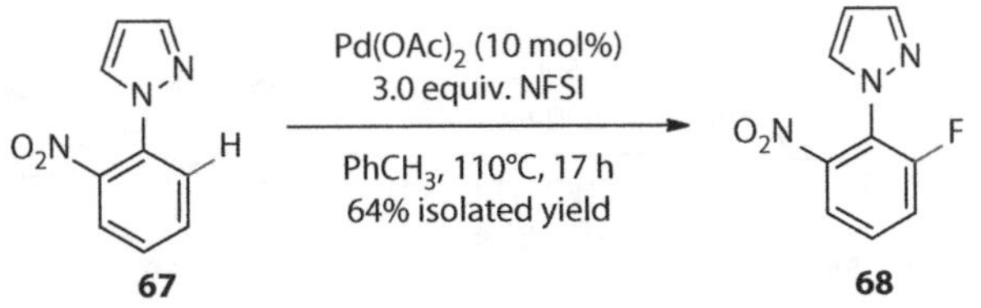

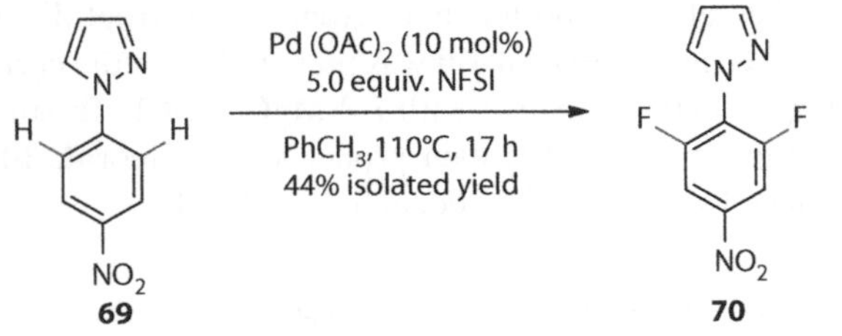

A large portion of this halogenation section has been devoted to electrophilic fluorination, which is justified considering the intense interest in the last decade. Electrophilic chlorination is ignored here since aryl chlorides are not as synthetically useful. However, important developments of electrophilic bromination and iodination during the last decade are summarized below since they are important substrates for transition metal-catalyzed cross-coupling reactions among other applications in organic chemistry.

A "greener" electrophilic bromination was reported to teach organic chemistry experiments for undergraduates.[39] The bromonium ion was generated from the reaction between NaBr and NaClO (bleach) in acetic acid and ethanol. To that end, acetanilide (**25**) was brominated to afford the *para*-brominated product **71** as the major isomer.

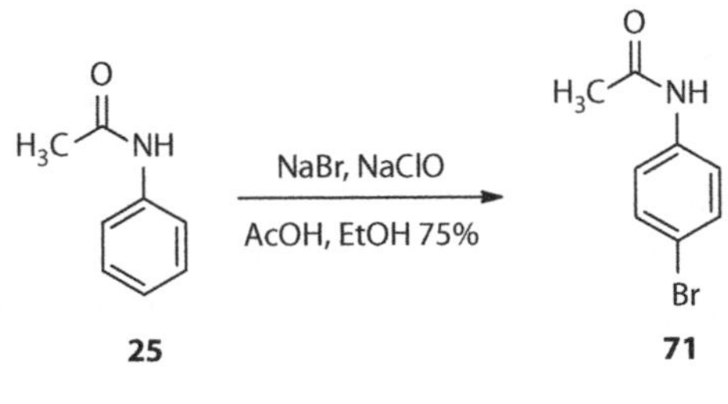

Applying the directing-group effects, we can predict the regiochemical outcome of EAS reactions. When there are three substituents on the benzene ring, the regiochemistry is harder to gauge. It is not obvious where the bromine would go on substrate **72** when it was treated with NBS. However, experimental data indicated bromination **73** was the near exclusive product.[40] Here the sterics were probably the dominating force for the regioselectivity.

As shown in the structures of vancomycin and 2,2′-bis(diphenylphosphino)-1,1′-binaphthyl (BINAP), biaryl atropisomers are isolable if the barrier to rotation about the single bonding linking the rings is high. Miller and coworkers discovered a novel tripeptide-catalyzed asymmetric bromination to carry out dynamic kinetic resolution of biaryl atropisomers.[41] In case of racemic bisphenol (±)-**75** as the substrate, electrophilic bromination with 3 equiv. of N-bromosuccinimide (NBS) took place exclusively on the phenol ring because it is more electron rich than the benzoic acid ring. With the aid of a catalytic amount of the chiral tripeptide **74**, nonracemic biaryl compound **76** was isolated with excellent optical yield and good chemical yield. Employing the same approach with minor variations on the peptide catalysts and bromonium source, enantioselective synthesis of biaryls has been extended to synthesize atropisomeric benzamides[42,43] and 3-arylquinazolin-4(3H)-ones[44] via atroposelective brominations.

In addition to palladium catalysts, copper(I) is able to facilitate C–H halogenation. For a substrate like **77**, controlling the mono- and di-selectivity is an issue encountered to accomplish mono-halogenation selectively. Through careful optimization, the best conditions to give mono-brominated product **79** in 82% yield, with

6% contamination of the dibrominated product **79**, when using 1 equiv. of CuBr and 0.5 equiv. of acetic acid as the promoter.[45]

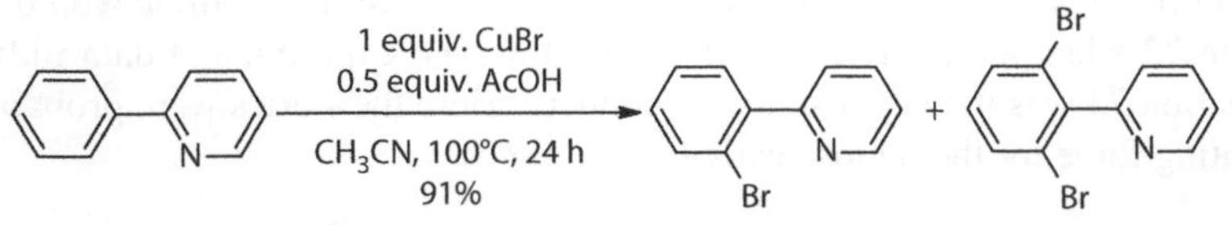

Pyrazolones have been employed as directing groups (DGs) for C–H halogenation as well.[46] For example, 3-methyl-1-phenyl-1*H*-pyrazol-5(4*H*)-one (**80**) was selectively brominated at the *ortho*-position to give **81** with the assistance of a palladium *N*-heterocyclic carbene (NHC) and using NBS as the bromonium source.

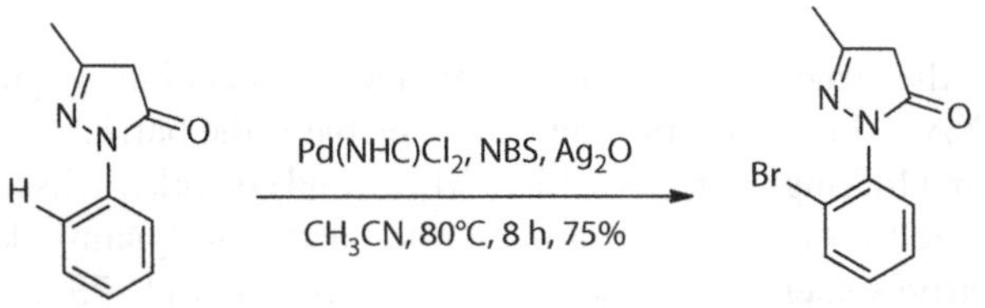

Bromination of heterocycles is interesting because the heteroatoms such as N, O, S change the regiochemical outcomes drastically. In case of benzo[*b*]thiophene-3-carboxylic acid (**82**), bromination takes place predominantly on the benzene ring because thiophene is an electron-deficient heteroaryl compound and an acid makes it even more so. Therefore, electrophilic bromination of **82** gave rise to 6-brominated product **83** and 5-brominated product **84** in a 1.5:1 ratio.[47]

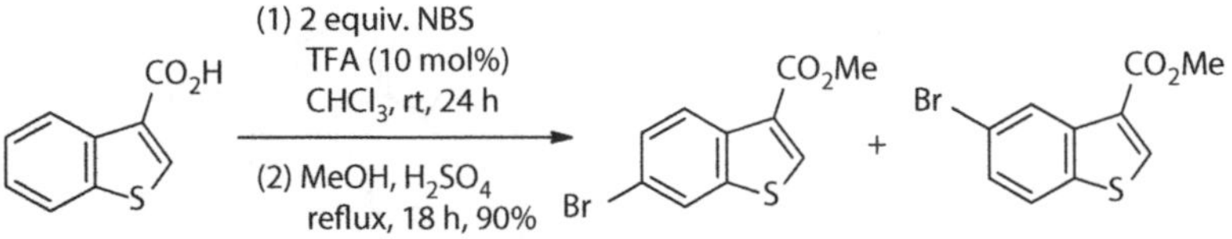

Bromination of the 1*N*-Boc protected pyrrole **85** using NBS gives the C2(α)-bromination product **86** predominantly.[48]

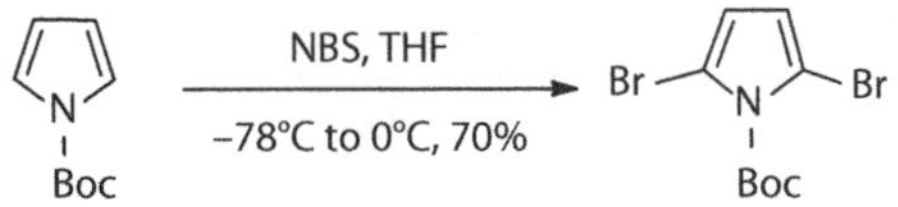

However, if the pyrrole ring is protected with a bulky protective group, the steric hindrance drives the substitution to the β-position. For example, triisopropylsilyl group (TIPS) protected pyrrole **87** was brominated with NBS, C3(β) positions were brominated predominantly to afford **88** and **89**, respectively, depending on the amount of NBS used.[49]

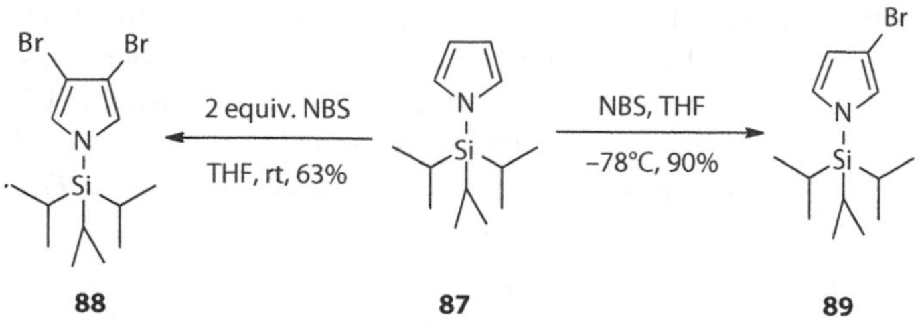

Since electrophilic fluorination and bromination are often applicable to electrophilic iodination, here I will focus on mostly C–H aromatic iodination below.

As early as 2007, Sanford described a regioselective palladium-catalyzed C–H halogenation using chelating groups as the DGs.[50] In addition to pyridine, O-methyl oxime on substrate 90 was a powerful DG for regioselective iodination to give iodobenzene 91. In the same vein, oxazoline on substrate 92 was equally efficient at directing the regioselective iodination to make iodobenzene 93.

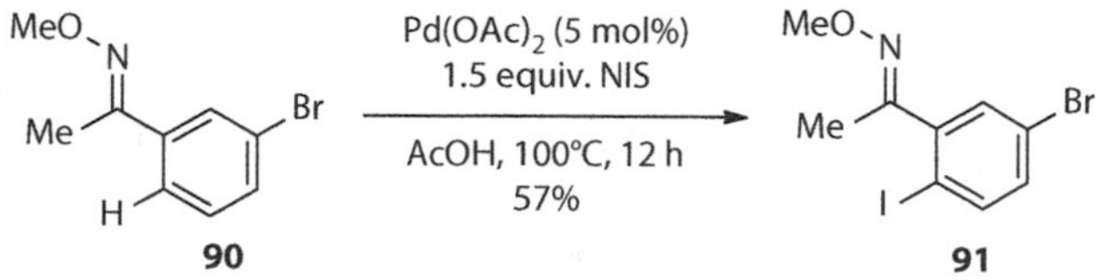

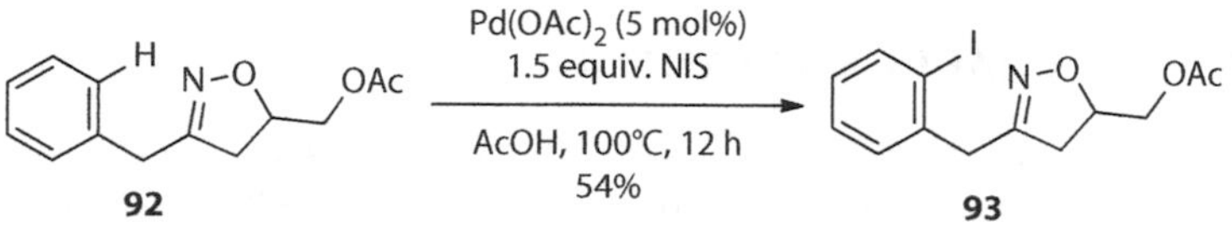

More than a dozen DGs have been found to be effective in directing C–H activation. One of them, amonotetrazole was reported to direct ortho-selective halogenation of arenes.[51] The distal distance between the key N atom on the tetrazole from the H to be four-bond lengths, making the putative intermediate involving the palladium atom a seven-membered ring. As exemplified by substrate 94, palladium-catalyzed ortho-iodination took place on the fluorobenzene ring at the bottom, not the one on top, to give 95.

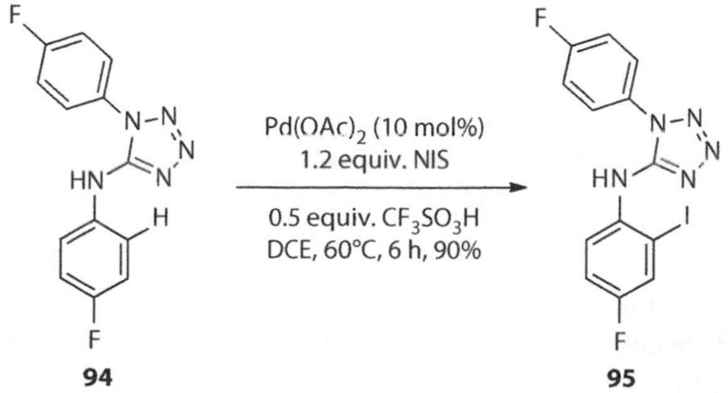

Benzo[d]thiazole was discovered to be a DG for C–H activation albeit an unremovable one. *ortho*-Bromination of 2-arylbenzo[d]thiazoles was achieved using *N*-halo-succinimide (NXS) as a halogen source and [Rh(Cp*Cl$_2$)]$_2$ as the catalyst.[52] Using *N*-iodosuccinamide (NIS) as the iodonium source, substrate **96** was iodinated at the *ortho*-position selectively to deliver iodoarene **97**.

4a,8a-Azaboranaphthalene (**98**) is an electronic isostere of naphthalene. When **98** was treated with the electrophilic *N*-iodosuccinamide (NIS) in the presence of the Lewis acid AlCl$_3$, the iodinated product **99** was obtained.[53] It could be further chlorinated by treating **99** with *N*-chlorosuccinamide (NCS) with AlCl$_3$ catalyst to produce dihalogenated 4a,8a-azaboranaphthalene **100**.

5.4   FRIEDEL–CRAFTS ALKYLATION

During the last decade, a book[54] on *Catalytic Asymmetric Friedel–Crafts Alkylations* and several reviews (*vide infra*) on Friedel–Crafts alkylations have been published.

For the Friedel–Crafts alkylation, the alkylation reagents include activated alkenes, carbonyl compounds, imines, and epoxides. The catalysts include Lewis acids such as AlCl$_3$, FeCl$_3$, BF$_3$, ZnCl$_2$, and TiCl$_4$; Brønsted acids such as HF, H$_2$SO$_4$, and H$_3$PO$_4$; and acidic oxide catalysts of, for example, silica-alumina type and cation-exchange resins. The three major types of *asymmetric* Friedel–Crafts alkylation were reviewed in 2008 by Poulsen and Jørgensen.[55]

The first type of the Friedel–Crafts alkylation uses *activated alkenes*, often in the form of α,β-unsaturated carbonyls, as the alkylating agents. In this area, the bidentate Cu(II)-bisoxazolines (BOX) have proven to be successful catalysts for asymmetric synthesis. In case of the Friedel–Crafts alkylation, when indole **101** is the substrate, diethyl arylidene malonates **102** are the alkylation reagents, and chiral copper complex $Cu(OTf)_2$-*t*-Bu-BOX (**103**) is the catalyst, the catalytic enantioselective Friedel–Crafts alkylation proceeded in excellent yield (45%–99%) and moderate enantiomeric excess (46%–69% ee) to afford alkylated indoles **104**.[56]

Organocatalysts have also been employed to catalyze enantioselective Friedel–Crafts alkylation when activated alkenes are used as the alkylating agents. Employing imidazolidinone **107** as the organocatalyst, MacMillan carried out the enantioselective Friedel–Crafts alkylation of indoles **105** using enals **106** as the alkylation agent to produce alkylated indoles **108**.[57]

The second category of asymmetric Friedel–Crafts alkylation involves *carbonyl compounds* as the alkylated agents. Taking advantage of the same bidentate complex $Cu(OTf)_2$-*t*-Bu-BOX (**103**) as the chiral catalyst, the reaction between electron-rich arene 2,4-dimethoxybenzen (**109**) and trifluoropyruvate (**110**) gave rise to adduct **111** in 58% yield and 86% ee.[58] The Cu(II) served as a bidentate ligand binding to the substrates and rendering the stereoselectivities.

Again, when carbonyl compounds are employed as the alkylating agents, organocatalysts have been employed to catalyze enantioselective Friedel–Crafts alkylation. In 2005, Török's large group carried out an enantioselective Friedel–Crafts alkylation of indoles (**108**) with aldehydes **109** ($R^1 = -CO_2Et$, Aryl–).[59] The organocatalysts that they selected were simple cinchona alkaloids such as **110**, where PHN stands for phenanthrene. The reaction afforded adduct **111** in 60%–96% yield and 82%–93% ee.

108   109   110 (10 mol%)   111

The third type of asymmetric Friedel–Crafts alkylation uses *imine compounds* as the alkylated agents. Once again, "copper leads the way."[55] For the asymmetric Friedel–Crafts alkylation of indole **108** with imine **112**, the complex of CuPF$_6$– Tol-BINAP (**113**) was utilized as the chiral catalyst to render adduct **114** in 67%– 89% yield and 78%–97% ee.[60]

108   112   113 (1–5 mol%)   114

After screening, an optically active thiourea derivative **115** was chosen as the organocatalyst for the asymmetric Friedel–Crafts alkylation using imine compounds generated *in situ* as the alkylated agents.[61] *N*-acyliminium ions were formed from the reaction between indoles **116** and alkyl aldehyde R^{1}CHO, with the subsequent enantioselective acyl-Pictet–Spengler reaction, producing tetrahydro-β-carboline derivatives **117** in 65%–81% yield and 85%–95% ee.[61]

115

116

(1) R^{1}CHO, MS or Na$_2$SO$_4$

(2) AcCl, 2,6-lutidine
115 (5–10 mol%)

117

Also in the arena of asymmetric Friedel–Crafts alkylation, Yu et al. in 2009 reviewed the nascent field of using chiral Brønsted acids as the catalysts.[62]

In one example of Friedel–Crafts alkylation of indoles **108**, enamine **118** served as a surrogate of imine because they are isomers.[63] With the aid of the Brønsted acid in the form of chiral phosphoric acid **119**, the reaction between **108** and **118** gave rise to adduct **120** in good to excellent yields and excellent ee. For the same methodology, if the enamine **121** was used as the alkylating agent, adduct **122** with a tertiary chiral center was assembled enantioselectively (94%–99% yield and 73%–97% ee).[63]

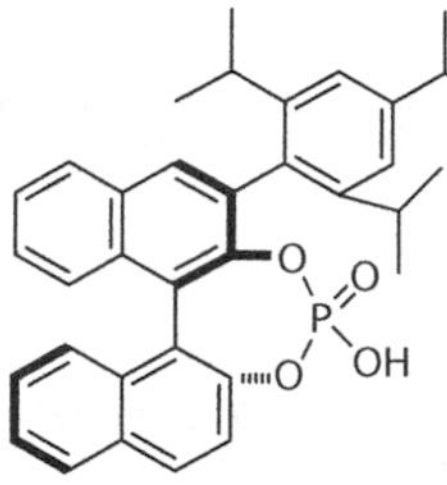

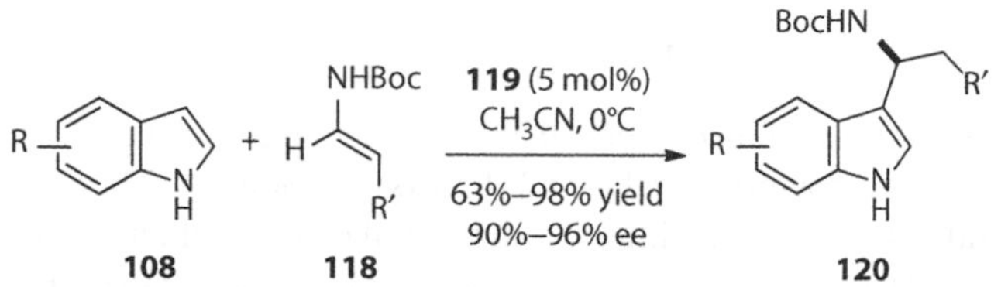

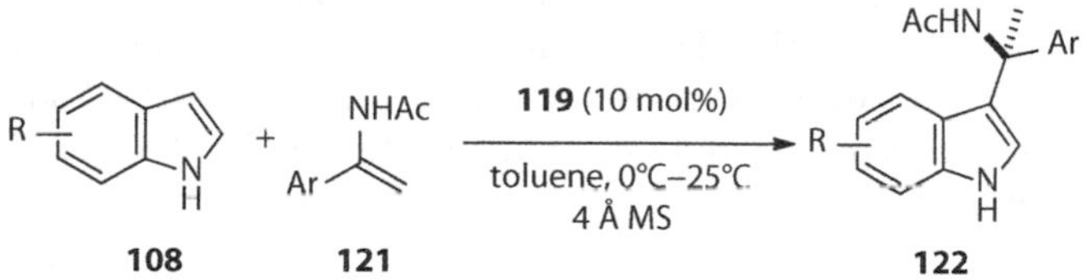

Alkylation of indoles almost always takes place on the C3 position. By blocking the C3 position with R^1 and using 4,7-dihydroindoles **123** as the substrates, You and coworkers achieved the asymmetric Friedel–Crafts alkylation of substrates **123** with enamine **124** to prepare the 2-substituted indoles **126** smoothly in good yields with up to >99% ee.[64]

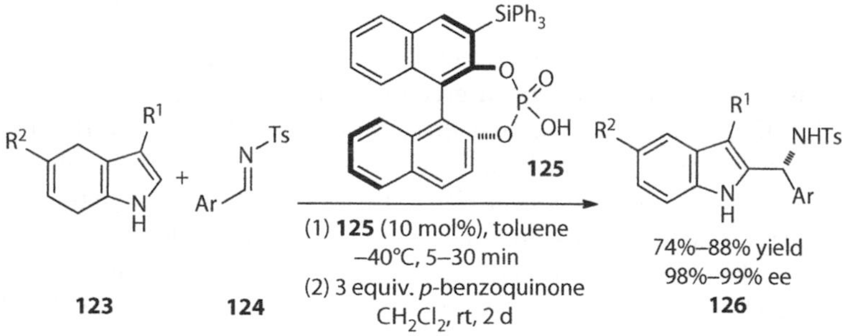

In addition to chiral phosphoric acids, chiral thioureas are another key class of Brønsted acids employed in the asymmetric Friedel–Crafts alkylation. In addition to thiourea **115** that catalyzed the acyl-Pictet–Spengler reaction of **116** to make tetrahydro-β-carboline derivatives **117**, thiourea **127** was also found to be an effective chiral Brønsted acid catalyst.[65] Here nitroolefin **129** was the active alkene serving as the alkylating agent. The Friedel–Crafts alkylation of indoles **128** with nitroolefin

129 proceeded to deliver adduct **130** in 35%–88% yield and 71%–89% ee. This could be considered as a conjugate addition (1,4-addition) as well.

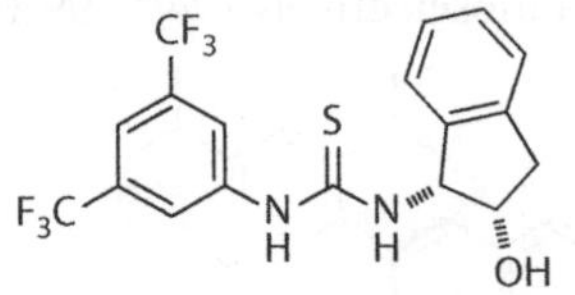

Another review also published in 2010 by de Figueiredo and colleagues focused on "organocatalyzed asymmetric Friedel–Crafts reactions."[66] There are five categories of chiral catalysts. In addition to the popular MacMillan's imidazolidinone catalysts (e.g., **107**), there are also cinchona alkaloid derivatives (e.g., **110**), thioureas such as **115** and **127**, chiral phosphoric acid (**119** and **125**), as well as diarylprolinol ethers as exemplified by **131**.[67] The Friedel–Crafts alkylation of 1-naphthols **132** with α,β-unsaturated aldehydes (enals **106′**) took place in a Michael addition fashion, followed by cyclization to produce chromanes **133** in 63%–93% yield with dr of 2:1 to 7:2 and er of 86:14 to 99:1.

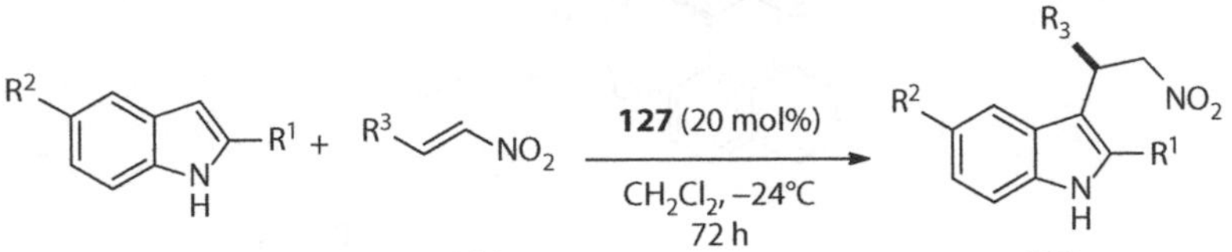

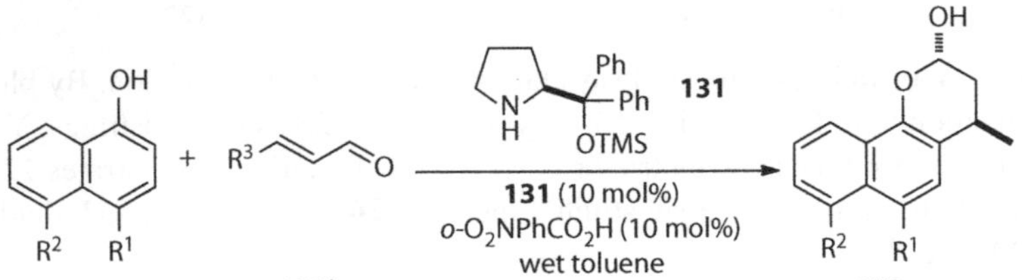

In addition to focusing on asymmetric synthesis, a review by Rueping and Nachtsheim in 2011 also summarized green chemistry of the asymmetric Friedel–Crafts alkylation reactions.[68] The first diastereoselective Friedel–Crafts alkylation reactions were developed by Bach and coworkers in 2005.[69,70] When substrate **134** (99% ee) was exposed to Lewis acid HBF_4, the Friedel–Crafts alkylation of arene Ar–H gave rise to adduct **136** with remarkable *syn*-diastereoselectivities up to 94:6 dr. The stereochemical outcome may be rationalized by the steric blockage by the bulky *tert*-butyl group as shown in benzyl carbocation intermediate **135**. In addition to the *tert*-butyl group, other regio-dominating groups also include nitro-, cyano-, or hydroxyl groups. The case in point of the nitro group is shown with substrate **137**.[71] Under the influence of catalytic $Bi(OTf)_3$, benzylation of silyl enol ether **138** with **137** led to the *anti*-product **139** with excellent diastereoselectivity. To be fair, the mechanism is closer to the S_N1 pathway than the Friedel–Crafts alkylation for the two reactions shown below because the reaction centers were not on the benzene rings.

134 135 136

137 138 139

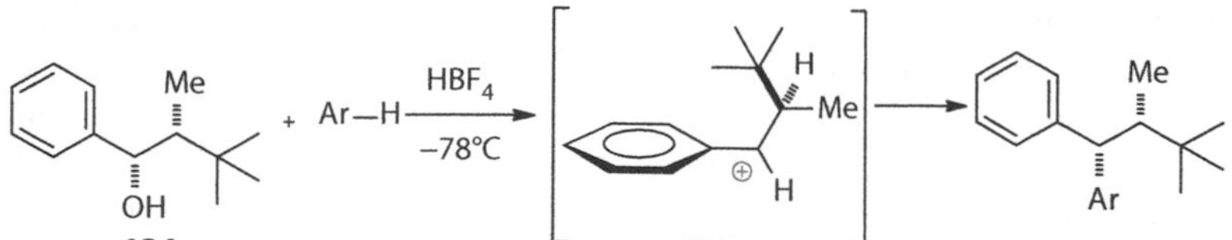

Now we move on to conventional Friedel–Crafts alkylations not involving chiral centers. The Friedel–Crafts alkylation with benzyl alcohols has proven to be an efficient approach for preparing 1,1-diarylalkanes. Using the venerable $FeCl_3$ as the Lewis acid catalyst, *o*-xylene (**140**) was benzylated with benzyl alcohol **141** nearly exclusively on the *para*-position to give adduct **142**.[72] The intramolecular version works as well for the Friedel–Crafts alkylations using benzyl alcohols as the alkylating agents as exemplified by transformation **143** → **144**.[73]

140 141 142

143 144

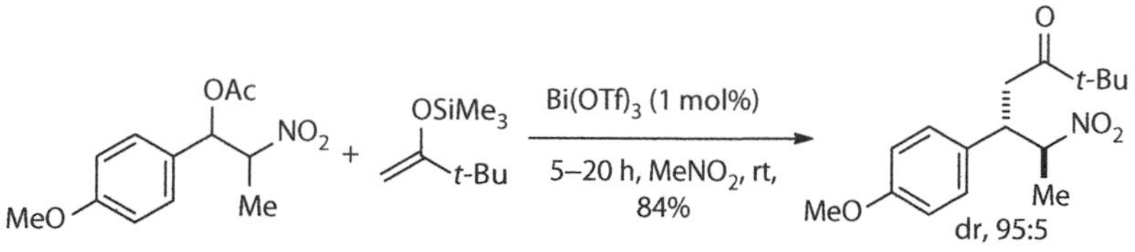

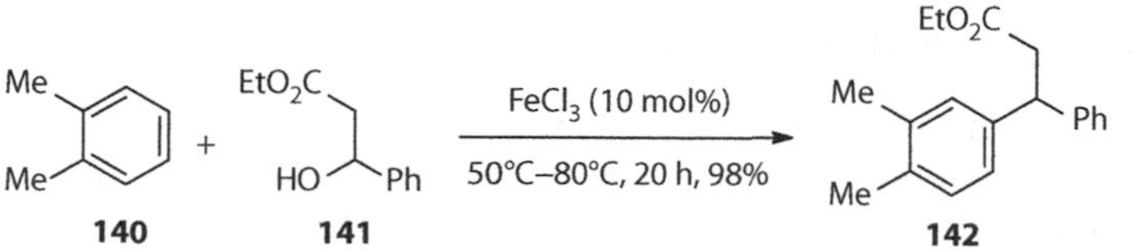

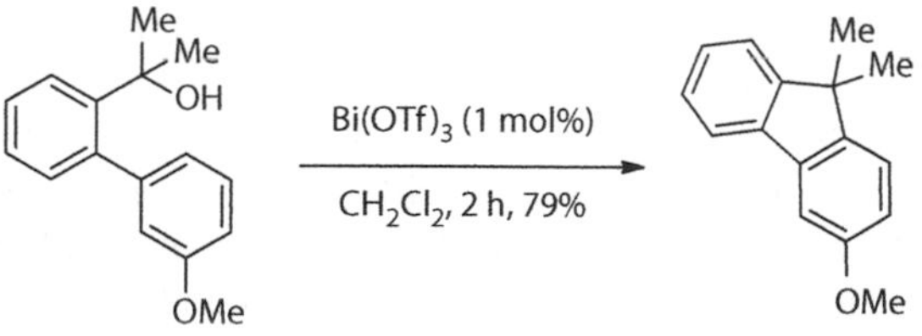

The synthesis of β-alkylpyrroles were reviewed in 2011.[74] A typical approach employing the Friedel–Crafts alkylation is showcased by transformation **145** → **146**.[75]

145 146

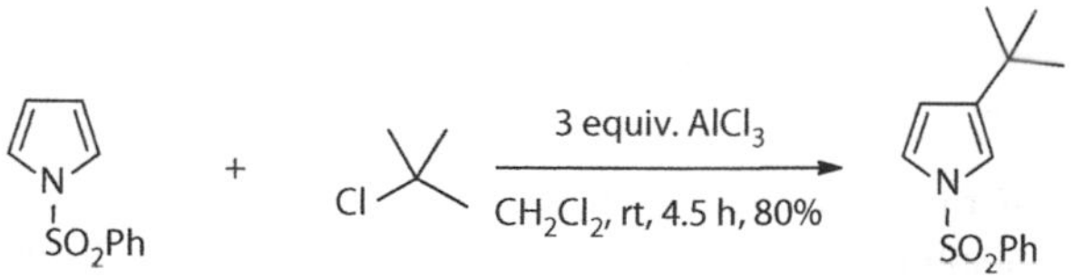

Klumpp reported a synthesis of functionalized 2-oxindole using Friedel–Crafts alkylation reaction.[76] When benzene and 3-hydroxyisatin **147** were exposed to a catalytic amount of triflic acid, two different outcomes were observed depending on the reaction conditions. When the reaction was carried out at rt, the Friedel–Crafts alkylation took place twice to afford adduct **148**. But if the reaction was done at 0°C, the Friedel–Crafts alkylation took place only once to deliver adduct **149**.

5.5 FRIEDEL–CRAFTS ACYLATION

Generally speaking, the Friedel–Crafts acylation is cleaner than the Friedel–Crafts alkylation since acylation does not have the issue of carbocation rearrangements often seen for alkyl cationic intermediates. On Friedel–Crafts acylation, a book titled *Advances in Friedel–Crafts Acylation Reactions: Catalytic and Green Processes* was published in 2009.[77] And a review on *Use of Solid Catalysts in Friedel–Crafts Acylation Reactions* was published in 2006.[78]

Normally, carboxylic acids are not active enough for the Friedel–Crafts reaction. Treatment of carboxylic acids with cyanuric chloride in the presence of pyridine was discovered to yield the corresponding acid chloride smoothly. Addition of aluminum chloride to the reaction mixture then provided a mild and efficient Friedel–Crafts acylation.[79] An example of the intramolecular version is shown by the transformation **150** → **151** although the intermolecular version worked just as well.

Photo-Friedel–Crafts acylation is known and is considered green chemistry. In one example, 1,4-naphthoquinone **152** in an ionic liquid as the solvent was acylated with butyraldehyde under irradiation to produce adduct **153**.[80] The reaction was carried out at room temperature.

152 + **153**

Meldrum's acid (2,2-dimethyl-1,3-dioxane-4,6-dione) is a versatile synthon and its synthetic utility was reviewed in 2010.[81] In 2009, Fillon and Fishlock[82] described a scandium triflate-catalyzed intramolecular Friedel–Crafts acylation of Meldrum's acid derivative **154** to give rise to ketone **155**. A mechanism was proposed by invoking the intermediacy of a ketene.

154 **155**

Friedel–Crafts reaction is usually performed with carboxylic acid chlorides or anhydrides because they are the most reactive among all carboxylic acid derivatives. Generally esters are not active enough to be the acylating agents. But under the right circumstances, esters may be activated enough to take part in the Friedel–Crafts reaction. For example, esters **156** were activated to intermediate **157** when exposed to 10 mol% of InBr$_3$ and dimethylchlorosilane with the concurrent evolution of the alkanes.[83] Under the influence of InBr$_3$, intermediate **157** was active enough to undergo Friedel–Crafts acylation with arene to afford product **158**.

156 **157** **158**

Meanwhile, α-acetoxybutendioic anhydride (**160**) is a unique ester. The anhydride functional group could serve as an activating group since it has two points to chelate with the Lewis acid catalyst.[84] The methodology also works when the anhydride moiety was replaced with a bis-ester functional group such as the Meldrum's acid derivative **154**.

159 **160** **161**

Among all carboxylic acid derivatives, amides are the least reactive. However, Klumpp and coworkers found that a series of amides capable of providing aromatic ketones in good yields.[85] When amide **161** was exposed to Brønsted super acid triflic acid, intramolecular Friedel–Crafts acylation occurred to afford 1-indanone **162**. A mechanism involving diminished C–N resonance through superelectrophilic activation is shown in transition state **163** and subsequent cleavage to acyl cations.

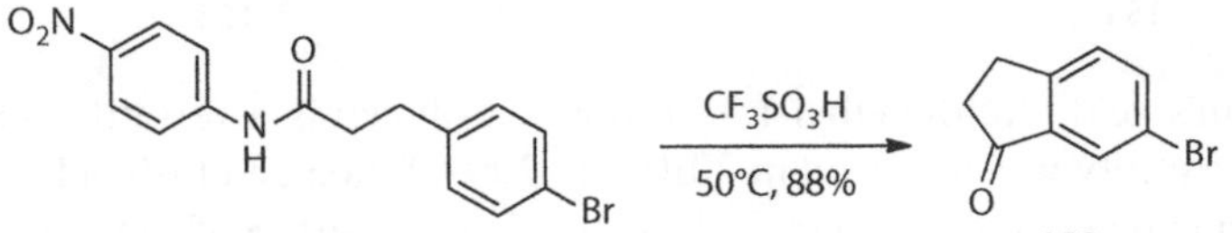

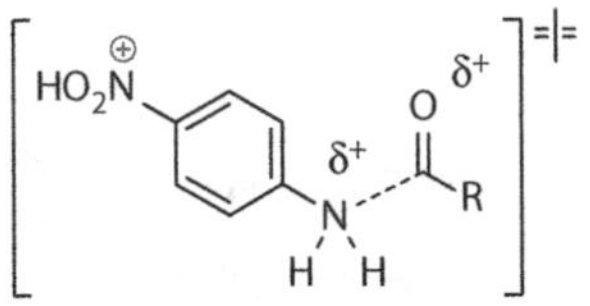

Acid chlorides can be activated using a simple iodide source such as KI or LiI to undergo nucleophilic attack from a variety of relatively weak nucleophiles.[86] The Friedel–Crafts acylation of hindered phenol **164** with pivaloyl chloride took place at the *para*-position to produce ketone **165**. However, if the *para*-position was blocked, then the same reaction conditions would have given the pivaloyl ester instead.

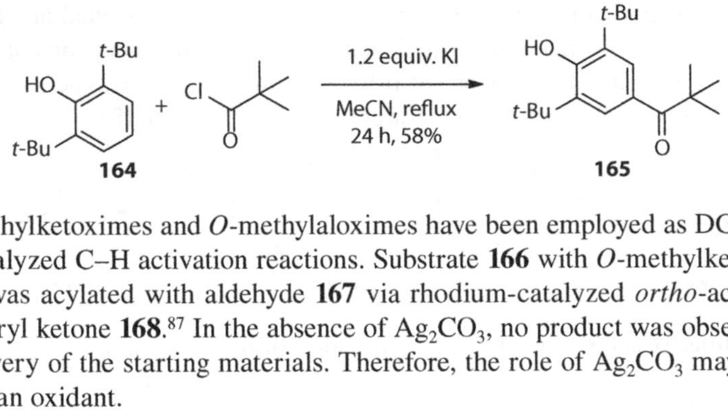

O-Methylketoximes and *O*-methylaloximes have been employed as DGs for rhodium-catalyzed C–H activation reactions. Substrate **166** with *O*-methylketoxime as the DG was acylated with aldehyde **167** via rhodium-catalyzed *ortho*-acylation to make diaryl ketone **168**.[87] In the absence of Ag$_2$CO$_3$, no product was observed with full recovery of the starting materials. Therefore, the role of Ag$_2$CO$_3$ may be more than just an oxidant.

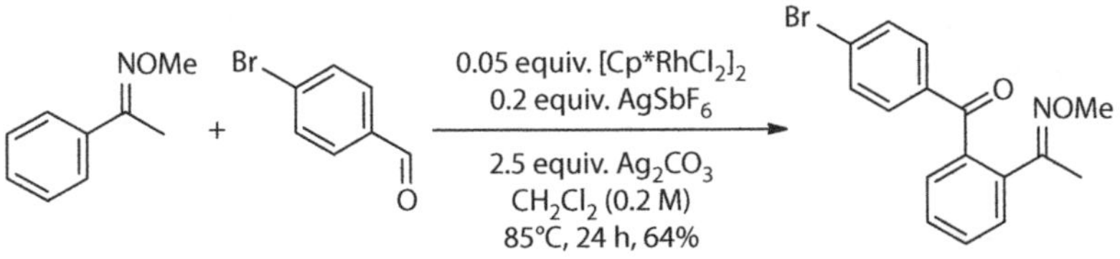

O-Methylketoximes have also been employed as DGs for palladium-catalyzed C–H activation reactions. In addition to aldehydes being the acylation agents, *alcohols* may be acylation agents if they are oxidized during the reaction. Substrate **169** used *O*-methylketoxime as the DG and underwent a palladium-catalyzed *ortho*-C–H activation followed by acylation from benzyl alcohol **170** after oxidation to afford acylation product **171**.[88] The net outcome is the same as a Friedel–Crafts acylation reaction.

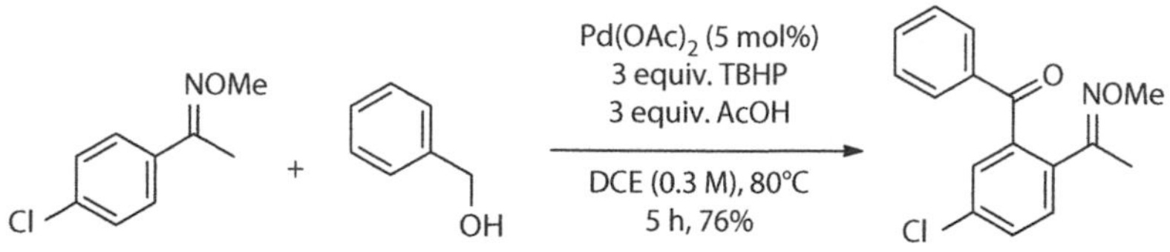

Oh et al. took advantage of two sequential Friedel–Crafts reactions to build 4-substituted 2-naphthols.[89] At first, the Friedel–Crafts acylation of phenylacetyl chloride **172** with alkyne **173** gave rise to β-chlorovinyl ketone **174**, a versatile synthetic precursor in its own right. A tandem Friedel–Crafts alkylation of **174** then assembled 4-substituted 2-naphthols **175**. For a specific example of phenylacetyl chloride (**172′**), its reaction with terminal alkyne **173′** proceeded smoothly to provide 4-substituted 2-naphthols **175′** in 81% yield.

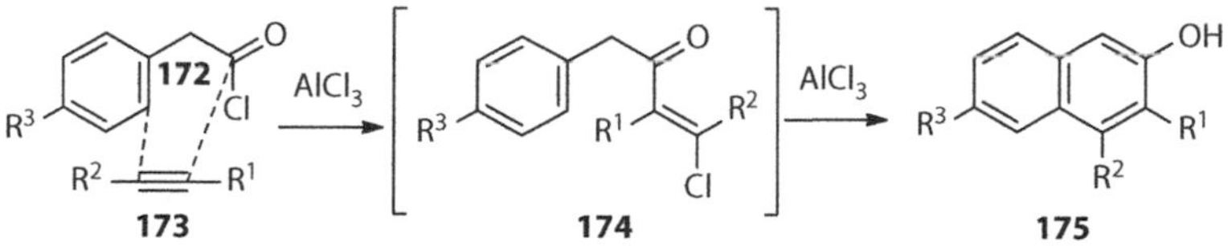

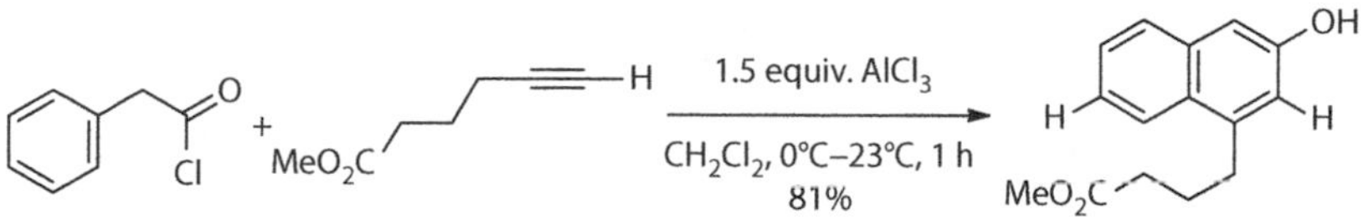

Without protection of the NH group on indoles **176**, a Friedel–Crafts acylation with propionic anhydride was achieved with regioselectivity favoring the C-3 position as shown in **177**.[90] The catalyst was ytterbium triflate and the solvent was an ionic liquid [BMI]BF$_4$ (1-butyl-3-methylimidazolium tetrafluroborate) thus the reaction is considered green chemistry.

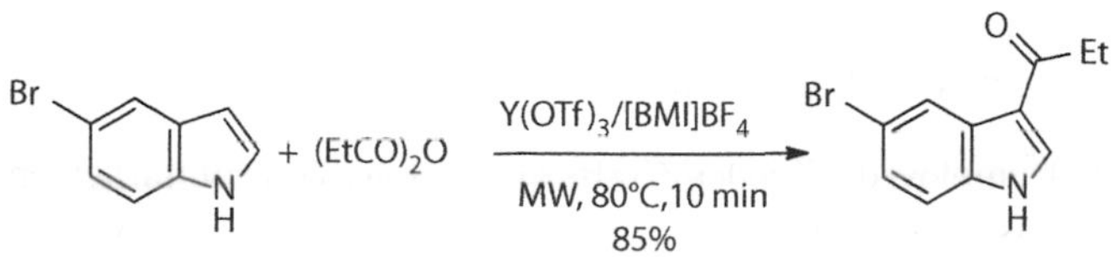

Normally, the Friedel–Crafts reactions are catalyzed by either Lewis acids or Brønsted acids. Aubé's group observed that 1,1,1,3,3,3-hexafluoro-2-propanol (HFIP)

could serve as a solvent as well as a catalyst for the Friedel–Crafts acylation.[91] In a way, one could view HFIP as a Brønsted acid because the six fluorine atoms makes it much more acidic than the corresponding isopropanol. The intramolecular version is showcased by transformation **178** → **179**.

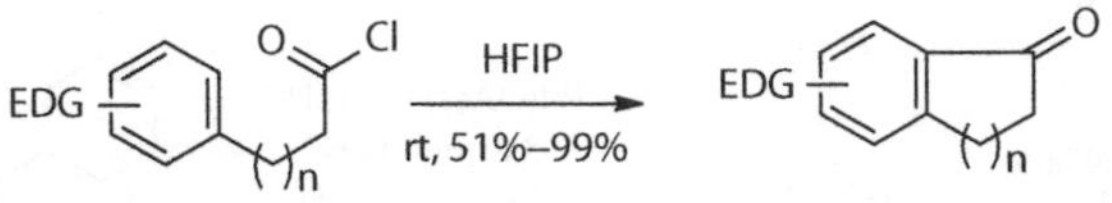

Copper-catalyzed aerobic intramolecular oxidative acylation of benzoylacetone derivative **180** afforded naphthoquinone **181** bearing two acyl groups at positions *ortho* to each other on the benzene ring.[92] In contrast, when ethyl ketone **182** was employed as the substrate, 1,3-indandione **183** was produced. The proposed mechanism is similar to that of the Friedel–Crafts acylation reaction because they all involve electrophilic attack of a carbonyl by the benzene ring.

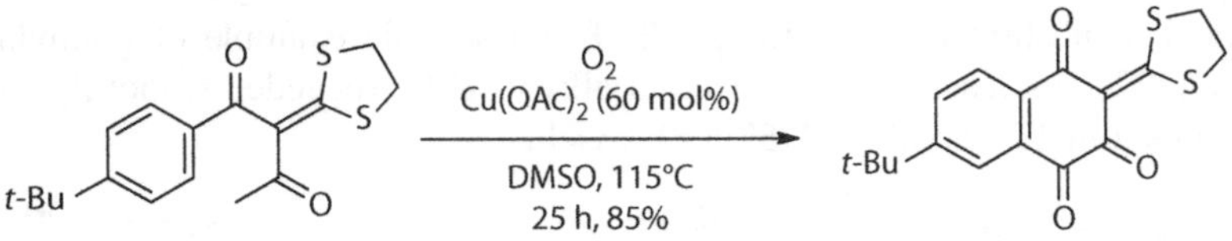

5.6 APPLICATIONS OF FRIEDEL–CRAFTS REACTION ON TOTAL SYNTHESIS

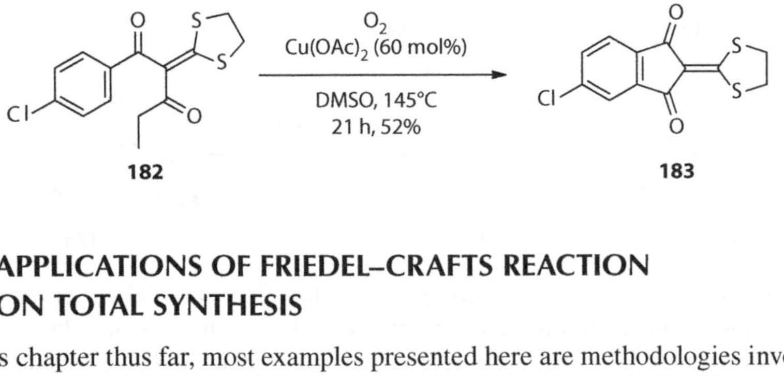

For this chapter thus far, most examples presented here are methodologies involving the Friedel–Crafts acylation. Now, some applications of the Friedel–Crafts alkylation and acylation to total synthesis are summarized below.

5.6.1 FRIEDEL–CRAFTS ALKYLATION

Fukuyama et al. employed a Friedel–Crafts alkylation as one of the key operations in the total synthesis of (+)-haplophytine.[93] A rather sophisticated alkylating agent **185** was prepared by the treatment of indole derivative **185** with NIS. Then subsequent Friedel–Crafts alkylation of aniline **186** took place with the facilitation of AgOTf at the *para*-position of the amine group, leading to adduct **187**.

184 185 186 187

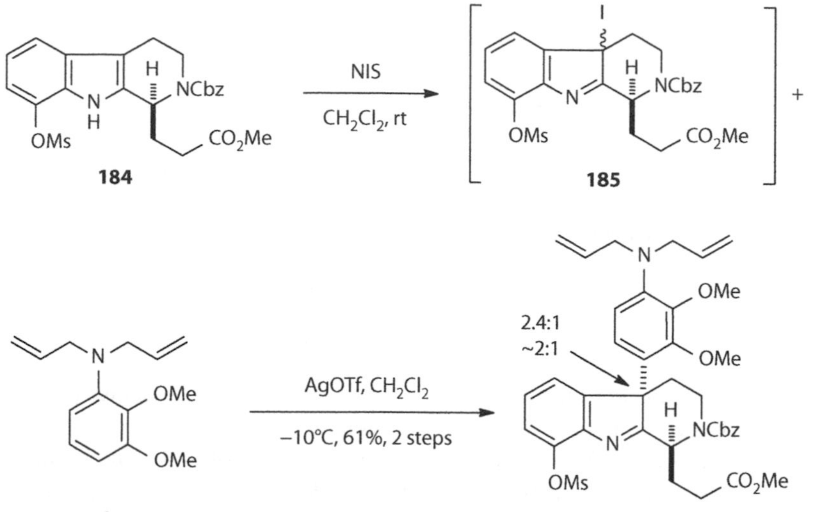

Tokuyama, Fukuyama and their colleagues again used the Friedel–Crafts alkylation as one of the key operations in the total synthesis of dictyodendrin A and B.[94] Under similar conditions for the transformation **186** → **187**, bromoindole **188** was alkylated at the C-2 position with *para*-methoxyphenylethyl bromide (**189**) smoothly under mild condition using AgOTf at −78°C to assemble adduct **190**.

188 189 190

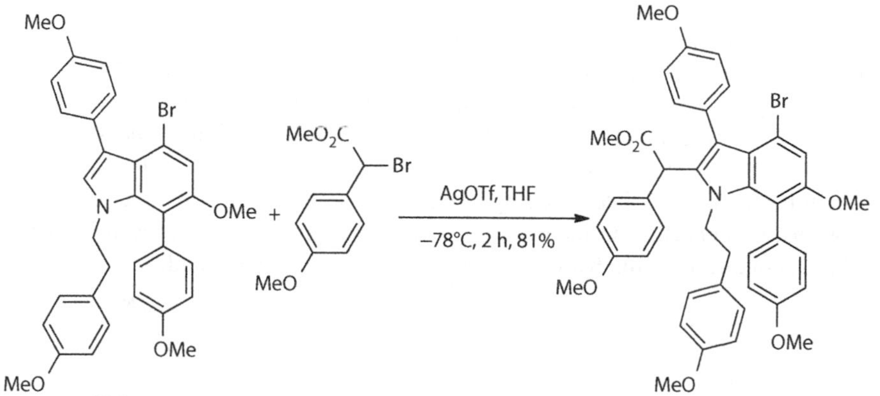

In an effort toward the total synthesis of microstegiol, Green et al. carried out the Friedel–Crafts alkylation of naphtholic acetate **191** with propargyl alcohol–$CO_2(CO)_6$ complex **192**.[95] The reaction was catalyzed by a boron Lewis acid. Interestingly, intermediate **193** was later manipulated to serve as a substrate for a Friedel–Crafts acylation to construct a seven-membered ring.

191 + **192**

0.7 equiv. Bu₂BOTf
$\xrightarrow{\text{CH}_2\text{Cl}_2, 0.5\text{ h}, 0°\text{C}}$
90%

193

In the total synthesis of laetevirenol A, Heo and group took advantage of an intramolecular Friedel–Crafts alkylation as the penultimate step.[96] It was proposed that the olefin intermediate from dehydration of tertiary alcohol under the influence of tosylic acid assembled the *trans*-cyclized product **194**, which was subsequently globally demethylated to deliver laetevirenol A.

p-TsOH, toluene

120°C, 18 h, 43%

194

In the total synthesis of mersicarpine, a cationic cyclization approach was adopted.[97] Under the influence of Lewis acid AlCl₃, cyclic carbamate **195** decomposed to expose a carbocation, which underwent an intramolecular Friedel–Crafts alkylation onto the indole ring to afford lactam **196**. Apparently the strategical installation of the chlorine at the C-3 position was critical to stabilize the indole ring. Without it, the intermediate was very labile to acidic conditions. Although chlorine is not on the mersicarpine molecule, it was later conveniently removed by acidic oxidation conditions using hydrogen peroxide in trifluoroacetic acid.

AlCl₃, CH₂Cl₂

rt, 86%

195   **196**

In an attempt to achieve a one-pot biomimetic total synthesis of an indole alkaloid yuechukene, Hong's group accomplished a Friedel–Crafts alkylation with a sterically encumbered α-alkyl enal **197**.[98] After extensive screening of a variety of Brønsted acids, solvents and reaction conditions, the organocatalytic alkylation of indole gave the desired product **198** in only 25% yield using quinine derivative **199** as the organocatalyst and phosphoric acid (*S*)-**200** as the additive.

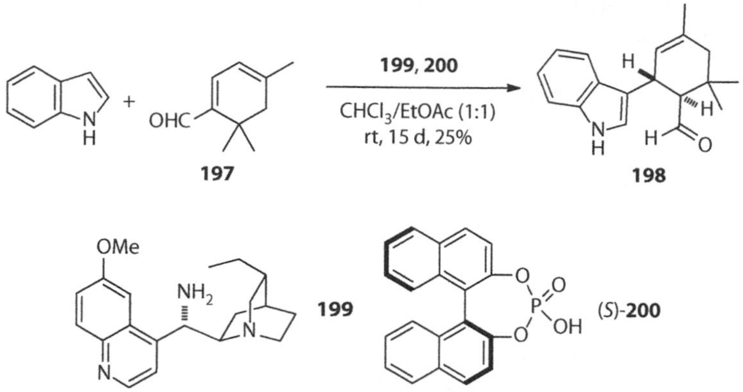

In a total synthesis of pisiferin (**202**), Xie's group took advantage of a Friedel–Crafts alkylation as one of the five key operations.[99] Substrate **201** was used as a racemate because the alcohol would lose chirality at the end anyhow. When it was exposed to 3 equiv. of BF₃·OEt₂, pisiferin (**202**) was obtained in 60% yield presumably via the intermediacy of the corresponding allylic cation.

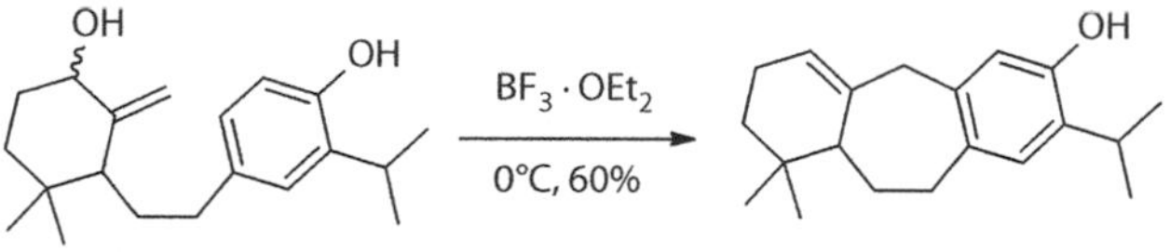

5.6.2 FRIEDEL–CRAFTS ACYLATION

As mentioned before, Friedel–Crafts acylation reactions are often superior to the alkylation reactions because the former do not undergo carbocationic rearrangement sometimes encountered by the latter. So it is not surprising that there are more synthetic applications of the former reactions.

In studies toward the total synthesis of lactonamycin, Barrett and colleagues assembled carboxylic acid **203** using a Negishi coupling reaction.[100] Exhaustive optimization of the Friedel–Crafts acylation of **203** resulted in conditions including zinc chloride and the Ghosez reagent (1-chloro-*N,N*-2-trimethyl-1-propylenamine) to give tetracycle **204** in excellent yield. This is a good example that sometimes hard work pays off.

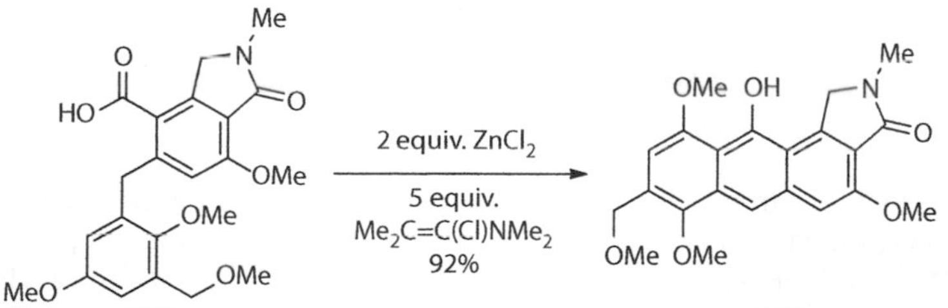

In 2008, Hecht et al. finished the total synthesis of the topopyrones as a new class of topoisomerase I poisons.[101] After a few false starts, they succeeded in carrying out a titanium-mediated *ortho*-directed Friedel–Crafts acylation of phenol **205** and installing the required acyl chain at the *ortho*-position to afford quinone **206**.

205   **206**

For a formal total synthesis of (±)-taiwanianiaquinol B and (±)-dichroanone, She et al. applied a domino-Friedel–Crafts acylation/alkylation sequence as the key operation to install the core structure.[102] Optimization of the reaction between anisole and carboxylic acid **207** to give adduct **208** in 75% yield arrived at the best conditions using methanesulfonic acid and P_2O_5, a powerful dehydrating agent. Fascinatingly, the reaction between anisole and linear carboxylic acid **209** also led to **208** under similar condition with a yield of 71%.

207   **208**

209

Sarpong et al. took a Ga(III)-catalyzed cycloisomerization approach to make (±)-icetexone and (±)-epi-icetexone.[102] In one of the key operations, carboxylic acid **210** was converted to the corresponding acid chloride, and the ensuing intramolecular Friedel–Crafts acylation was catalyzed with aluminum chloride to give indanone **211**. The benzene ring on **211** is substituted on all six positions.

210   **211**

Due to their electron-rich nature, pyrroles and indoles are useful substrates for the Friedel–Crafts reaction. Bull et al. discovered that 1,5-diazabicyclo[4.3.0]non-5-ene (DBN) could serve as a nucleophilic catalyst for the Friedel–Crafts acylation of pyrroles and indoles with acid chlorides.[104] Mechanistically, DBU attacks the acid chloride functional group and expels the chloride, forming a more reactive N-acyl-DBN intermediate, which was actually detected by the mass spectroscopy (MS). Thus, catalysis by DBN, pyrrole ester **212** was acylated with 1.2 equiv. of p-toluoyl chloride to give the α-acylated product **213**, which upon hydrolysis produced anti-inflammatory drug Tolmetin in 78% isolated yield.

An intramolecular Friedel–Crafts acylation of chiral carboxylic acid **214** was accomplished using chlorosulfonic acid to afford tetralone **215**, which was converted to (+)-sertraline after a reductive amination.[105]

Strong Lewis acid BBr_3 is an efficient demethylation agent as well as a catalyst for the Friedel–Crafts acylation. In a total synthesis of diptolidonesin G (**217**), substrate **216** was exposed to 30 equiv. BBr_3, which effected a global demethylation followed a Friedel–Crafts acylation to deliver the natural product **217** in an impressive 92% yield.[106]

In the total synthesis of sandresolide B and amphilectolide, Trauner and cowork-ers used the intramolecular Friedel–Crafts acylation as one of the key operations.[107] After meticulous optimizations, the only conditions that led to desired product entailed activation of acid **218** with trifluoroacetic anhydride (TFAA) followed by gentle heating with zinc chloride. Short reaction times and stoichiometric zinc chloride were crucial to the ring closure to afford seven-membered product **219**. Although the absolute configuration at the chiral center α- to the carbonyl group was incorrect, it was epimerized using 1,8-diazabicycloundec-7-ene (DBU) as a base.

218   **219**

5.7 MISCELLANEOUS EAS REACTIONS

In the presence of acid promoters, trifluoromethylthiolation of phenols takes place via
the EAS mechanism. For instance, trifluoromethylthiolation of 2,3,5-trimethylphenol
(**220**) gave rise to the *para*-substituted product **221** using (*N*-trifluoromethylsulfanyl)
aniline as the reagent as triflic acid as the promoter.[108]

220   **221**

Triflic acid as a Brønsted acid was also the acid promoter for the an intramolecular cyclization.[109] When arene-ynamide **222** was treated with 1.2 equiv. triflic
acid, cyclization took place to deliver 3*H*-pyrrolo[2,3-*c*]quinazoline **223** via the EAS
mechanism as well. The intermediate is supposed to be a highly reactive ketenimin-
ium ion. This methodology was developed to the total synthesis of marine natural
products.

222   **223**

Pappo et al. developed an EAS of aldehydes, followed by reduction *in situ* with
triethylsilane.[110] The methodology is highly chemo- and regioselective in introducing
primary alkyl substituents into aromatic compounds. The net outcome is alkylation, which could serve as a two-step alternative of a Friedel–Crafts acylation. This
method may be showcased by 17β-estradiol (**224**) as the substrate. The EAS of iso-
butyraldehyde was catalyzed by ethanethiol to give the alcohol intermediate, which
was reduced *in situ* to give alkylated product **225** in 67% yield. The same reaction
also worked for benzaldehyde with a yield of 85%.

224 + isopropyl CHO → EtSH then Et₃SiH, Cu(OTf) (2.5 mol%), CF₃CH₂OH, rt to 50°C, 67% → **225**

An investigation into the mechanism of the Pechmann condensation reaction using NMR suggested that EAS was the key step for both of the possible pathways.[111] To reconcile the fact that intermediates **229** and **230** and NMR data, the authors concluded the Pechmann condensation reaction probably first proceeds to form tertiary alcohol **229** via an EAS. An intramolecular transesterification then affords intermediate **230**, which undergoes dehydration to deliver coumarin **228**.

226 + **227** → Cat. → **228**

226 + 227 → EAS → **229**

Transesterification → **230** → −H₂O → **228**

In summary, we have reviewed the advances during the last decade of EAS. Halogenation and Friedel–Crafts reactions continue to play important roles in organic synthesis as evidenced by their application in total synthesis.

REFERENCES

1. Taylor, R. *Electrophilic Aromatic Substitution*, Wiley: Chichester, 1990.
2. Ingold, C. K. *Structure and Mechanism in Organic Chemistry*, 2nd ed.; Cornell University Press: Ithaca, New York, 1986.
3. Xu, X. F.; Zilberg, S.; Haas, Y. *J. Phys. Chem. A* **2010**, *114*, 4924–4933.
4. Lenoir, D. *Angew. Chem. Int. Ed.* **2003**, *42*, 854–857.
5. Hartshorn, S. R. *Chem. Soc. Rev.* **1974**, *3*, 167–192.
6. Myhre, P. C.; Owen, G. S.; James, L. L. *J. Am. Chem. Soc.* **1968**, *90*, 2115–2123.
7. Sun, X. *Curr. Org. Chem.* **2014**, *18*, 3066–3077.
8. Jones-Wilson, T. M.; Burtch, E. A. *J. Chem. Educ.* **2005**, *82*, 616–617.
9. Wieder, M. J.; Barrows, R. *J. Chem. Educ.* **2008**, *85*, 549–551.

10. Sánchez-Viesca, F.; Gómez Gómez, M. R.; Berros, M. *J. Chem. Educ.* **2011**, *88*, 944–946.
11. Garabatos-Perera, J. R.; Rotstein, B. H.; Thompson, A. *J. Org. Chem.* **2007**, *72*, 7382–7385.
12. Begunov, R. S.; Sokolov, A. A.; Belova, V. O.; Fakhrutdinov, A. N.; Shashkov, A.S.; Fedyanin, I. V. *Tetrahedron Lett.* **2015**, *56*, 5701–5704.
13. Shopsowitz, K.; Lelj, F.; MacLachlan, M. J. *J. Org. Chem.* **2011**, *76*, 1285–1294.
14. Balz, G.; Schiemann, G. *Ber. Dtsch. Chem. Ges.* **1927**, *60*, 1186–1190.
15. Grakauskas, V. *J. Org. Chem.* **1970**, *35*, 723–728.
16. Hyman, H. H.; Shaw, M. J.; Filler, R. *J. Am. Chem. Soc.* **1970**, *92*, 6498–6502.
17. Hehre, W.J.; Hiberty, P. C. *J. Am. Chem. Soc.* **1974**, *96*, 7163–7165.
18. Cacace, F.; Giacomello, P.; Wolf, A. P. *J. Am. Chem. Soc.* **1980**, *102*, 3511–3515.
19. Moilliet, J. S. *J. Fluorine Chem.* **2001**, *109*, 13–17.
20. Alric, J. P.; Marquet, B.; Billard, T.; Langlois, B. R. *J. Fluorine Chem.* **2005**, *126*, 661–667.
21. Fifolt, M. J.; Olczak, R. T.; Mundhenke, R. F.; Bieron, J. F. *J. Org. Chem.* **1985**, *50*, 4576–4582.
22. Venturini, F.; Navarrini, W.; Famulari, A.; Sansotera, M.; Dardani, P.; Tortelli, V. *J. Fluorine Chem.* **2012**, *140*, 43–48.
23. Banks, R. E.; Besheesh, M. K.; Fraenk, W.; Klapötke, T. M. *J. Fluorine Chem.* **2003**, *124*, 229–232.
24. Darwish, A. D.; Avent, A. G.; Abdul-Sada, A.; Gol'dt, I. V.; Hitchcock, P. B.; Kuvytchko, I. V.; Taylor, R. *Chem. Eur. J.* **2004**, *10*, 4523–4531.
25. Hiller, A.; Patt, J. T.; Steinbach, J. *J. Organomet. Chem.* **2006**, *691*, 3737–3742.
26. Borodkin, G. I.; Zaikin, P. A.; Shubin, V. G. *Tetrahedron Lett.* **2015**, *47*, 2639–2642.
27. Bluck, G. W.; Carter, N. B.; Smith, S. C.; Turnbull, M. D. *J. Fluorine Chem.* **2004**, *125*, 1873–1877.
28. Heravi, M. R. P. *J. Fluorine Chem.* **2008**, *129*, 217–221.
29. Borodkin, G. I.; Shubin, V. G. *Russ. Chem. Rev.* **2010**, *79*, 259–283.
30. Teare, H.; Robins, E. G.; Kirjavainen, A.; Forsback, S.; Sandford, G.; Solin, O.; Luthra, S. K.; Gouverneur, V. *Angew. Chem. Int. Ed.* **2010**, *49*, 6821–6824.
31. Lee, E.; Kamlet, A. S.; Powers, D. C.; Neumann, C. N.; Boursalian, G. B.; Furuya, T.; Choi, D. C.; Hooker, J. M.; Ritter, T. *Science* **2011**, *334*, 639–642.
32. Gouverneur, V. *Nat. Chem.* **2012**, *4*, 152–154.
33. Pravst, I.; Stavber, S. *J. Fluorine Chem.* **2013**, *156*, 276–282.
34. Liu, P.; Gao, Y.; Gu, W.; Shen, Z.; Sun, P. *J. Org. Chem.* **2015**, *80*, 11559–11565.
35. Yamada, S.; Gavryushin, A.; Knochel, P. *Angew. Chem. Int. Ed.* **2010**, *49*, 2215–2218.
36. Sorokin, V. I.; Pozharskii, A. F.; Ozeryanskii, V. A. *J. Fluorine Chem.* **2013**, *154*, 67–72.
37. Qiu, Z.; Quan, Y.; Xie, Z. *J. Am. Chem. Soc.* **2013**, *135*, 12192–12195.
38. Testa, C.; Roger, J.; Scheib, S.; Fleurat-Lessard, P.; Hierso, J.-C. *Adv. Synth. Catal.* **2015**, *357*, 2913–2923.
39. Cardinal, P.; Greer, B.; Luong, H.; Tyagunova, Y. *J. Chem. Educ.* **2012**, *89*, 1061–1063.
40. Li, H.-J.; Wu, Y.-C.; Dai, J.-H.; Song, Y.; Cheng, R.; Qiao, Y. *Molecules* **2014**, *19*, 3401–3416.
41. Gustafson, J. L.; Lim, D.; Miller, S. J. *Science* **2010**, *328*, 1251–1255.
42. Gustafson, J. L.; Lim, D.; Barrett, K. T.; Miller, S. J. *Angew. Chem. Int. Ed.* **2011**, *50*, 5125–5129.
43. Barrett, K. T.; Miller, S. J. *J. Am. Chem. Soc.* **2013**, *135*, 2963–2966.
44. Diener, M. E.; Metrano, A. J.; Kusano, S.; Miller, S. J. *J. Am. Chem. Soc.* **2015**, *137*, 12369–12377.
45. Du, Z.-J.; Gao, L.-X.; Lin, Y.-J.; Han, F.-S. *ChemCatChem* **2014**, *6*, 123–126.

46. Purohit, V. B.; Karad, S. C.; Patel, K. H.; Raval, D. K. *Catal. Sci. Technol.* **2015**, *5*, 3113–3118.
47. Liger, F.; Pellet-Rostaing, S.; Popwycz, F.; Lemaire, M. *Tetrahedron Lett.* **2011**, *52*, 3736–3739.
48. Fürstner, A.; Krause, H.; Thiel, O. R. *Tetrahedron* **2002**, *58*, 6373–6380.
49. Franco, F.; Greenhouse, R.; Muchowski, J. *J. Org. Chem.* **1982**, *47*, 1682–1688.
50. Kalyani, D.; Dick, A. R.; Anani, W. Q.; Sanford, M. S. *Org. Lett.* **2006**, *8*, 2523–2526.
51. Sadhu, P.; Alla, S. K.; Punniyamurthy, T. *J. Org. Chem.* **2013**, *78*, 6104–6111.
52. Ding, Q.; Zhou, X.; Pu, S.; Cao, B. *Tetrahedron* **2015**, *71*, 2376–2381.
53. Sun, F.; Lv, L.; Huang, M.; Zhou, Z.; Fang, X. *Org. Lett.* **2014**, *17*, 5024–5027.
54. Bandini, M.; Umani-Ronchi, A. (Eds.); *Catalytic Asymmetric Friedel–Crafts Alkylations*, Wiley-VCH: Weinheim, 2009.
55. Poulsen, T. B.; Jorgensen, K. A. *Chem. Rev.* **2008**, *108*, 2903–2915.
56. Zhuang, W.; Hansen, T.; Jørgensen, K. A. *Chem. Commun.* **2001**, 347–348.
57. Austin, J. F.; MacMillan, D. W. C. *J. Am. Chem. Soc.* **2002**, *124*, 1172–1173.
58. Gathergood, N.; Zhuang, W.; Jørgensen, K. A. *J. Am. Chem. Soc.* **2000**, *122*, 12517–12522.
59. Török, B.; Abid, M.; London, G.; Esquibel, J.; Török, M.; Mhadgut, S. C.; Yan, P.; Prakech, G. K. S. *Angew. Chem. Int. Ed.* **2005**, *44*, 3086–3089.
60. Liu, C.; Widenhoefer, R. A. *Org. Lett.* **2007**, *9*, 1935–1938.
61. Jia, Y.-X.; Zhong, J.; Zhu, S.-F.; Zhang, C.-M.; Zhou, Q.-L. *Angew. Chem. Int. Ed.* **2007**, *46*, 5565–5567.
62. You, S.-L.; Cai, Q.; Zeng, M. *Chem. Soc. Rev.* **2009**, *38*, 2190–2201.
64. Kang, Q.; Zheng, X.-J.; You, S.-L. *Chem. Eur. J.* **2008**, *14*, 3539–3542.
63. Herrera, R. P.; Sgarzani, V.; Bernardi, L.; Ricci, A. *Angew. Chem. Int. Ed.* **2005**, *44*, 6576–6579.
65. Hong, L.; Wang, L.; Sun, W.; Wong, K.; Wang, R. *J. Org. Chem.* **2009**, *74*, 6881–6884.
66. Terrasson, V.; Marcia de Figueiredo, R.; Campagne, J. M. *Eur. J. Org. Chem.* **2010**, 2635–2655.
67. Hong, L.; Liu, C.; Sun, W.; Wong, K.; Wang, R. *Org. Lett.* **2009**, *11*, 2177–2180.
68. Rueping, M.; Nachtsheim, B. J. *Beilstein J. Org. Chem.* **2011**, *6*, 1–24.
69. Mühlthau, F.; Schuster, O.; Bach, T. *J. Am. Chem. Soc.* **2005**, *127*, 9348–9349.
70. Mühlthau, F.; Stadler, D.; Goeppert, A.; Olah, G. A.; Prakash, G. K. S.; Bach, T. *J. Am. Chem. Soc.* **2006**, *128*, 9669–9675.
71. Rubenbauer, P.; Bach, T. *Tetrahedron Lett.* **2008**, *49*, 1305–1309.
72. Iovel, I.; Mertins, K.; Kischel, J.; Zapf, A.; Beller, M. *Angew. Chem. Int. Ed.* **2005**, *44*, 3913–3917.
73. Rueping, M.; Nachtsheim, B. J.; Ieawsuwan, W. *Adv. Synth. Catal.* **2006**, *348*, 1033–1037.
74. Tsuchimoto, T. *Chem. Eur. J.* **2011**, *17*, 4064–4075.
75. Mueller-Westerhoff, U. T.; Swiegers, G. F. *Synth. Commun.* **1994**, *24*, 1389–1393.
76. Gasonoo, M.; Klumpp, D. A. *Tetrahedron Lett.* **2015**, *56*, 4737–4739.
77. Sartori, G. (Ed.); *Advances in Friedel–Crafts Acylation Reactions: Catalytic and Green Processes*, CRC Press: Boca Raton, Florida, 2009.
78. Sartori, G.; Maggi, R. *Chem. Rev.* **2006**, *106*, 1077–1104.
79. Kangani, C. O.; Day, B. W. *Org. Lett.* **2008**, *10*, 2645–2648.
80. Murphy, B.; Goodrich, P.; Hardacre, C.; Oelgemöller, M. *Green Chem.* **2009**, *11*, 1867–1870.
81. Dumas, A. M.; Fishlock, D. *Acc. Chem. Res.* **2010**, *43*, 440–454.
82. Fillion, E.; Fishlock, D. *Tetrahedron* **2009**, *65*, 6682–6695.
83. Nishimoto, Y.; Babu, S. A.; Yasuda, M.; Baba, A. *J. Org. Chem.* **2008**, *73*, 9465–9468.
84. Chavan, S. P.; Garai, S.; Dutta, A. K.; Pal, S. *Eur. J. Org. Chem.* **2012**, 6841–6845.

85. Raja, E. K.; DeSchepper, D. J.; Nilsson Lill, S. O.; Klumpp, D. A. *J. Org. Chem.* **2012**, *77*, 5788–5793.
86. Wakeham, R. J.; Taylor, J. E.; Bull, S. D.; Morris, J. A.; Williams, J. M. J. *Org. Lett.* **2013**, *15*, 702–705.
87. Yang, Y.; Zhou, B.; Li, Y. *Adv. Synth. Catal.* **2012**, *354*, 2916–2920.
88. Sharma, S.; Kim, M.; Park, J.; Kim, M.; Kwak, J. H.; Jung, Y. H.; Oh, J. S.; Lee, Y.; Kim, I. S. *Eur. J. Org. Chem.* **2013**, 6656–6665.
89. Kim, H. Y.; Oh, K. *Org. Lett.* **2014**, *16*, 5934–5936.
90. Tran, P. H.; Tran, H. N.; Hansen, P. E.; Do, M. H. N.; Le, T. N. *Molecules* **2015**, *20*, 19605–19619.
91. Motiwala, H. F.; Vekariya, R. H.; Aubé, J. *Org. Lett.* **2015**, *17*, 5484–5487.
92. Luo, H.; Pan, L.; Xu, X.; Liu, Q. *J. Org. Chem.* **2015**, *80*, 8282–8289.
93. Ueda, H.; Satoh, H.; Matsumoto, K.; Sugimoto, K.; Fukuyama, T.; Tokuyama, H. *Angew. Chem. Int. Ed.* **2009**, *48*, 7600–7603.
94. Okano, K.; Fujiwara, H.; Noji, T.; Fukuyama, T.; Tokuyama, H. *Angew. Chem. Int. Ed.* **2010**, *49*, 5925–5929.
95. Taj, R. A.; Green, J. R. *J. Org. Chem.* **2010**, *75*, 8258–8270.
96. Choi, Y. L.; Kim, B. T.; Heo, J.-N. *J. Org. Chem.* **2012**, *77*, 8762–8767.
97. Lv, Z.; Li, Z.; Liang, G. *Org. Lett.* **2014**, *16*, 1349–1351.
98. Dange, N. S.; Hong, B.-C.; Lee, G.-H. *RSC Adv.* **2014**, *4*, 59706–59715.
99. Li, Y.; Li, L.; Guo, Y.; Xie, Z. *Tetrahedron* **2015**, *71*, 9282–9286.
100. Wehlan, H.; Jezek, E.; Lebrasseur, N.; Pave, G.; Roulland, E.; White, A. J. P.; Burrows, J. N.; Barrett, A. G. M. *J. Org. Chem.* **2006**, *71*, 8151–8158.
101. Elban, M. A.; Hecht, S. M. *J. Org. Chem.* **2008**, *73*, 785–793.
102. Tang, S.; Xu, Y.; He, J.; He, Y.; Zheng, J.; Pan, X.; She, X. *Org. Lett.* **2008**, *10*, 1855–1858.
103. de Jesus Cortez, F.; Sarpong, R. *Org. Lett.* **2010**, *12*, 1428–1431.
104. Taylor, J. E.; Jones, M. D.; Williams, J. M. J.; Bull, S. D. *Org. Lett.* **2010**, *12*, 5740–5743.
105. Roesner, S.; Casatejada, J. M.; Elford, T.G.; Sonawane, R. P.; Aggarwal, V. K. *Org. Lett.* **2011**, *13*, 5740–5743.
106. Kim, K.-S.; Kim, I.-Y. *Org. Lett.* **2010**, *12*, 5314–5317.
107. Chen, I. T.; Baitinger, I.; Schreyer, L.; Trauner, D. *Org. Lett.* **2014**, *16*, 166–169.
108. Jereb, M.; Gosak, K. *Org. Biomol. Chem.* **2015**, *13*, 3103–3115.
109. Yamaoka, Y.; Yoshida, T.; Shinozaki, M.; Yamada, K.-I.; Takasu, K. *J. Org. Chem.* **2015**, *80*, 957–964.
110. Parnes, R.; Pappo, D. *Org. Lett.* **2015**, *17*, 2924–2927.
111. Tyndall, S. Wong, K. F.; Van Alstine-Parris, M. A. *J. Org. Chem.* **2015**, *80*, 8951–8953.

6 Rearrangement and Fragmentation Reactions

Hannah Payne, Sharon Molnar, and Micheal Fultz

CONTENTS

6.1   INTRODUCTION

For the undergraduate organic student, rearrangement and fragmentation reactions are not as commonplace as substitution, addition, or elimination reactions. Be it much to their surprise, they do indeed explore rearrangement–fragmentation reactions early on in their organic chemistry journey. "Rearrangement" reactions, in practicality, describe organic reactions involving either a one-step or multiple-step atom (or large fragment) migration.[1]

Generally, rearrangements take place when carbocations are present. A case in point are Wagner–Meerwein rearrangements associated with [1,2]-hydride, alkyl, or aryl shifts.[2] The driving force, of course, for these rearrangements is the creation of a more energetically stable carbocation trending $3° > 2° > 1°$.[3]

(a)

(b)

A 2° carbocation is more energetically favored than a 1°. Therefore, the positive charge on the 1° position is eliminated via a hydride shift.[3]

In a similar fashion, alkyl and aryl shifts occur to create a more energetically cc favored carbocation.[3]

Most synthetic chemists would agree that reducing the number of necessary steps to achieve a desired product increases efficiencies and yields, while reducing material cost and man-hours. In this vein, Wang utilized the aza-Cope–Mannich cyclization to construct the D ring within (±)-cycloclavine (see structure in Section 6.7.2).[4] The synthesis of this 3,4-fused indole skeleton, across 14–17 steps, has posed a significant challenge to chemists. The utilization of the aza-Cope–Mannich cyclization allowed Wang et al. to devise a 7-step synthetic strategy that increased the overall percent yield to 27% versus reported low yields (0.2%, 1.2%, and 2.3%).[5,6]

The average undergraduate organic student is aware of the fragmentation that occurs during mass spectrometry, where a given compound is exposed to high ion frequency. The following literature, however, indicates that fragmentations occur in reactions to create target compounds.

6.2 REARRANGEMENTS

Rearrangement reactions are a broad class of organic reactions in which a molecule's carbon skeleton is "rearranged" to afford a structural isomer of the parent compound. In essence, these reactions involve the migration of an atom from one location to another in the carbon skeleton architecture. Two key rearrangements are the [1,2]-rearrangements and pericyclic reactions.[7]

6.3 [1,2]-REARRANGEMENTS

[1,2]-Rearrangements involve the movement of a substituent from one atom to another across two adjacent atoms.[4] The most readily known examples include the Wagner–Meerwein rearrangement, as mentioned above, and the [1,2]-Wittig rearrangement. The [1,2]-Wittig rearrangement transposes aryl alkyl ethers, in the presence of a stoichiometric amount of a strong base, to the corresponding secondary or tertiary alcohols.[8]

• 1,2-Migration of substituent R from carbon 2 to carbon 3.[5]

6.4   PERICYCLIC REACTIONS

In the case of pericyclic reactions, a transition state molecule, possessing cyclic geometry, occurs in a concerted fashion.[9]

6.5   FRAGMENTATIONS

As per their name, fragmentation reactions generate fragments of the originating compound via carbon–carbon bond breakage. The structural feature that allows for fragmentation is the presence of a carbon β wherein an electron deficiency may develop and accommodate carbocationic character.[10] This type of reaction has a higher frequency of occurrence when the γ atom is a heteroatom containing an unshared electron pair. The presence of an unshared electron pair is critical in the stabilization of the newly formed cationic center.[9] Either heterolytic or homolytic hydrocarbon fragmentation may occur, however, each is governed by its own set of rules.[11–13] Heterolytic fragmentation, for instance, is governed by "polarity alternation"[11] and the prediction of the product generated is based upon the concepts of consonance/dissonance, conjoinment/disconjoinment, and resonance.[10,11] Homolytic hydrocarbon fragmentation, on the other hand, is less reliably predicted and the energy differences are less directing.[13]

The ease with which hydrocarbon fragmentation occurs may be enhanced by altering the charge distribution on the homopolar carbon backbone.[11–13] A favorable charge distribution is established by introducing polar functional groups in such a manner as to create consonant/dissonant patterns.[12] The following examples demonstrate hydrocarbons that either fragment well or resist fragmentation altogether.

d = OH, OR, OCOR, NRR', NR(COR), SH, SR, F, Cl, Br, I

Fragmentation reactions are either concerted or stepwise in nature. Regardless of the fashion in which the fragmentation occurs, it is perfunctory for each member of the atomic chain to engage in electron pair exchange.[12]

Concerted mechanistic pathways have the added restraints of molecular geometry due to the need of continuous overlap of the participating orbitals.[10] Discussion on the most well-known fragmentation reactions will follow later in this chapter.

For the purpose of this chapter rearrangements that involve electron deficient carbons (resulting from insufficient valence shell electrons or induction) will be considered.

6.6 REARRANGEMENTS

6.6.1 CLAISEN REARRANGEMENTS

The Claisen rearrangement is one of the better-known reactions in the rearrangement category. Discovered by L. Claisen in 1912,[14] this rearrangement refers to the thermal [3,3]-sigmatropic rearrangement of ally vinyl ethers **1** to the corresponding γ,δ-unsaturated carbonyl compound **2**.[8]

Under the broad umbrella of Claisen rearrangements, there are multiple subclasses which undergo concerted pericyclic reactions. A brief overview of the aza-Claisen, Johnson–Claisen, Claisen–Ireland, and Eschenmoser–Claisen variations is provided in the following sections.

6.6.1.1 Aza-Claisen Rearrangement

The thermal [3,3]-sigmatropic rearrangement of *N*-allyl enamines, such as **3** to the corresponding imines **4**, occurs through a concerted process. This rearrangement usually takes place via a chair-like transition state in which the substituents are arranged in quasi-equatorial positions.[5,14,15]

After solidifying their understanding of the ether-directed, stereoselective aza-Claisen rearrangement, Jamieson and Sutherland were able to execute the process in achieving efficient synthesis of cyclic natural products containing *erythro*-hydroxyl and -amino functional groups.[16] Jameson and Sutherland chose to target the synthesis of piperidine alkaloid, α-conhydrine **5**. Up to this point, the biologically active compound (originally isolated from the poisonous plant *Conium maculatum* L in 1856) posed a synthetic challenge.[17] In general, compounds possessing a 2-(1-hydroxyalkyl)-piperidine unit (e.g., the aforementioned targeted α-conhydrine **5**) are of interest owing to their antiviral and antitumor properties.[18,19] Jamieson and Sutherland's quest for the conhydrine began with the production of allylic trichloroamines. The allylic trichloroacetimidate **6** was prepared using a catalytic amount of 1,8-diazabicyclo[5.4.0]undec-7-ene (DBU) and trichloroacetonitrile. Aza-Clasen rearrangement of **6** in tetrahydrofuran (THF), using bis-(acetonitrile)palladium(II) chloride (10 mol%) as the catalyst, generated erythro- and threo-allylic trichloroamines **7** and **8** in 52% yield over the two steps in a 12:1 ratio. Exchanging the THF for a noncoordinating solvent, such as toluene, increases the yield (55%) and boosts the stereochemical outcome (16:1).

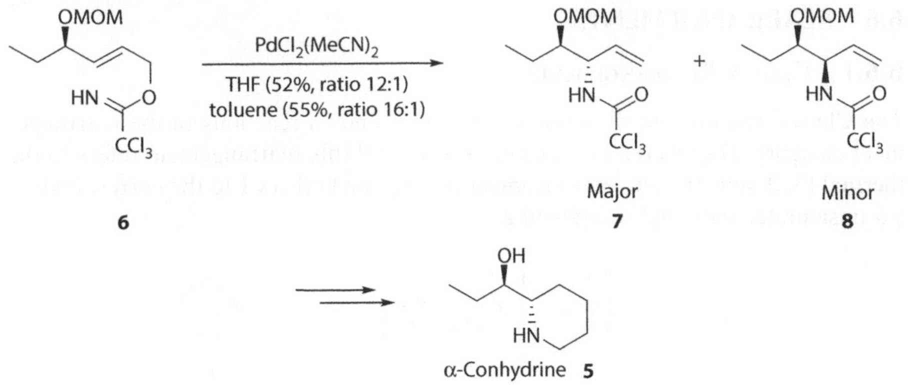

α-Conhydrine **5**

6.6.1.2 Johnson–Claisen Rearrangement

When allylic alcohols **9** are heated in the presence of excess triethyl orthoacetate under weakly acidic conditions, a ketene acetal intermediate **10** is created.[8,20] This intermediate undergoes a facile [3.3]-sigmatropic rearrangement to afford γ,δ-unsaturated esters **11**.[8]

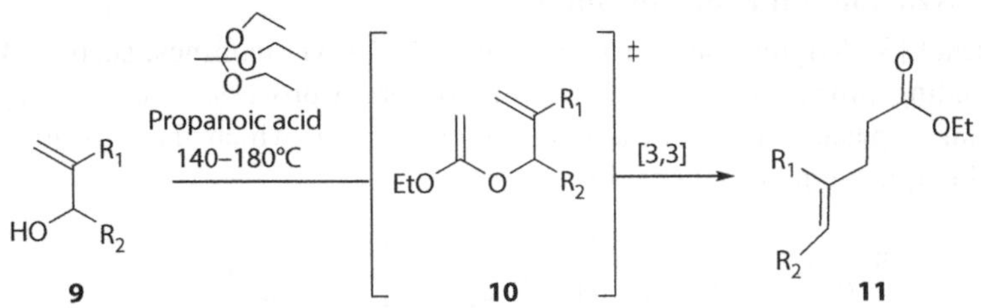

Johnson–Claisen rearrangement is an efficient diastereoselective route for the creation of quaternary stereocenters of the C-3 position of cyclic lactams of γ-hydroxy-α,β-unsaturated lactams, such as physostigmine **12**. Compound **13** was subjected to standard Johnson–Claisen rearrangement reaction conditions (triethyl orthoacetate, propionic acid, reflux), which produced compound **14** in 97% yield and >99% enantiomeric purity.[21]

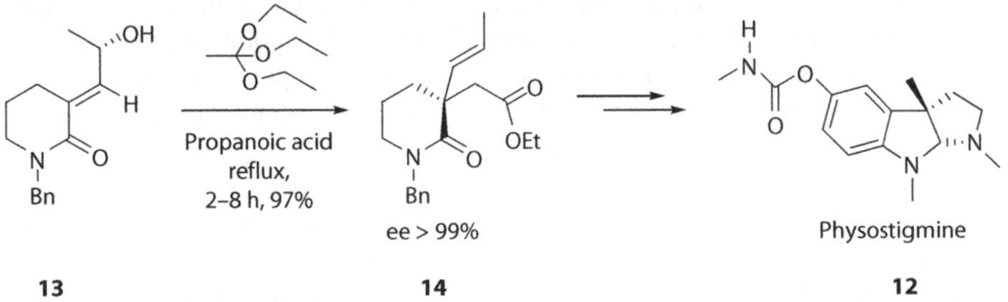

6.6.1.3 Claisen–Ireland Rearrangement

Standard Claisen–Ireland rearrangements convert allylic esters **15** to acids through O-trialkylsilylketene acetals such as **16** will undergo [3,3]-sigmatropic rearrangement to afford γ,δ-unsaturated carboxylic acids **17**.[8,22]

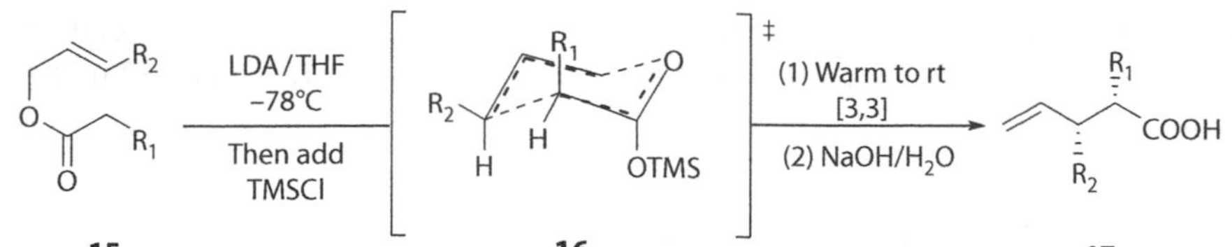

15   **16**   **17**

Williams utilized the Claisen–Ireland rearrangement for the stereo-controlled total synthesis of **18** (4-hydroxydictyolactone) which established the carbon backbone asymmetry of the contiguous C2, C3, C10 stereotriad. This was accomplished by synthesizing the nonracemic Ireland–Claisen precursor **19**, followed by the addition of LDA (lithium diisopropylamide) into a cold reaction mixture containing TMSCl (trimethylsilyl chloride), and Et₃N. Once heated to 70°C, the intermediate product E (*O*)-trimethylsilyl ketene acetal **20** is converted to the carboxylic acid **21** (dr 94:6) and isolated in high yield.[23]

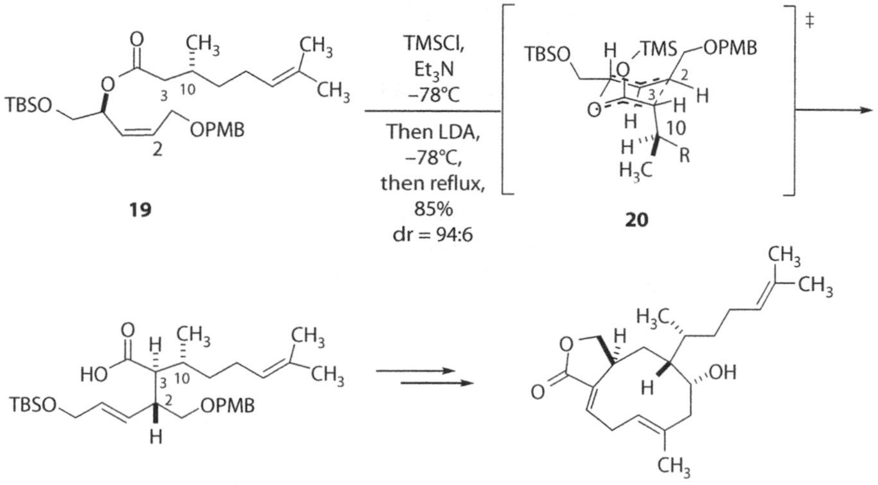

19   **20**

21   4-hydroxydictyolactone **18**

6.6.1.4   Eschenmoser–Claisen

When allylic or benzylic alcohols **22** are heated in *N,N*-dimethylacetamide dimethyl acetal in xylenes, the intermediate **23** undergoes [3,3]-sigmatropic rearrangement to afford γ,δ-unsaturated amides **24**.[8,24]

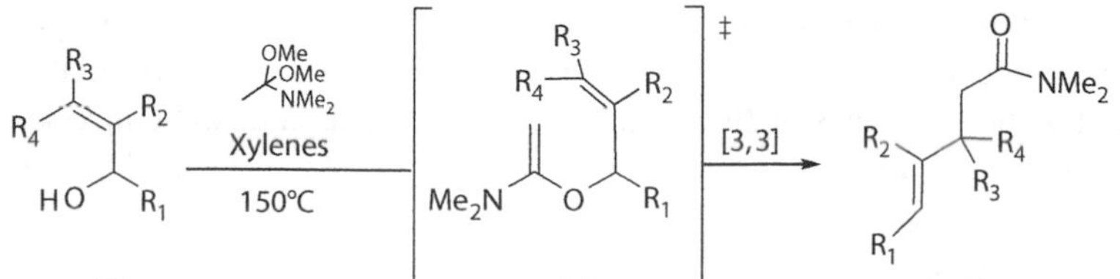

22   **23**   **24**

An excellent example of the Eshenmoser–Claisen rearrangement is the synthesis of the tricyclic compound aplykurodinone **25**. This process begins by taking the crude intermediate **26**, mixing it with *N,N*-dimethylacetamide dimethyl acetal and microwaving in a sealed reaction vessel to produce γ,δ-unsaturated amide **27** in 90% yield.[25]

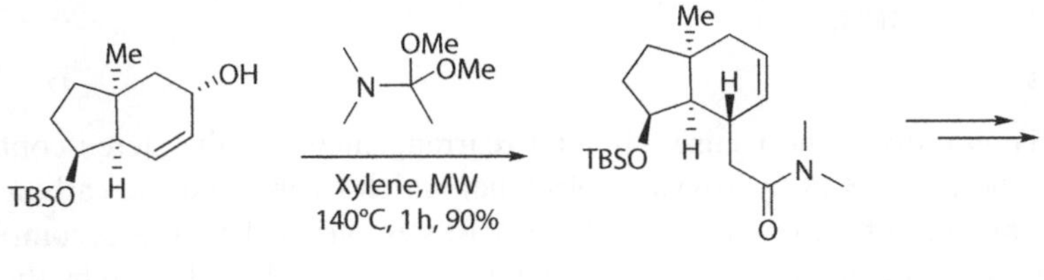

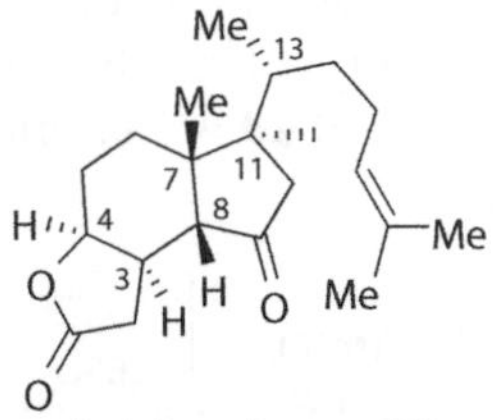

Aplykurodinone **25**

6.7 COPE REARRANGEMENTS

The Cope rearrangement was discovered in 1940 when A. C. Cope observed that ethyl (1-methylpropylidene)-cyanoacetate **28** rearranged into isomeric ethyl (1-methylpropenyl)-allylcyanoacetate **29** upon distillation.[26]

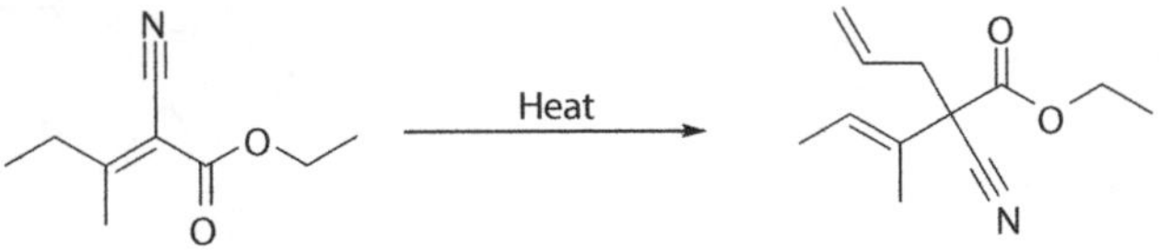

A shared attribute of the Claisen and Cope rearrangements is their multiple subclasses that all undergo concerted pericyclic reactions. The anionic oxy-Cope and aza-Cope subclasses are discussed below.

6.7.1 ANIONIC OXY-COPE

This type of rearrangement is mechanistically identical to the oxy-Cope, yet it occurs at an accelerated rate of 10^{10}–10^{17}. This rate acceleration is attributable to the treatment of the 1,5-dien-3-ol **30** with a base, such as potassium *tert*-butoxide, which provides an oxide transition state **31** to reveal aldehydes **32** as the final product.

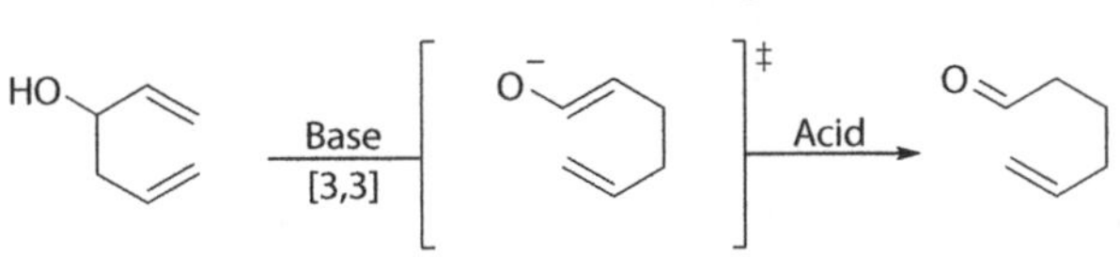

Mulinanes (33) are a class of diterpenoids that are biologically active natural compounds.[27] They can be synthesized by employing the anionic oxy-Cope rearrangement to construct the relative configuration of the C8 stererocenter.[28] Addition of potassium hydride to a solution of alcohol 34 and 18-crown-6 affords the desired aldehyde 35 as a major product (73% yield) due to the anionic oxy-Cope's high stereoconvergency.[28]

6.7.2 Aza-Cope

When nitrogen is incorporated within the 1,5 π bond system, the [3,3]-sigmatropic rearrangement is referred to as aza-Cope. Variants of this Cope subclass include 1-aza, 2-aza (36), 3-aza, and 1,3-, 2,3-, 2,5-, 3,4-diaza-Cope rearrangements.[8]

Wang utilized the aza-Cope–Mannich cyclization to construct the D ring within (±)-cycloclavine 39, which consists of a complex 3,4-fused polycyclic molecular architecture.[4] The synthesis of this 3,4-fused indole skeleton has posed a daunting challenge to chemists given its low yield (0.2%, 1.2%, and 2.3%) across 14–17 steps.[5,6] By using the aza-Cope–Mannich cyclization, Wang devised a synthetic strategy, employing the aza-Cope–Mannich cyclization, and achieved a seven step synthetic approach with an impressive 27% overall yield. The aza-Cope–Mannich tandem reaction is initiated by a cationic 2-aza-Cope rearrangement (2-azonia-[3,3]-sigmatropic rearrangement) followed by the Mannich reaction. The desired aldehyde 44 is obtained in high yield (83%) from aldehyde 40 when subjected to aza-Cope–Mannich conditions ($FeCl_3$, CH_2Cl_2, reflux 45°C, 0.2 h) with 2-hydroxyhomopropargyl tosylamine 41.

40   **41**   **42**   **43**

44   (±)-Cycloclavine

39

6.8   KETONE AND KETENE FORMATION REARRANGEMENTS

6.8.1   CARROLL REARRANGEMENT

The Carroll rearrangement is defined by the [3,3]-sigmatropic rearrangement of allylic β-keto esters (**45**) to γ,δ-unsaturated ketones **46**.[7]

45   **46**

Abe determined that this particular rearrangement would be beneficial in synthesizing the *trans*-bicyclic enone of spirocurcasone **47**. Via aldol condensation, 2-oxobutyl derivatives **50/51** are produced by the Carroll rearrangement of β-keto allyl ester **48**. This, in turn, leads to the formation of the desired tricyclic diterpenoid final product, spirocurcasone **47**.[29] Efficient synthesis of related tricyclic diterpenoid curcusones is highly sought after due to their antiproliferative activity toward mouse lymphoma L5178Y cell lines.[30] Bringing together the β-keto allyl ester **48** with LDA in cyclopentyl methyl ether (CPME) at −78°C, results in the lithium dienolate intermediate **49**. The lithium dienolate intermediate is then refluxed in CPME for 20 h. During this time, a rearrangement takes place of the enolate anion from the C1 to the C2 position, which is followed by decarboxylation. The process of rearrangement initiates mainly from the opposite face of the diethyl acetal group at C3. The final bulk product is a mixture containing the target compound, α-2-oxobutyl **50** (15%) and the nondesirable β-2-oxobutyl product **51** (46%). Ensuing treatments of the isolated **50** render the target spirocurcasone derivative.

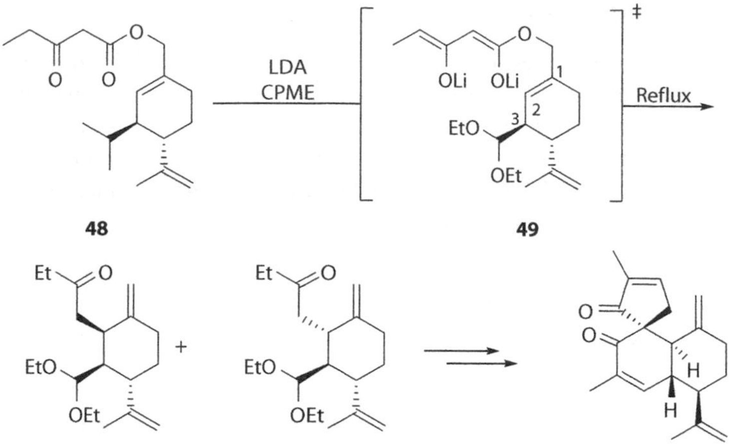

48   LDA CPME   **49**   Reflux

50 (15%)   **51** (46%)   Spirocurcasone **47**

6.8.2   RUPE REARRANGEMENT

When a tertiary propargylic alcohol **52** undergoes acid-catalyzed dehydration, a rearrangement takes place that results in the [1,2]-shift of the hydroxyl group. Subsequently, the corresponding α,β-unsaturated ketone **53** is created.[8,31]

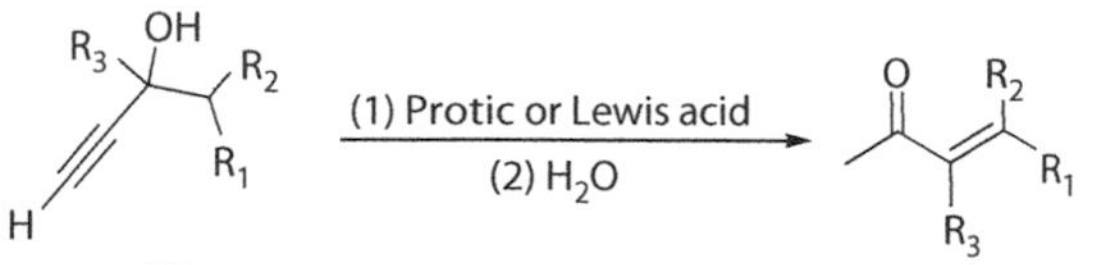

52   (1) Protic or Lewis acid (2) H_2O **53**

The Rupe rearrangement, in tandem with Donnelly–Farrell cyclization, can result in the conversion of **54** to **55**. This is noteworthy due to **55** being an essential intermediate for the synthesis of 4,5,5-trimethyl-5,6-dihydrobenzo[c][2,7]naphthyridine, **56**.[32] This debrominated analog of marine alkaloid veranamine (which shows potential as an antidepressant)[33] can be synthesized in three steps with a 38% overall yield.

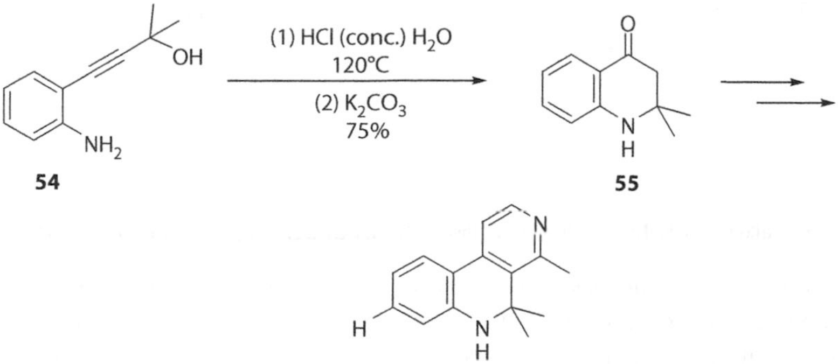

54   (1) HCl (conc.) H_2O 120°C (2) K_2CO_3 75%   **55**

4,5,5-trimethyl-5,6-
dihydrobenzo[c][2,7]naphthyridine **56**

6.8.3 SAKURAI REACTION

The Neber rearrangement results in the formation of an azirine (**37**) when *O*-acylated ketoximes **58** are treated with a base. Subsequent transformations of the azirine ring results in a α-amino ketone **59** or ketals.[34] This rearrangement is critical not only for creating intermediates to synthesize heterocycles, but for introducing α-amino ketones into natural products.[7] Unlike the Beckmann rearrangement, this process creates nonstereospecific amino ketones.

The synthesis of motualevic acids A–F, and (E) & (Z) geometrical isomers of antazirines **60**, is accomplished utilizing the Neber rearrangement in the later stages of the synthesis. Tosylated compound **61** is catalyzed with quinidine in toluene at 0°C in order to induce the chirality that is present in (E)-antarzirine. The desired (R)-(E)-antazirine (ent-**60**) was produced in high yield (85%) and 76% ee.[35]

6.8.4 BAKER–VENKATARAMAN REARRANGEMENT

β-Diketones are important synthetic intermediates in the production of chromones, flavones, coumarins, and isoflavones.[8] This rearrangement takes place when an aromatic *ortho*-acyloxyketone **62** is treated with a base, which catalyzes a rearrangement that yields the corresponding aromatic β-diketones **63**.

R_1 = alkyl, aryl, NH_2; R_2 = alkyl, aryl; base = KOH, KOt-Bu, NaH, Na Metal, KH, C_5H_5N

A method for the synthesis of the natural product platachromone B **64** and related compounds has been developed and takes advantage of the Baker–Venkataraman rearrangement. Base-catalyzed Baker–Venkataraman rearrangement of acrylate derivatives **65a–c** will yield (2Z,4E)-5-aryl-3-hydroxy-1-[2-hydroxy-4,6-dimethoxy-3-(3-methylbut-2-en-1-yl)phenyl]penta-2,4-dien-1-ones **66a–c**.[36]

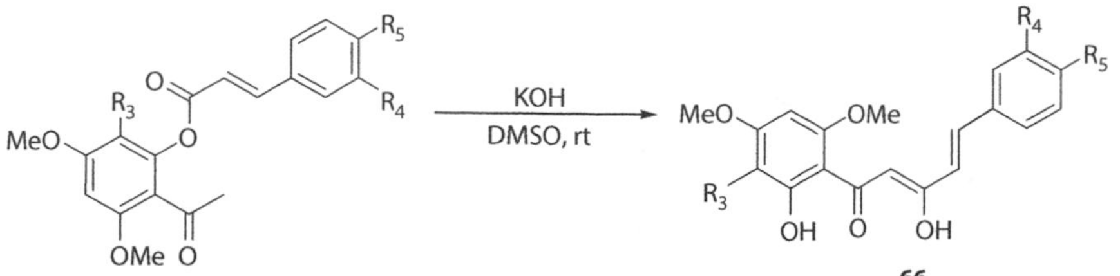

65

a: $R_3 = CH_2CHCMe_2$, $R_4 = R_5 = H$
b: $R_3 = CH_2CHCMe_2$, $R_4 = H$, $R_5 = OMe$
c: $R_3 = CH_2CHCMe_2$, $R_4 = R_5 = OMe$

66

a: $R_3 = CH_2CHCMe_2$, $R_4 = R_5 = H$
b: $R_3 = CH_2CHCMe_2$, $R_4 = H$, $R_5 = OMe$
c: $R_3 = CH_2CHCMe_2$, $R_4 = R_5 = OMe$

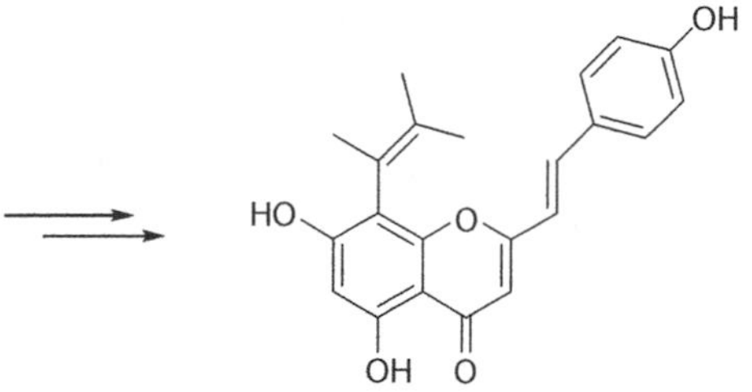

Platachrome B **64**

6.8.5 FERRIER REARRANGEMENT

In general, the Ferrier rearrangement refers to the Lewis acid promoted rearrangement of unsaturated carbohydrates.[37] Depending on the starting material, 1,2-glycal (Type I) or exocyclic enol ether **67** (Type II), either a 2,3-unsaturated glycosyl compound or highly substituted cyclohexanone (such as **68**) is created, respectively.[8]

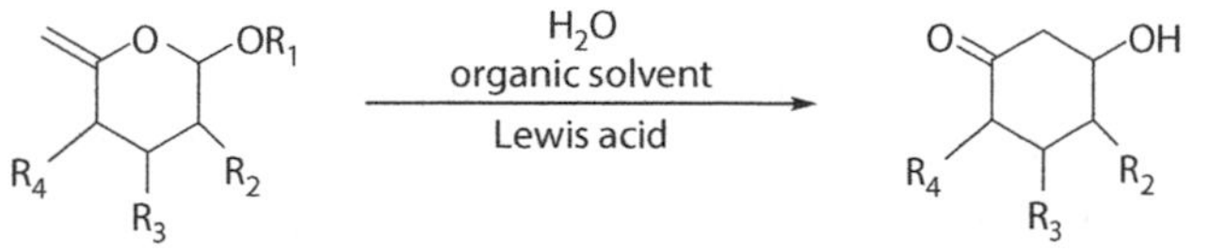

R_3 = alkyl; R_{4-5} = O-alkyl, O-acyl

The Ferrier rearrangement can be utilized to convert a sugar into a ketone, as shown by Seth for the preparation of fluoro cyclohexenyl nucleic acid (F-CeNA).[38] F-CeNA **69** has the potential to be used in antisense therapeutics, which is a specific manner of treating genetic disorders. Fluorine substitution improves the hybridization of CeNA-modified oligonucleotides in a three prong approach by (1) increasing the strength of Watson–Crick base pairing; (2) creation of a conformationally more flexible CeNa ring versus the hexitol ring, and (3) the production of extreme half-chair conformations of CeNA (which resemble the conformational states of the C2′-endo and C3′-endo sugar puckers of the furanose ring in DNA and RNA). Seth transformed the sugar **70** into the carbocycle **71** via the Ferrier rearrangement by introducing **70** to catalytic PdCl$_2$.

F-CeNA thymine
phosphoramidite **69**

70 **71**

6.8.6 TIFFENEAU–DEMJANOV REARRANGEMENT

The Tiffeneau–Demjanov rearrangement is regarded as a variant of the pinacol rearrangement, and can be carried out on four- to eight-membered rings.[39] The percent yields of ring-enlarged products, although better than the Demjanov rearrangement, decrease with increasing ring size.[39–41] Cycloalkanone **72** are created when β-aminoalcohols **73** are combined with nitrous acid. The cycloalkanone **72**, however, has a tendency to undergo a ring expansion.[39–41]

73 **72**

There has been a heightened interest in the synthesis of Merrilactone A **76** due to its potential as a nonpeptide neurotrophic factor that can be used for neurodegenerative diseases.[42] However, its complex compact sesquiterpene architecture, which features five rings and seven chiral centers (five of these chiral centers are contiguous fully substituted carbon atoms) has made its synthesis quite a feat. The Tiffeneau–Demjanov rearrangement can be utilized to construct one of the five rings. In order to transform the cyclobutanone **74** to the required cyclopentanone **75** with regioselective ring enlargement, the least substituted methylene carbon underwent preferential migration. Ethyl diazoacetate in the presence of $BF_3 \cdot Et_2O$ yielded the cyclopentanone **75** in very good yield (88%).[43]

Merrilactone A
76

74 **75**

6.9 ALDEHYDE (OR KETONE) FORMATION REARRANGEMENTS

6.9.1 FRIES REARRANGEMENT

Phenolic esters (**77**) in the presence of a Lewis, Brønsted, or solid acid results in a rearrangement to the corresponding *ortho*-(**78**) or *para*-(**79**) substituted phenolic aldehydes (or ketones).[8,44] While this rearrangement allows for the preparation of acyl phenols, the harsh reaction conditions that the rearrangement requires can only accommodate esters with stable acyl components.[7] Another concern is the presence of heavily substituted aromatic or acyl components. A high degree of substitution can invoke steric constraints that will result in an overall decrease in the yield.[1]

Lewis, Brønsted
or solid acid
R_1 = alkyl, —OR, —NR$_2$, -aryl
R_2 = H, alkyl, aryl

77   **78**   **79**

The Fries rearrangement allows for the introduction of the terminal α-substituted butenolide in the total synthesis of muricadienin **80**.[45] This compound is sought after since it is the unsaturated putative precursor in the biosynthesis of *trans-* and *cis-*solamin. In essence, subsequent *O*-acylation with fatty acid **81** was followed by the *in situ* Fries rearrangement triggered by DMAP. The resulting tricarbonyl intermediate **82** was then directly reduced with NaBH$_3$CN in acetic acid to afford α-alylated butenolide **83** in an excellent yield of 98% over three chemical transformations.[46]

DCC, DMAP,
DIPEA, DCM,
0°C, -rt, 12 h

NaBH$_3$CN,
AcOH,10°C
-rt,12h

98%
(over 2
steps)

82

81

83

(+)-Muricadienin
80

6.9.2   MEYER–SCHUSTER REARRANGEMENT

The Meyer–Schuster rearrangement refers to the acid-catalyzed isomerization of secondary and tertiary propargylic alcohols (**84**) through a [1,3]-shift of the hydroxyl group to the corresponding α,β-unsaturated aldehyde or ketone **85**.[8]

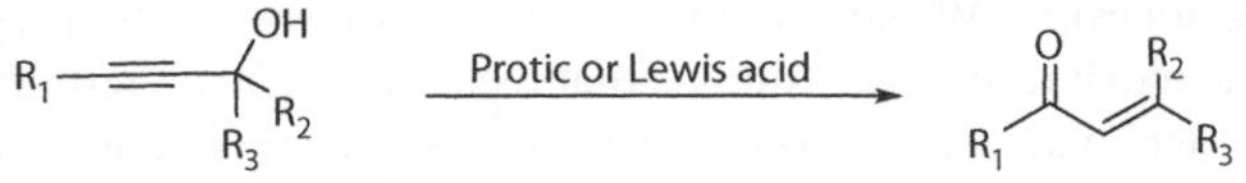

84 → **85**

R_1 = H, alkyl, aryl; R_{2-3} = H, aryl, or alkyl with no H atoms adjacent to α-carbon

This rearrangement was utilized by Beretta to construct an enone moiety for the synthesis of B_1 and L_1 prostanoids (**88**).[47] Cyclopentanoid phytoprostanes (PhytoPs) are a group of bioactive compounds that perform various functions, such as immunomodulatory activity on dendritic cells, as well as antiinflammatory and apoptotic properties. With this being said, the enone moiety of **87** can be synthesized by subjecting the propargylic alcohol **86** to Nolan's Au(I) dinuclear catalyst [(IPrAu)$_2$(u-OH)] BF_4 in MeOH − H_2O (1:1).[48] The result is a smooth transformation to enone **87** in 76% isolated yield with ≥95% E-diastereoselectivity.

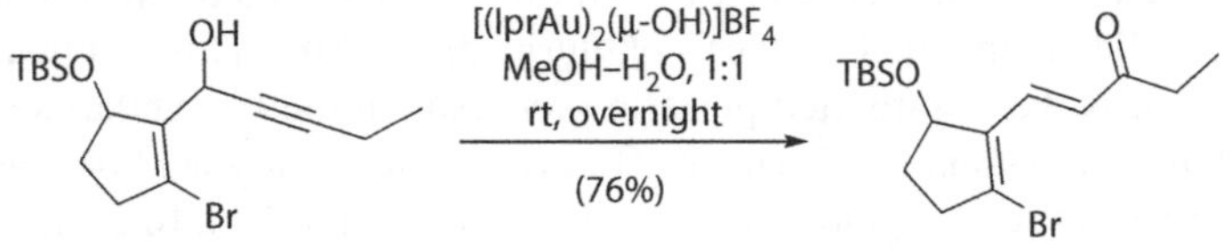

86 → **87**

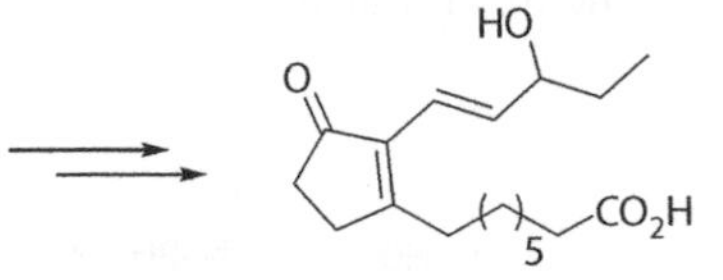

[(IprAu)$_2$(μ-OH)]BF$_4$
MeOH–H$_2$O, 1:1
rt, overnight

(76%)

88 (RS)-16-L$_1$-PhytoP

6.9.3 PINACOL REARRANGEMENT

When vicinal diols (**89**) are treated with a Lewis or protic acid they undergo an [1,2]-alkyl, aryl, or hydride shift to afford an aldehyde **91** or ketone **93** through a carbocation intermediate (**90,92**).[7] The reaction can be highly regioselective, which is determined by the relative migratory aptitudes of the substituents attached to the carbon adjacent to the carbocation center. The relative migratory aptitudes correlate with their ability to stabilize a positive charge, such that the following trend is observed: aryl ~ H ~ vinyl ~ t-Bu >> cyclopropyl > 2° alkyl > 1° alkyl.[49,50]

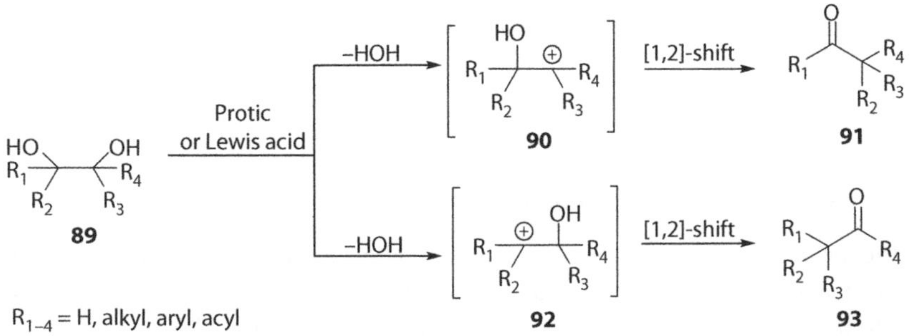

R_{1-4} = H, alkyl, aryl, acyl

Ingenane diterpenes possess important biological activity ranging from anticancer to anti-HIV.[51] The development of a synthetic route for diterpenoid (+)-ingenol **94**, was a challenging process due to the challenging scaffold design of the compound. The reaction that leads to **96** that is used for the complete synthesis of ingenol **94** had to be carefully tweaked in order to reach completion. Treating alcohol **95** with $BF_3 \cdot Et_2O$ in CH_2Cl_2 and quenched at $-78°C$ resulted in a recovery of only starting material. When the reaction was quenched with saturated aqueous $NaHCO_3$ at $-78°C$, a low yield of the desired product **96** was obtained. Upon further investigation, it was noted that the temperature of the reaction quench was essential to getting a high yield of desired product (80%) with the following conditions 1:1 $MeOH/Et_3N$ at $-40°C$.[52]

95

$$BF_3 \cdot Et_2O$$
$$CH_2Cl_2$$
$$-78°C \text{ to } -40°C$$
$$\text{then } MeOH/Et_3N$$
$$80\%$$

96

(+)-Ingenol
94

6.9.4 SEMIPINACOL REARRANGEMENT

Similar to the pinacol rearrangement, when 2-heterosubstituted alcohols (**97**) are subjected to mild acidic conditions they undergo an [1,2]-alkyl, aryl, or hydride shift **98** to afford an aldehyde or ketone **99**.[7,53] The semipinacol rearrangement is almost exclusively utilized in complex molecule synthesis due to its predictability and mild reaction conditions.[8]

$R_{1-4} = H$, alkyl, aryl, acyl; $X = Cl$, Br, I, SR, OTS, OMs, N_2
Mild conditions = $LiClO_4/THF/CaCO_3$, Et_3Al/CH_2Cl_2, etc.

One such complex molecule synthesis is that of peribysin E and its analogs, as developed by Handore and Reddy.[54] Peribysin E **100** exhibits potent cell adhesion inhibitory activity and therefore serves as a potential candidate as an anticancer and antiinflammatory agent. Epoxy alcohol **101**, underwent a semipinacol-type rearrangement to afford the peribysin E analog **102**.

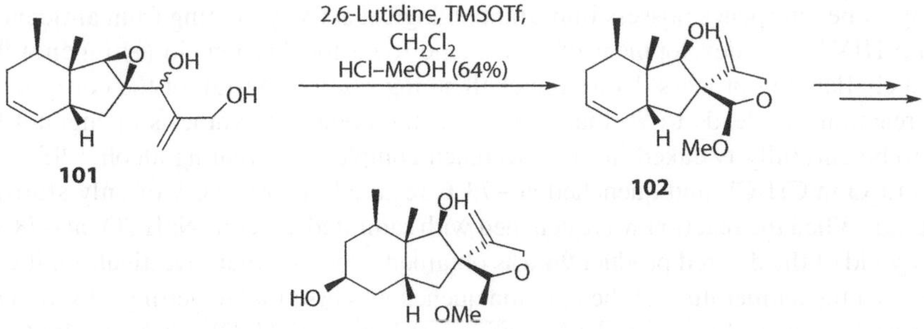

PeribysinE
100

6.10 CARBOXYLIC ACID FORMATION REARRANGEMENTS

6.10.1 FAVORSKII REARRANGEMENT

The Favorskii rearrangement is useful for highly branched carboxylic acid synthesis, where an α-halo ketones possesses at least one α-hydrogen.[55,56] Treatment of such starting materials with a base, in the presence of a nucleophile, induces skeletal rearrangement via a cyclopropane intermediate. The creation of the new carboxylic acid, or carboxylic acid derivative, is regio- and steroselective. Given this consideration, the Favorskii rearrangement is sensitive to structural factors and reaction environment.

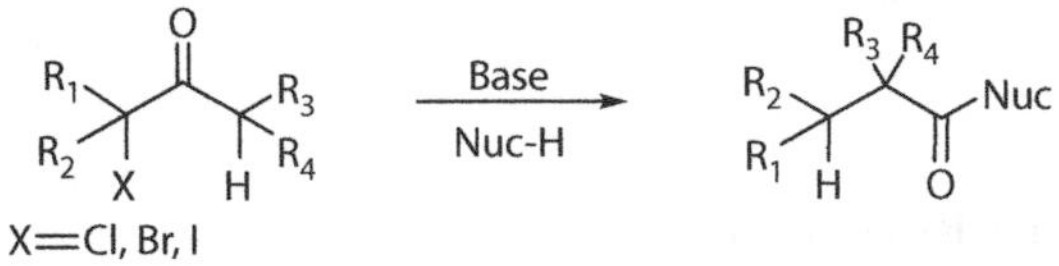

The synthetic optimization of nitroxides emanates has garnered interest after it was determined that this class of compounds ameliorates the toxic effects of radiation during cancer therapy.[57,58] The Favorskii rearrangement of 3,5-dibromo-4-oxo-2,2,6,6-tetramethylpiperidine **103** was performed in the presence of piperidine to form the piperidine derivative **104** directly in 89% yield.[59]

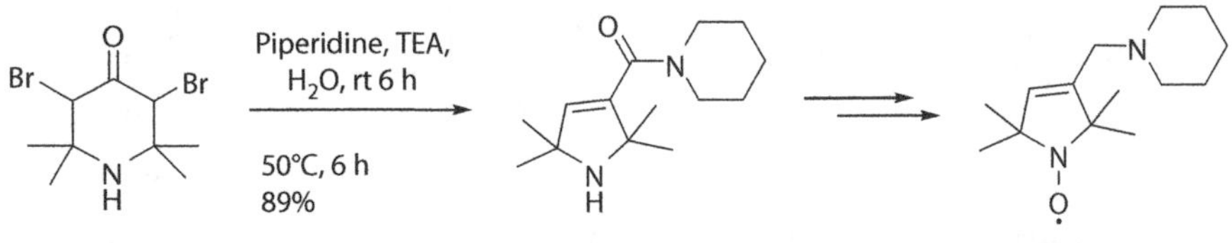

6.10.2 HOMO-FAVORSKII REARRANGEMENT

Similar to the Favorskii rearrangement, the homo-Favorskii occurs when a β-halo ketone **105** is treated with a base in the presence of a nucleophile. Carboxylic acids or carboxylic acid derivatives **107** are formed through cyclobutanone intermediates like **106**.[8,60]

105 **106** **107**

108 **109** **110**

111 **110** Kelsoene **112**

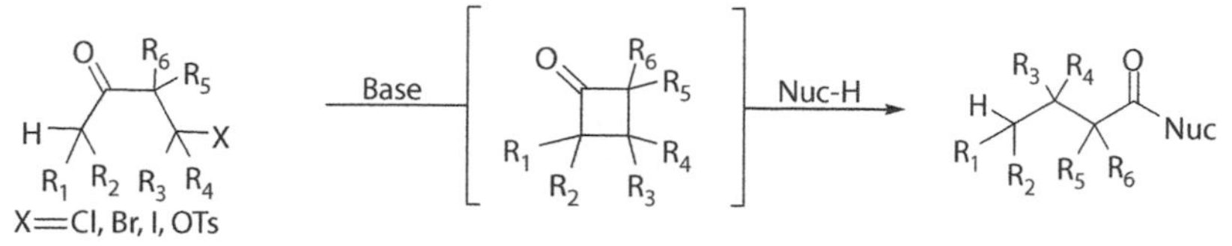

The synthesis of a sesquiterpene with a unique tricycle decane skeleton, (+)-kelsoene (**112**),[40] involves a base-catalyzed reaction of γ-keto tosylate **108**. This process initiates a homo-Favorskii rearrangement to cyclobutanone **109**. Treatment of γ-keto tosylate with excess *t*-BuOK at room temperature results in rapid (<2 min) formation of a 5:4 mixture of two cyclobutanes, **109** and **110** (combined yield 95%). The exposure of this mixture to *p*-TsOH in trifluoroethanol (at 0°C for 4 h) induced the clean isomerization into a roughly 1:1, separable mixture (90%) of two cyclobutanones **109** and **110**. However, only cyclobutanone **109** underwent the rearrangement to cyclobutanone **111**, while the other intermediate remains unchanged.[61]

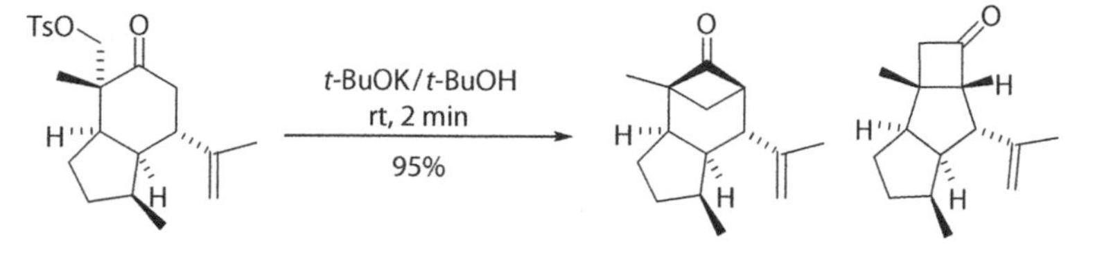

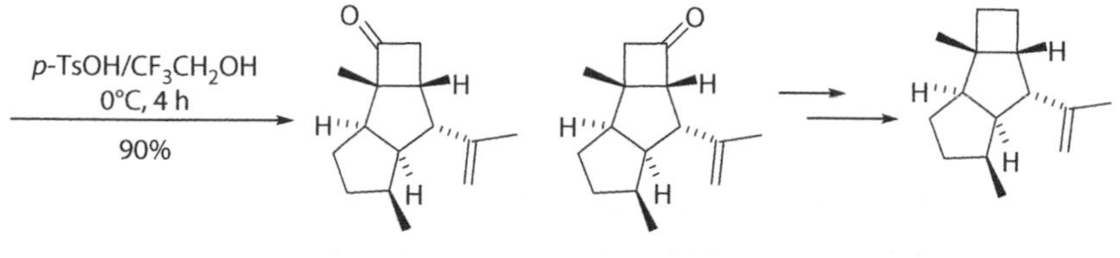

6.10.3 Quasi-Favorskii Rearrangement

There are α-halo ketones **113** that, due to their structural arrangement, undergo what is referred to as the quasi-Favorskii rearrangement. These α-halo ketones lack α′ position hydrogen atoms or are bicyclic with an α′-hydrogen atom at bridgehead carbon. In either case, just as with the Favorskii rearrangement, skeletal rearrangement occurs when the compound is treated with a nucleophile to yield a carboxylic acid or their derivative **114**.[8,62]

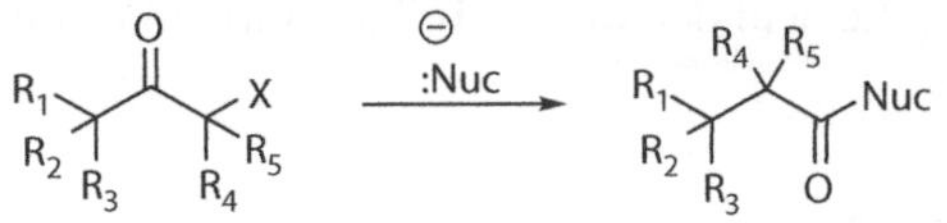

113

X=Cl, Br, I
α-haloketone
(no α' hydrogen)

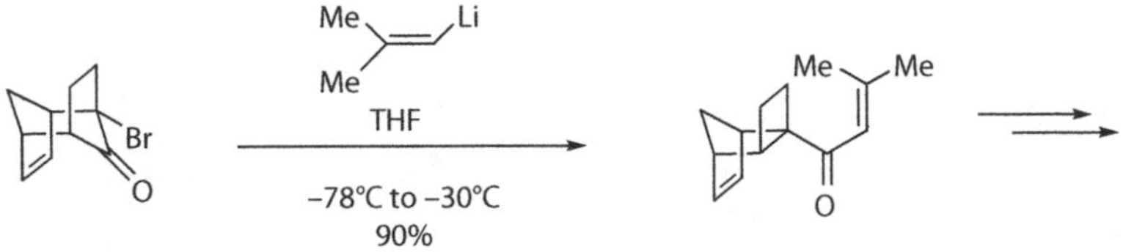

114

THF

−78°C to −30°C
90%

115

116

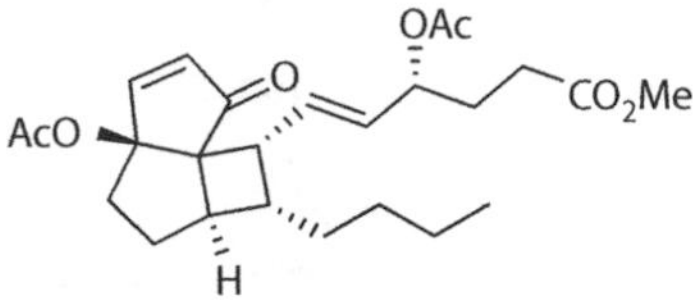

Tricycloclavulone **117**

A natural product synthesis that benefits from the quasi-Favorskii rearrangement is that of prostanoid tricycloclavulone **117**. The carboxylic core of this compound is generated from its [4 + 3]-cycloadduct. This intermediate adduct was achieved via the reaction of 2,5-difromocyclopentane with cyclopentadiene to provide ketone **115**. The adduct then underwent quasi-Favorskii rearrangement to provide ketone **116**, followed by a sequence of ring-opening and ring-closing metathesis (RCM) to yield the targeted natural product tricycloclavulone.[63]

6.10.4 BENZILIC ACID REARRANGEMENT

The Benzilic acid rearrangement refers to the rearrangement of α-diketones **118** to salts of α-hydroxy acids in the presence of a base.[64] The desired final product is achieved after the salt is acidified.[8]

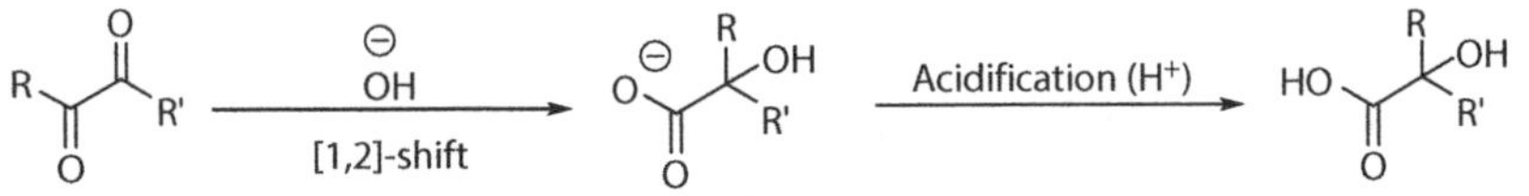

118   **119**   **120**

Taiwaniaquinoids are rearranged diterpenoids that have significant bioactivity,[8,46,65] such as aromatase inhibition or cytotoxicity to human oral epidermoid carcinoma KB cells.[66] The total synthesis of (−)-taiwaniaquinone H (**123**) from abietane **121a/122b** was partly accomplished by treating the hydroxydione **121a** and **121b** with lithium bis(trimethylsilyl)amide (LHMDS) to yield the hydrofluorene derivative **122a/122b**.[66]

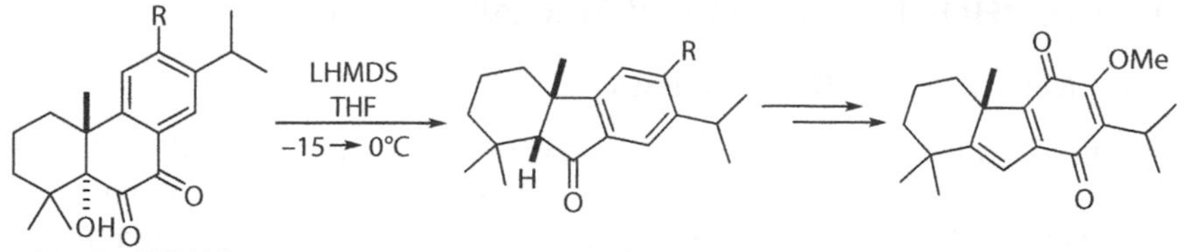
LHMDS
THF
−15→0°C

121a: R = H (90%)
121b: R = OMe (91%)

122a: R = H (65%)
122b: R = OMe (6%)

(−)-Taiwaniaquinone H **123**
(63% over 3 steps)

6.10.5 Baeyer–Villiger Rearrangement

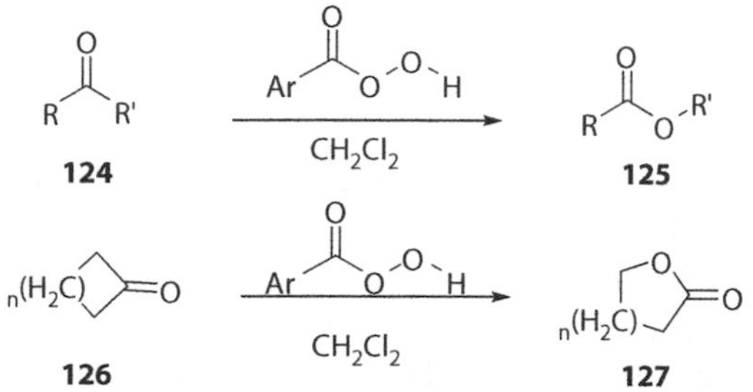
CH_2Cl_2
CH_2Cl_2

124 **125** **126** **127**

The Baeyer–Villiger rearrangement, also known as the Baeyer–Villiger oxidation, refers to the transformation of ketones **124** into esters **125** and cyclic ketones **126** into lactones **127** or hydroxy acids by peroxyacids.[8,67]

The Baeyer–Villiger rearrangement is one of the main features for the synthesis of carbazole alkaloids **131a/131**.[68,69] These compounds have a multitude of desirable therapeutic properties, such as: antitumor, antibiotic, antiviral, anti-HIV, antiinflammatory, antimalarial, psychotropic, antihistamine, antioxidative, and significant antituberculos activiites. When aromatic aldehydes **128** are placed in Baeyer–Villiger oxidation reaction conditions, the corresponding unisolable formate ester **129** is formed. *In situ* hydrolysis then transforms **129** to the known phenolic compounds **130** in 95%–96% yields.[68,69]

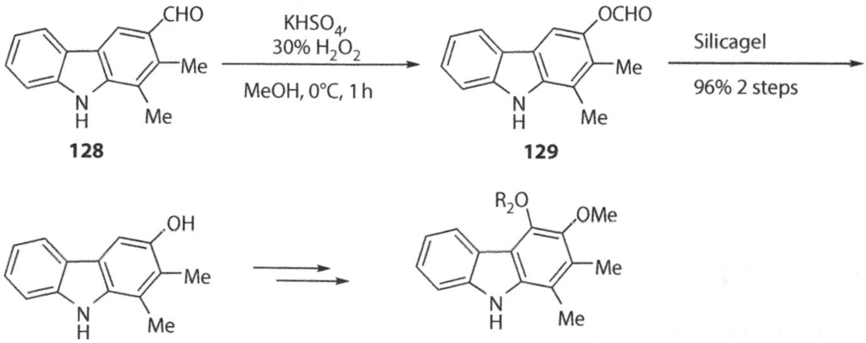
$KHSO_4$,
30% H_2O_2
MeOH, 0°C, 1 h
Silicagel
96% 2 steps

128 **129**

130 R = Me

Carbazomycin B (**131a**, R_2 = H)
Carbazomycin A (**131b**, R_2 = Me)

6.11 ALCOHOL FORMATION REARRANGEMENTS

6.11.1 MISLOW–EVANS REARRANGEMENT

Due to their ability to undergo reversible 1,3-transposition, allylic sulfoxides **132** can be transformed into allylic alcohols **134** and vice versa.[8] However, although this reaction is reversible, equilibrium lies largely to the left (it should be noted that the sulfonate is not detectable by NMR [nuclear magnetic resonance].)[70]

132   **133**   **134**

R_1 = alkyl, aryl; R_2 = alkyl, aryl, propargyl

Thiophile = PhSNa, P(OMe)$_3$, P(OEt)$_3$, P(NEt$_2$)$_3$, Et$_2$NH

The synthesis of the enantiomers of Gabosine D **138a** and E **138b** can be accomplished by converting allyl sulfide **135** prepared from (–)-quinic acid. This synthetic route begins with a Mislow–Evans rearrangement and subsequently follows a series of sequential reactions.[71] These natural products are of interest due to their antiprotozoal activity, DNA binding properties, and their ability to inhibit glyoxalase-I and glycosidases.[72,73] Oxidation of allyl sulfide **135** with *m*-CPBA (chloroperoxybenzoic acid) afforded a 1:1 mixture of sulfoxide **136**. After thermolysis of **136** in the presence of (EtO)$_3$P, the allyl alcohol **137** is produced as a single diastereomer in 98% yield.[71] A cascade of subsequent reactions creates the target compound in the form of **138** and **139**.

135   **136**

137   **138a** R = H (59%)
                                     138b R = Ac (64%)

6.12 AMINE FORMATION REARRANGEMENT

6.12.1 CURTIUS REARRANGEMENT

The Curtius rearrangement refers to the thermal decomposition (which is catalyzed by both protic and Lewis acids) of acyl azides **139** to the corresponding isocyanates

140.[74] These isocyanates can then be reacted with a multitude of reagents (e.g., water, amines, or alcohols) to form amines **141**, urea **142**, or carbamates **143**, respectively.[8,74]

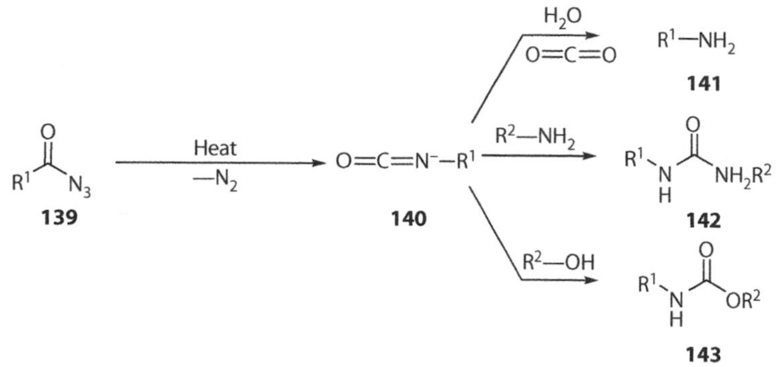

An example of how the Curtius rearrangement can be utilized to synthesize complex natural products is manifested in the total synthesis of aspeverin **147**.[75] Prenylated indole alkaloids, such as asperverin, exhibit important biological activities ranging from antibiotic and anthelmintic to potent cytotoxicity properties. The initial synthetic step was to convert the carboxylic acid to the acyl azide. The carbamate **146** was yielded when the acyl azide undergoes Curtis rearrangement in the presence of 2-(trimethylsilyl)ethanol. This carbamate was then converted to the natural product to complete the synthetic sequence.

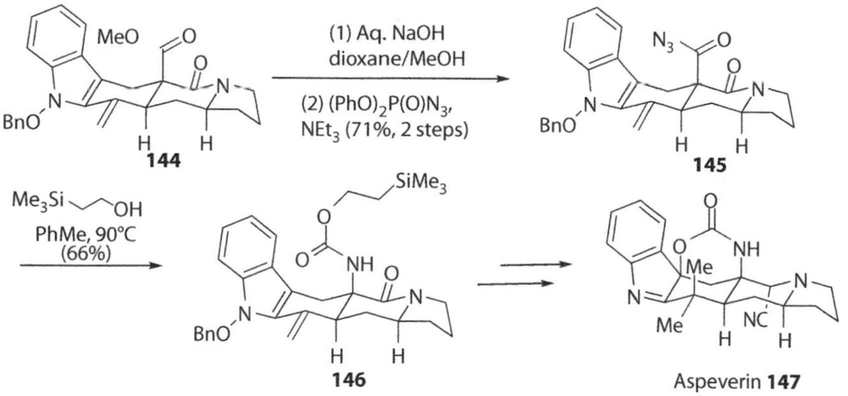

6.12.2 Lossen Rearrangement

When O-acyl hydroxamic acids **148** are converted to the corresponding isocyanate **149**, this is referred to as the Lossen rearrangement.[76] Nucleophilic attack of these isocyanate renders a variety of different products. It is of interest to note that the activation of the oxygen is a key mechanistic step since free hydroxamic acids do not undergo this particular rearrangement. Additionally, milder reaction conditions lends to Lossen rearrangements being preferentially chosen over the Hofmann and Curtius rearrangements.[8,76]

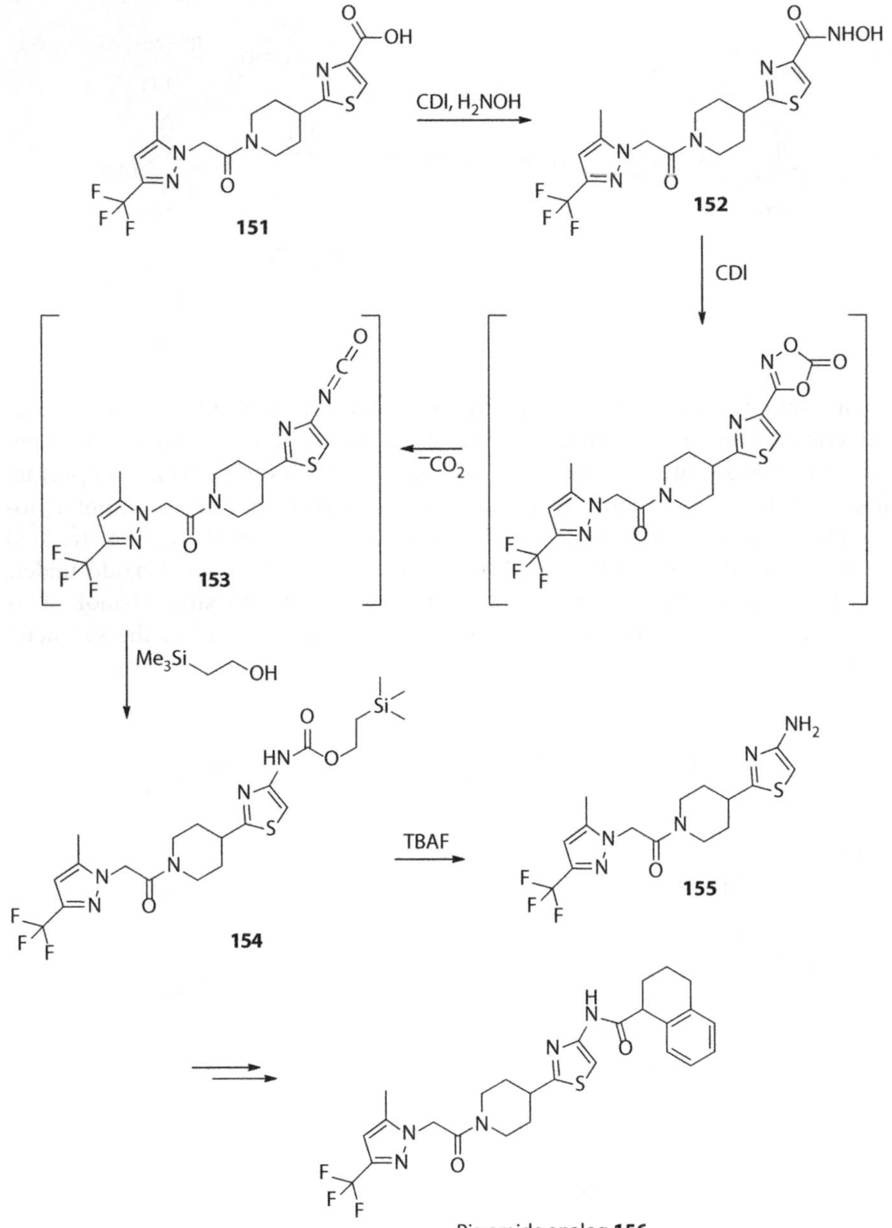

Bis-amide analog **156**

R_{1-2} = alkyl, aryl; R_3 = CO-alkyl, CO-aryl, Cl, SiR$_3$, C$_6$H$_3$(EWG)$_2$(O-aryl), PO$_2$R, SO$_2$R, C=NR(NHR); base: NaOH, KOH, DBU, (i-Pr)$_2$NEt; nucleophile: H$_2$O, ROH, RNH$_2$

A prime example of the Lossen rearrangement's functionality over other methods to generate isocynates can be displayed by Sulzer-Mosse's synthesis of N-thiazol-4-yl-salicylamides **156**.[77] This new family of compounds has the potential to be anti-oomycete agents. The Lossen approach was selected since the authors did not feel that Beckmann-type ketones were suitable building blocks, and chose to avoid any Curtius-type azide derivatives. The Lossen-type transformation proceeds seamlessly, with carbonyldiimidazole playing two key parts. First this reagent facilitates the formation of the hydroxamic acid under mild conditions. Second, as soon as **151** is formed, additional carbonyldiimidazole causes its cyclization to the dioxalone **152**. Subsequent loss of carbon dioxide delivers the isocynate **153**. This highly reactive species is then trapped with 2-(trimethylsilyl)ethanol to deliver the Teoc-protected amine **155**. Surprisingly, even given the complexity of carboxylic acid **151**, chemoseletivity is high in this process. The final step is the delivery of the thiazole-4-amine **154** by tetrabutyl-ammonium fluoride-promoted cleavage of the carbamate function.

6.12.3 HOFMANN REARRANGEMENT

The Hofmann rearrangement, also known as the Hofmann reaction, refers to the conversion of primary carboxamides **157** to the corresponding one-carbon shorter amines **158**.[78] This method does not allow for the presence of base sensitive groups in the amide, due to the basic reaction conditions. This rearrangement does give high yields for a variety of aliphatic and aromatic amides. In addition, complete retention of amine product's configuration is ensured if the starting amide is enantiopure at the alpha position.[12,78]

Oxycodone **161** is a semisynthetic analgesic that is clinically prescribed as a primary opioid for cancer pain management.[79,80] The construction of the morphinan skeleton **160** is possible via a Hofmann rearrangement/lactamization cascade sequence from amide **159**. In essence, a (diacetoxyiodo)benzene-mediated Hofmann rearrangement followed by hydrolysis of the resulting isocynate affords a primary amine. The ensuing spontaneous attack of lactone by the amine forms the desired lactam in good yield.[79]

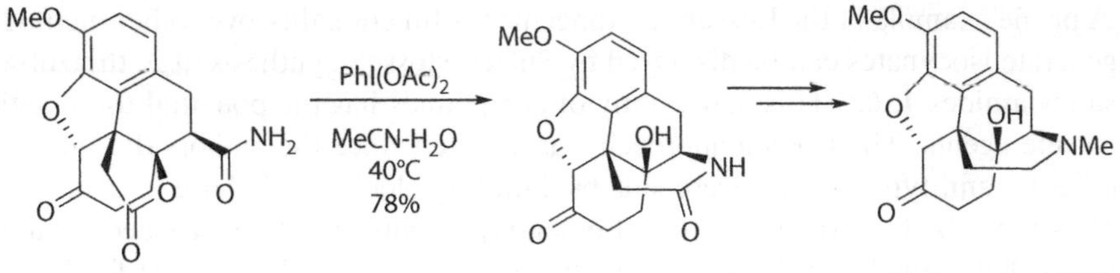

6.12.4 AZA-[2,3] WITTIG REARRANGEMENT

The nitrogen analog of the highly stereoselective rearrangement of α-metalated ethers to metal alkoxides is referred to as the aza-[2,3] Wittig rearrangement.[81] This rearrangement involves the isomerization of α-metalated tertiary amines **163** to skeletally rearranged metal amides **164**. With subsequent work-up, the homoallylic secondary amine **165** is then obtained.[8]

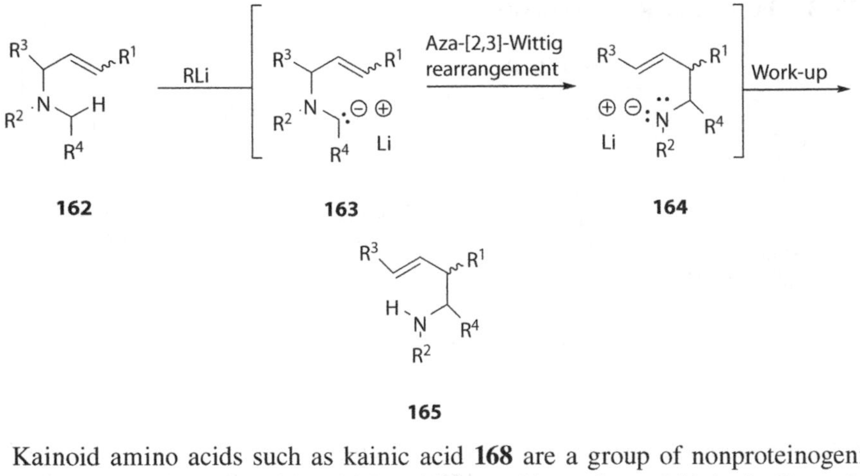

Kainoid amino acids such as kainic acid **168** are a group of nonproteinogenic pyrrolidine dicarboxylic acids which exhibit a wide variety of biologically active properties.[82] These compounds have been used as insecticides, anthelmintic agents, and most prominently neuroexcitatory agents. The route to creating the kainoid skeleton depends on the aza-[2,3]-Wittig sigmatropic rearrangement to efficiently install the relative stereochemistry between C2 and C3. This rearrangement began by utilizing LDA at −78°C to deprotonate **166** with warming to 0°C for 2 h to give the pivotal unnatural amino acid derivative in 78% yield.[83]

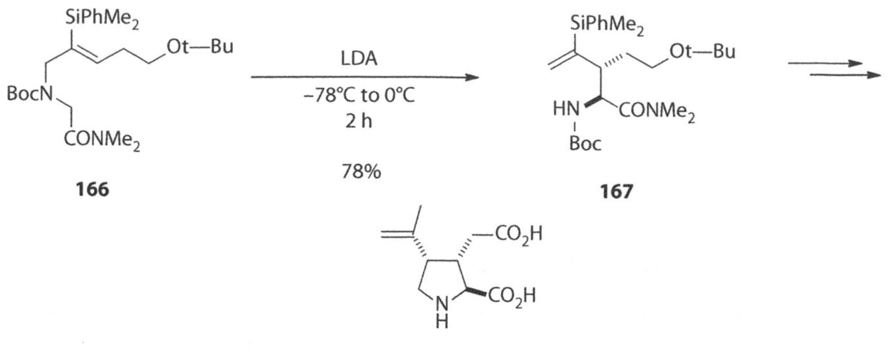

166 → LDA, −78°C to 0°C, 2 h, 78% → 167

(±)-kainic acid **168**

6.12.5 CIAMICIAN–PLANCHER REARRANGEMENT

The Ciamician–Plancher rearrangement refers to the 1,2-migratory shift of an alkyl or aryl group from position 2 to position 3 of indolenines, or vice versa, to form compounds originating from a more stable carbocation intermediate. As can be noticed from the following example, the rearrangement of 3,3-dimethylindole **169** goes through carbocation intermediates **170** and **171** to yield 2,3-dimethylindole **172**. The rearrangement requires high temperatures, and depends on the thermodynamic stability of rearranged product.[84]

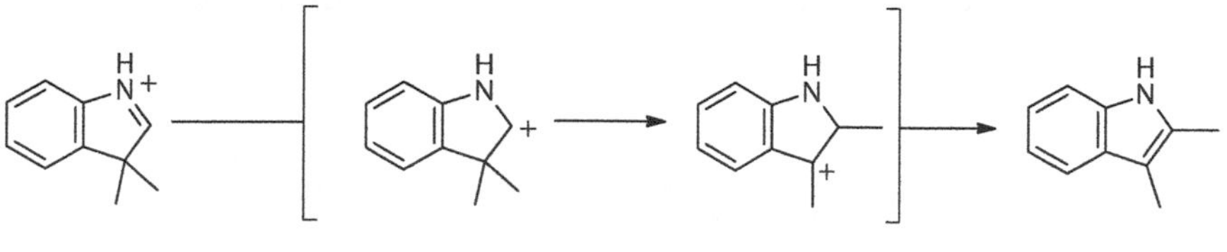

A notable use of this rearrangement is in the synthesis of tetrahydrocarbazoles. Tricyclic 1,2,3,5-tetrahydrocarbazoles have a wide range of biological applications, especially in the neurological arena. The enantioselective synthesis proceeds in a one-pot manner. Once the AgSbF$_6$ is added, the substrate becomes an activated complex. A subsequent S$_N$2-type attack of the indole C3 position gives the spirocylcic indolenine intermediate **175**. This intermediate then undergoes a 1,2-migratory shift to yield **176**.[85]

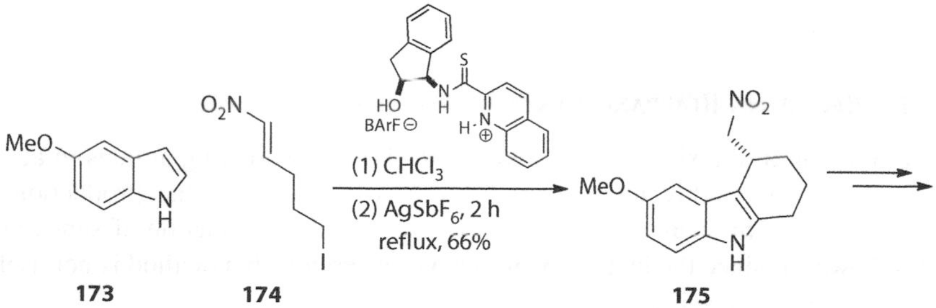

173   174   (1) CHCl$_3$ (2) AgSbF$_6$, 2 h   reflux, 66%   175

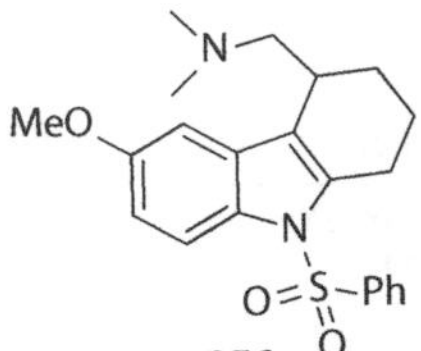

176

6.12.6 Meisenheimer Rearrangement

There are specific tertiary amine *N*-oxides **177** that will undergo a thermal rearrangement ([1,2],[2,3] in cyclic or acyclic systems) to the corresponding *O*-substituted-*N,N*-disubstituted hydroxylamines **178**.[86]

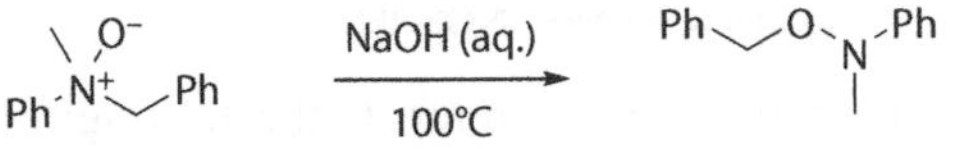

177   **178**

The enantiospecific total synthesis of the δ-lactonic marine natural product (+)-tanikolide (**181**) was achieved using [2,3]-Meisenheimer rearrangement as the key reaction. Under oxidative conditions, **179** smoothly rearranged to give **180** in 83% yield. The allylic *N*-oxide can actually direct the *m*-CPBA double-bond epoxidation to the syn position.[87]

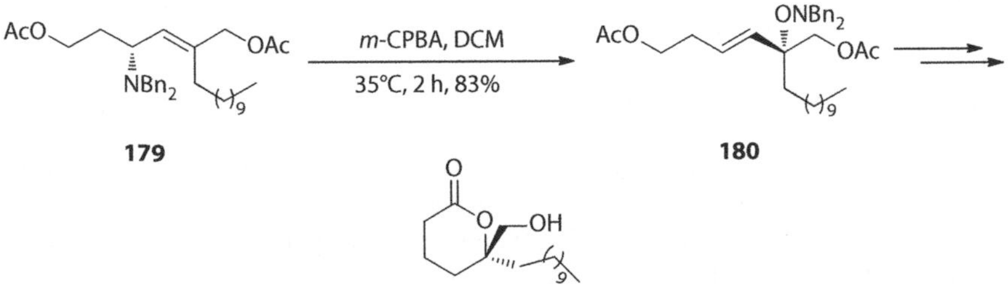

179   **180**

(+)-Tanikolide **181**

6.13 AMIDES

6.13.1 Beckmann Rearrangement

The conversion of aldoximes and ketoximes to the corresponding amides in acidic medium has been highly utilized in industry, especially for the production of ε-caprolactam.[88] This rearrangement can be used for the production of substituted amides, however, since the hydrogen atom never migrates, this method is not useful for the synthesis of unsubstituted amides.[89]

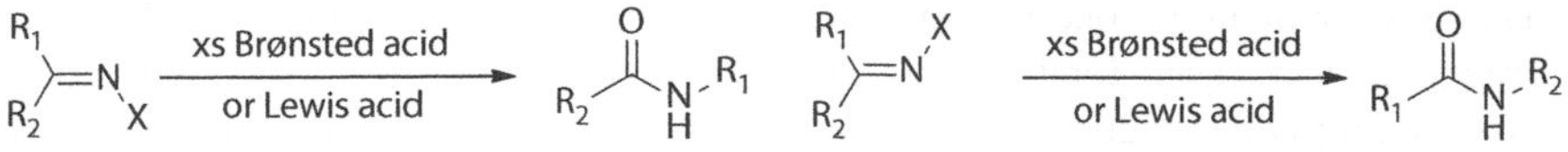

R$_1$ is anti to X

R$_2$ is anti to X

R$_{1-2}$ = alkyl, aryl, heteroaryl; X = OH, OTs, OMs, Cl

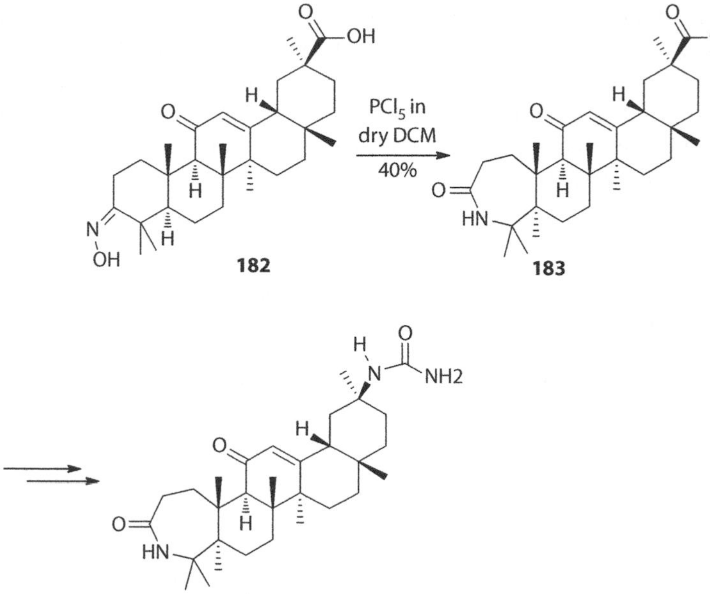

Urea derivative **184**

Inhibition of 11β-hydroxysteroid dehydrogenase (11β-HSD) type 1 and type 2 by glycyrrhetinic acid and hydroxamic acid derivatives has proven to be an effective treatment of metabolic diseases (11β-HSD1), chronic inflammatory diseases, and cancer (11β-HSD2).[90] The Beckman rearrangement of the 3-oxime **182** to a seven-membered ring **183**, and the rearrangement of the C-ring from 11-keto-12-ene to 12-keto-9(11)-ene can be utilized to create derivatives of glycyrrhetinic acid **184**.[90]

6.13.2 OVERMAN REARRANGEMENT

Overman described the facile thermal and mercuric ion catalyzed rearrangement of allylic trichloroacetimidates to afford the corresponding trichloroacetimidates in 1974.[91] Hence, this rearrangement performs 1,3-transpositions of alcohol and amine functionalities via the [3,3]-sigmatropic rearrangement of allylic trichloroacetimidates.

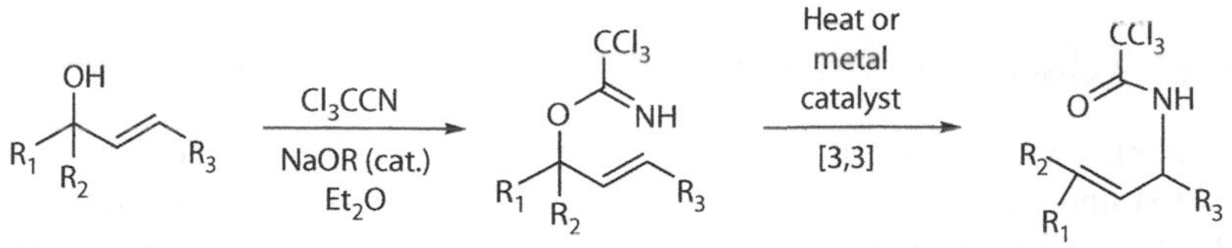

1°, 2°, or 3° allylic alcohol

R$_{1-3}$ = H, alkyl, aryl; metal catalyst: Hg(OCOCF$_3$)$_2$, Hg(NO$_3$)$_2$, Pd$^{(II)}$-salts

The development of one-pot multireaction processes involving the Overman rearrangement are used for the rapid and efficient synthesis of amino-substituted carboxylic and heterocyclic compounds. Before the Overman rearrangement could take place, a highly regioselective ring opening of the epoxide with allylmagnesium bromide and copper(I) salt was necessary to introduce the alkenyl side chain and generate the secondary alcohol **185** that would ultimately lead to the directing group. One-pot ether-directed Overman rearrangement leading to **188** followed by a RCM reaction process completed the cycloalkene. The phenanthridone skeleton could rapidly be prepared from the carboxylic amide **189** generated from the one-pot reaction, which can be used in the preparation of natural products such as (+)-7-deoxypancratistatin **190**. This compound has high *in vitro* antiproliferative activity against human tumor cell lines, and its analogs are also of interest.[92]

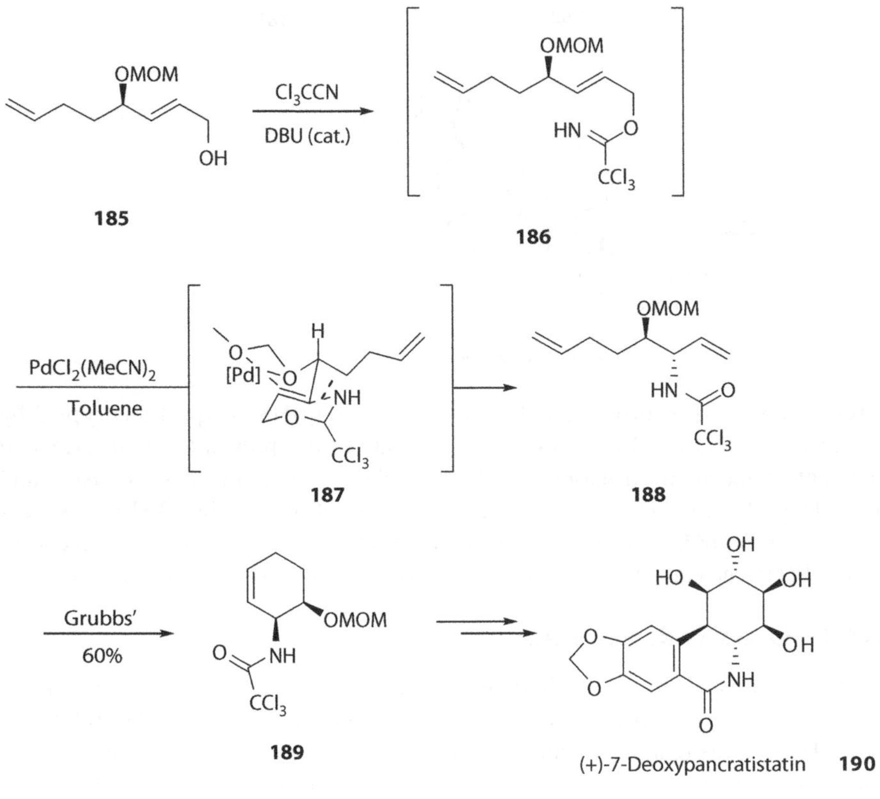

6.13.3 ANIONIC *ORTHO*-FRIES REARRANGEMENT

The anionic *ortho*-Fries rearrangement occurs when *ortho*-lithiated *O*-aryl carbamates **191** undergo a facile intermolecular [1,3]-acyl migration to furnish the substituted salicylamides **192** at room temperature.[93,94]

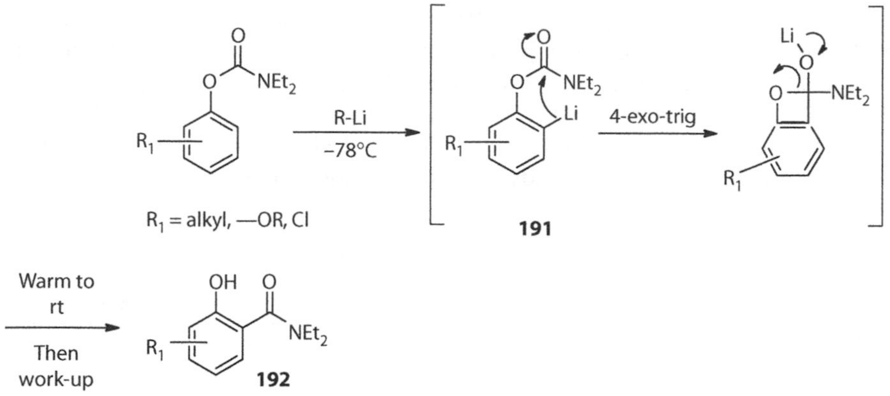

$R_1 = alkyl, \!—OR, Cl$

191

192

(+)-(R)-Concentricolide (**195**), a fungal metabolite of the Ascomycete *Daldinia concentrica*, has exhibited anti-HIV effects.[95] One of the key features for the synthesis of this compound includes an anionic *ortho*-Fries rearrangement to furnish 3-iodosalicylamide. The synthetic route begins with commercially available 2-iodophenol. The hydroxyl group is carbamated to create **193**, which is then transformed into 3-iodosalicylamide **194** via anionic *ortho*-Fries rearrangement.[96]

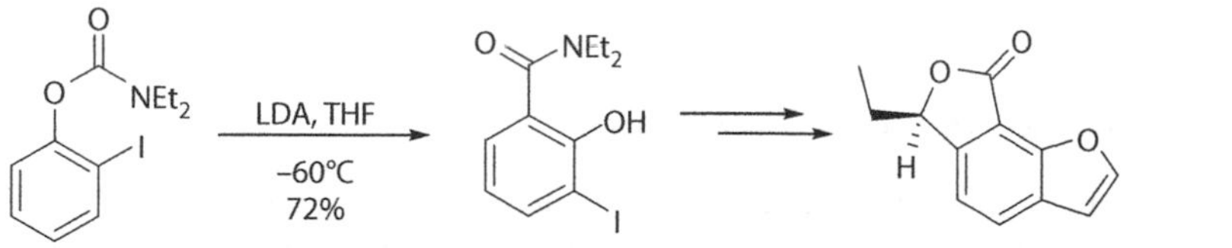

193   **194**   (+)-(R)-concentricolide **195**

6.14 HYDROCARBON REARRANGEMENTS

6.14.1 WAGNER–MEERWEIN REARRANGEMENT

The Wagner–Meerwein rearrangement pertains to the generation of a carbocation followed by the [1,2]-shift of an adjacent carbon–carbon bond to generate a new carbocation.[97] This rearrangement is used to describe all [1,2]-shifts of hydrogen, alkyl, and aryl groups.

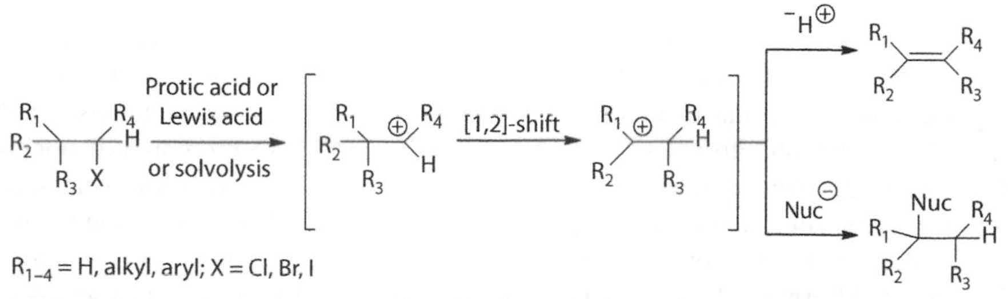

$R_{1–4} = H, alkyl, aryl; X = Cl, Br, I$

The synthesis of (–)-huperzine A (**199**), a potential candidate for treatment of Alzheimer disease, can be accomplished by utilizing the Wagner–Meerwein rearrangement to induce double-bond migration.[98] Ding developed a one-pot sequence by reacting alcohol **196** with SOCl₂ to yield putative carbocation **197**, the Wagner–Meerwein rearrangement followed to generate **198**. Through additional steps Huperzine A was completed to study its biological application.

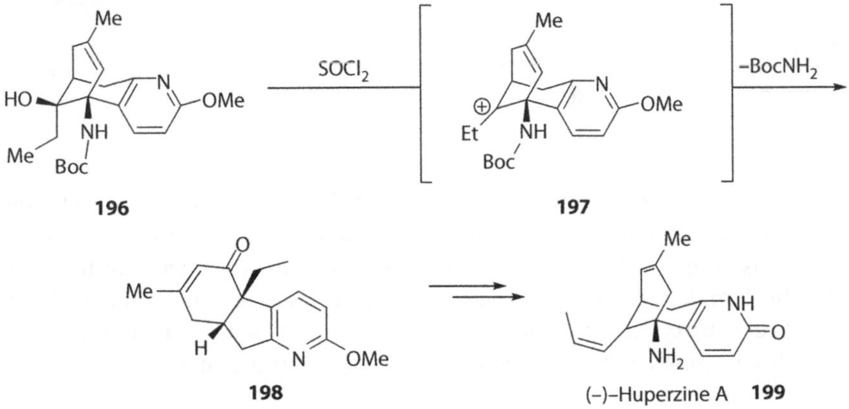

6.14.2   DIENONE–PHENOL REARRANGEMENT

Speaking to the migration of alkyl groups, Dienone–Phenol rearrangement relates to the acid- and base-catalyzed or photochemically induced migration of these groups in cyclohexadienones **200**.[99] This rearrangement is used widely for the production of highly substituted phenols **201**.[8]

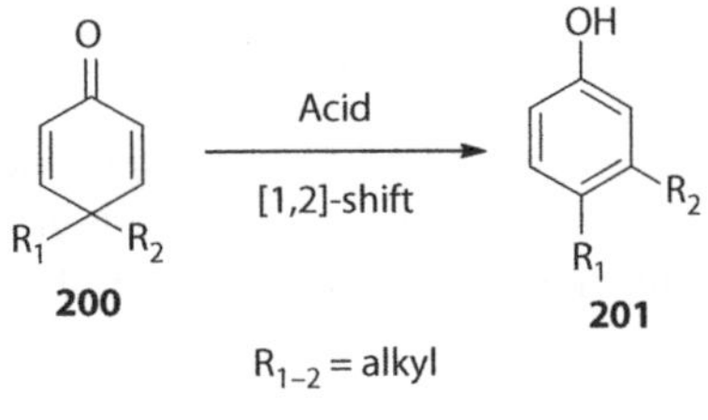

Many researchers have pursued the synthesis of 4,6-di-*tert*-butyl-2,2-dipentyl-2,3-dihydro-5-benzofuranol (BO-653 structure **205**) due to it being an antiatherogenic antioxidant that exhibits high radical scavenging activities against lipid peroxidation and inhibitory action of LDL oxidation. The base-promoted dienone–phenol rearrangement is invoked to efficiently construct the dihydrofuran moiety on the aromatic ring carrying the requisite functionalities without employing strong acid conditions. The hydroxy-dienone **202**, upon exposure to potassium *tert*-butoxide in dimethylformamide (DMF), brought about the selective 1,2-shift of an alkyl group to yield the phenols **203** and **204**.[100]

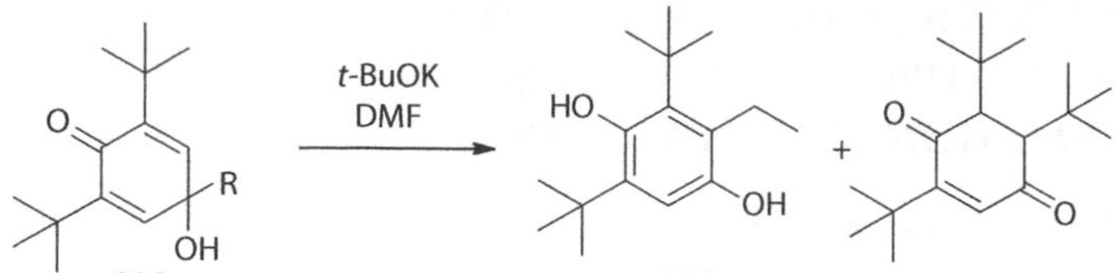

202 → **203** + **204**

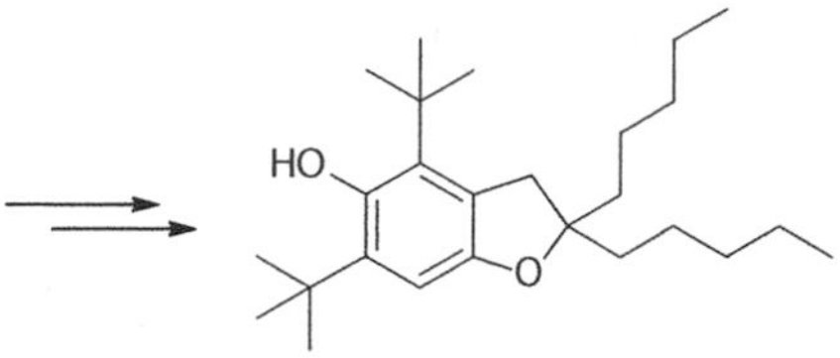

BO-653 **205**

6.14.3 RAMBURG–BACKLUND REARRANGEMENT

An α-halogenated sulfone subjected to a base will produce an alkene via episulfone intermediates according to the Ramburg–Backlund rearrangement.[101] This reaction is an optimal method for constructing 1,1- or 1,2-di, tri or tetra-substituted alkenes.

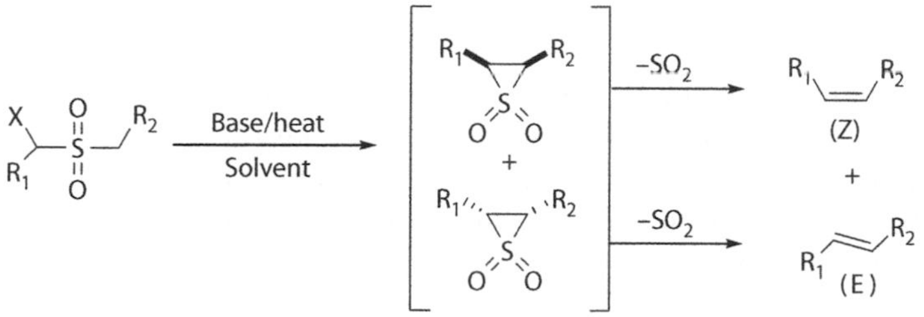

R_{1-2} = H, alkyl, aryl, heteroaryl, CO_2R; X = Cl, Br, I, OTs
Base: KOH, NaOH, KOt-Bu; solvent: THF, t-BuOH/DCM

C-Glycolipids such as **208** are of interest due to their potential as cell growth inhibitors. The Ramberg–Backlund rearrangement in combination with ionic hydrogenation allows for the synthesis of *C*-glycosides with high stereoselectivity at the anomeric center. Subjecting sulfone **206** to $C_2F_4Br_2$ in KOH/Al_2O_3 and refluxing affords the β-sulfone **207** in 60% yield.[102]

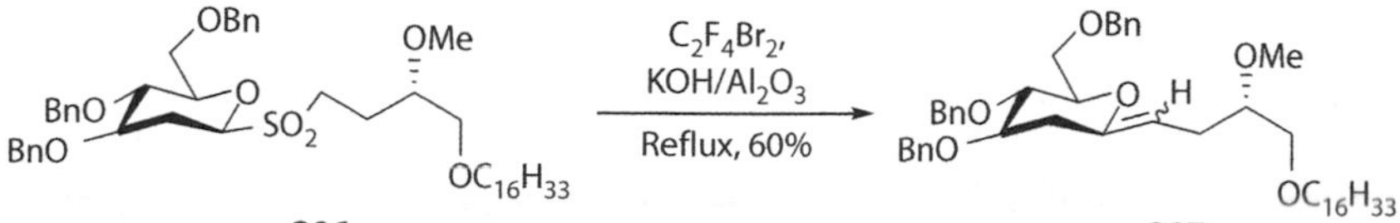

206 → **207**

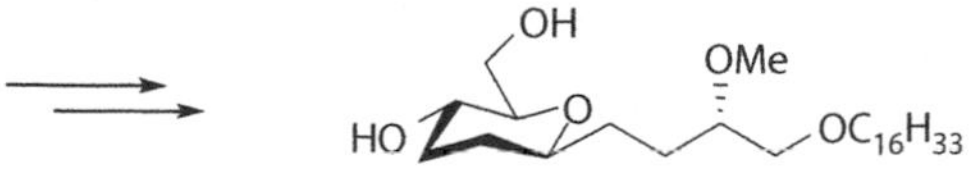

C-glycoside **208**

6.15 OXACYCLIC, CARBOCYCLIC, OXAZOLES, TETRAHYDROPYRAN, AND TETRAHYDROFURAN FORMATION REARRANGEMENTS

6.15.1 Petasis–Ferrier Rearrangement

Five-membered or six-membered enol acetals (**209** and **212**) will rearrange to the corresponding tetrahydrofurans (**211**) or tetrahydropyrans (**214**), respectively, when subjected to a Lewis acid. This rearrangement proceeds via an oxocarbenium ion.[103]

209 **210** **211**

212 **213** **214**

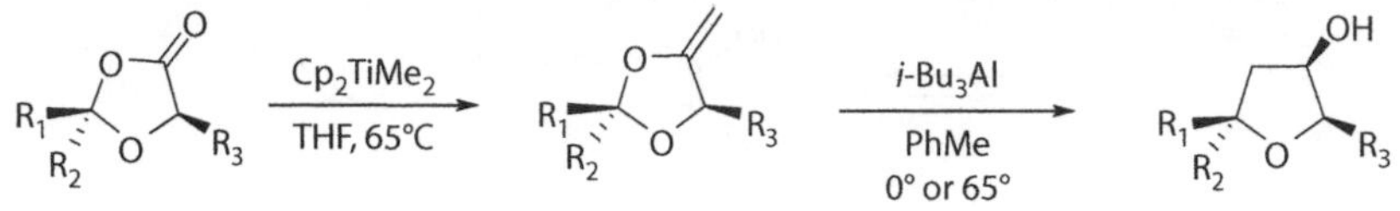

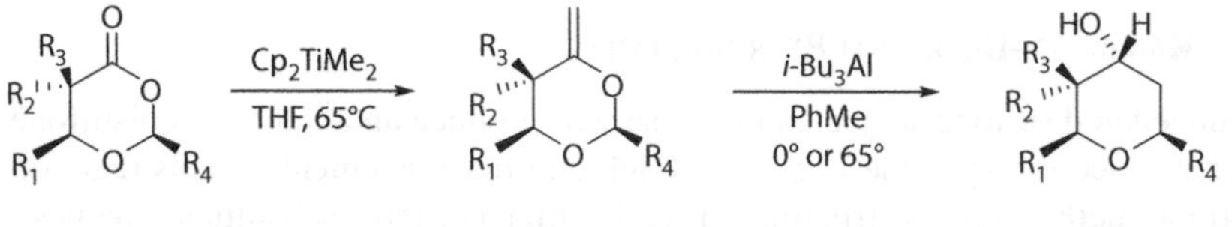

Petasis–Ferrier rearrangements of 4-(vinyloxy)-, 4-(propenyloxy)-, and 4-(isopropenyloxy)azetidin-2-ones to corresponding 4-(carbonylmethyl)azetidin-2-ones (**216**) is promoted by trimethylsilyl triflate. Compounds (**216**) possessing this architecture serve as intermediates (examples **219** and **220**) in the synthesis of carbapenem antibiotics. Compound **215** is alkylated with benzyl bromoacetate to yield **216**, which is subsequently subjected to the Petasis–Ferrier rearrangement to afford an inseparable mixture of **217** and **218** in a 7:1 ratio. These compounds may then be processed through a series of steps to produce the carbapenams **219** and **220** in a ratio of 7:1.[104]

215 **216**

219 **220**

56%, **217**: **218** = 7:1 Carbapenams (53%, 7:1)

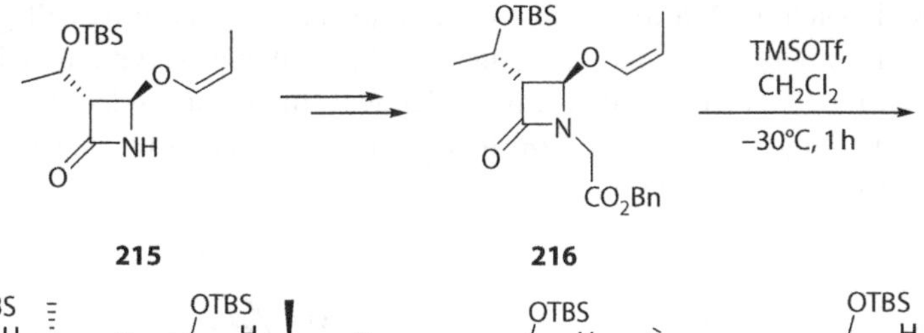

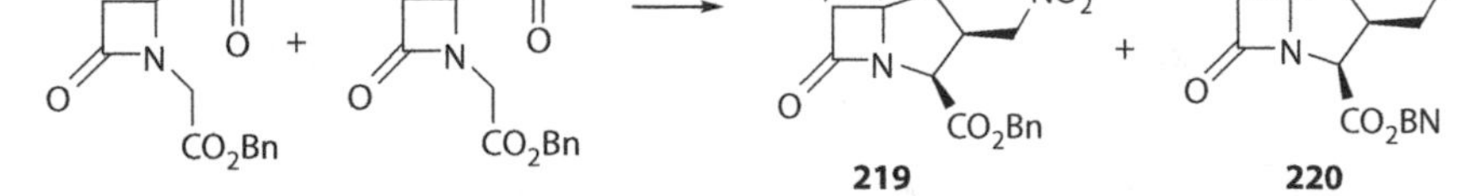

6.15.2 CORNFORTH

When 4-carbonyl substituted oxazoles are subjected to heat, they undergo rearrangement to their isomeric counterpart.[105] However, the extent of this rearrangement depends on the thermodynamic stability of the starting material versus the product.[106]

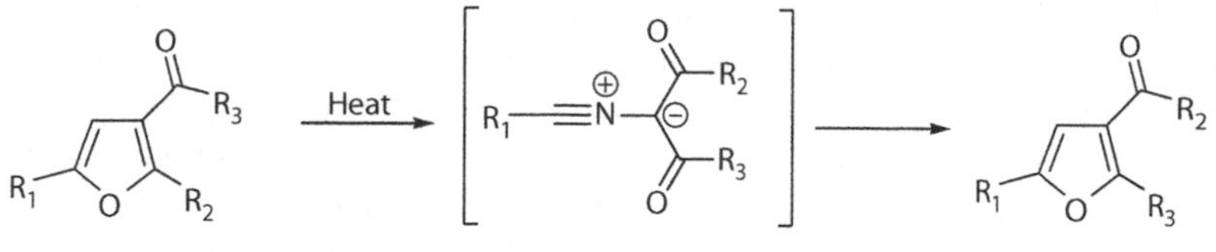

R_1 = alkyl; R_2 = alkyl or Ph; R_3 = Ph, Me, OMe, CO_2Et

5-Amino-oxazole containing structures have biological activity and therapeutic potential. For example, 4-(methoxycarbonyl)-2-(1-normon-2-yl)-5-piperidin-1-yloxaczole **221** is a pseudomonic acid-derived antibiotic and oxazolo-[5,4-d]pyrimidine is an inhibitor of ricin and shiga toxins.[107] The Cornforth rearrangement is an efficient method for rapid generation of these products. This is owing to the formal rearrangement which occurs upon heating 5-alkoxyoxazole-4-carboxamides (≥100°C for 17 h). This particular rearrangement is believed to proceed via an intermediate nitrile ylide. The Cornforth rearrangement is a critical step in converting amide **221** to the required ester **222**. This is achieved through thermal rearrangement conducted at 180°C to provide the advanced intermediate toward the target compound **223**.[108]

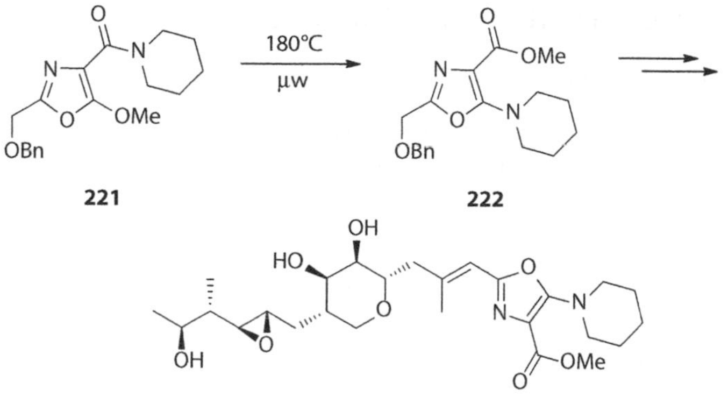

4-(methoxycarbonyl)-2-(1-normon-2-yl)-
5-piperid in-1-yloxaczole **223**

6.15.3 DIMROTH REARRANGEMENT

The Dimroth rearrangement refers to heterocyclic isomerization in which endocyclic, or exocyclic heteroatoms, and their attached substituents are translocated via a ring-opening–ring-closure sequence.[109]

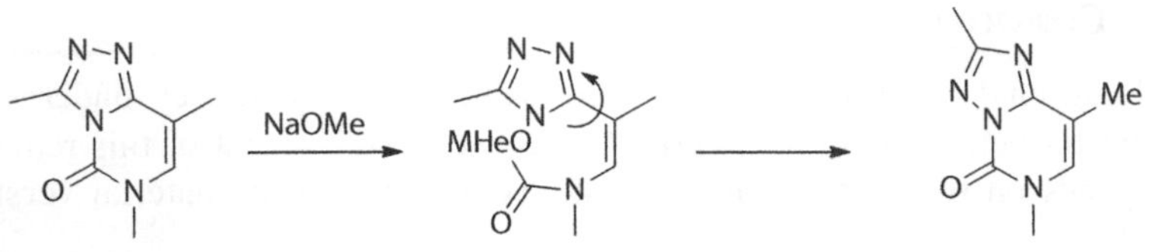

The Dimroth rearrangement is a synthetic option for the synthesize aza-analogs of purine. Variation of the substituents at positions 2, 5, and 7 of the 1,2,4-triazolo-1,3,5-triazine heterocycle system allows for the preparation of compounds with a wide spectrum of biological activity.[110] Some examples of this set of compounds' application areas: antidepressants, treatment for Parkinson disease, immunosupressors, antiinflammatory and antitumor agents, and antithymidine phosphorylase activity. The desired 2-aryl(hetaryl)-5-amino-1,2,4-triazolo[1,5-α]-1,3,5-triazin-7-ones **224** were obtained via oxidative cyclization of the corresponding arylidene(hetarylmethylidene)hydrazinyl-1,3,5-triazinones **225** with lead(IV) tetraacetate in acetic acid.

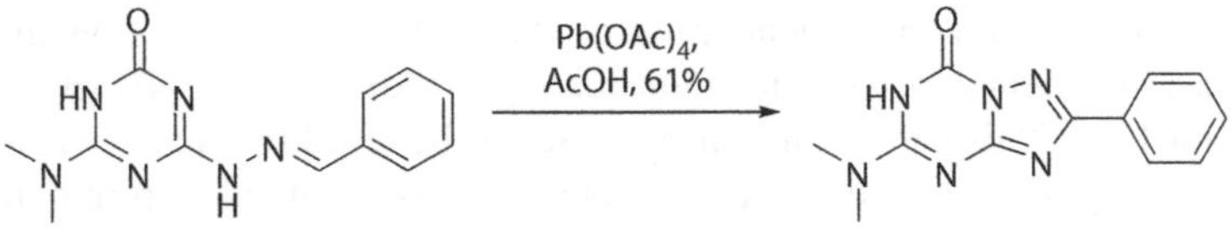

6.16 REARRANGEMENTS RESULTING IN LESS COMMON FUNCTIONAL GROUPS

6.16.1 SMILES REARRANGEMENT

This rearrangement covers intramolecular nucleophilic aromatic rearrangement, and requires activation via electron-withdrawing groups at the *ortho-* or *para*-positions.[8,111]

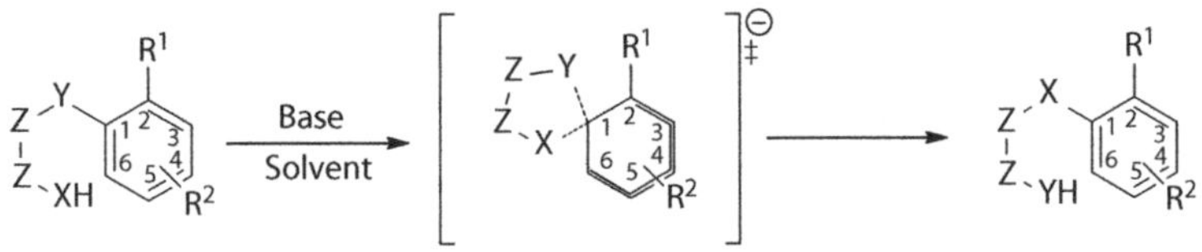

XH = NHCOR, CONH$_2$, SO$_2$NH$_2$, OH, NH$_2$, SH, SO$_2$H, CH$_3$;
Z = sp$_2$ or sp$_3$ hybridized substituted-or unsubstituted carbon, C = O, sp$_3$ nitrogen
Y = S, O, SO$_2$, CO$_2$, S=O, SO$_3$, I$^+$, P$^+$,
R$_1$ = EWG = NO$_2$, SO$_2$R, Cl; R$_2$ = alkyl, halogen, NO$_2$
Base: NaOH, KOH, RONa, RLi, K$_2$CO$_3$/DMSO

The Smiles rearrangement allows for convenient construction of a novel biaryl lactams, which are an important class of heterocyclic compounds that exhibit a plethora of biological properties. One specific example is the human A_3 adenosine receptor antagonist **228**. 2-bromo-5-(*o*-tolyl)pyrazolo[1,5-a]pyridol[3,2-e]pyrazin-4(5H)-one **227** is transformed from 3-bromo-1-(3-chloropyridin-2-yl)-*N*-(*o*-tolyl)-1*H*-pyrazole-5-carboxamide **226** via intramolecular Smiles rearrangement cyclization.[112]

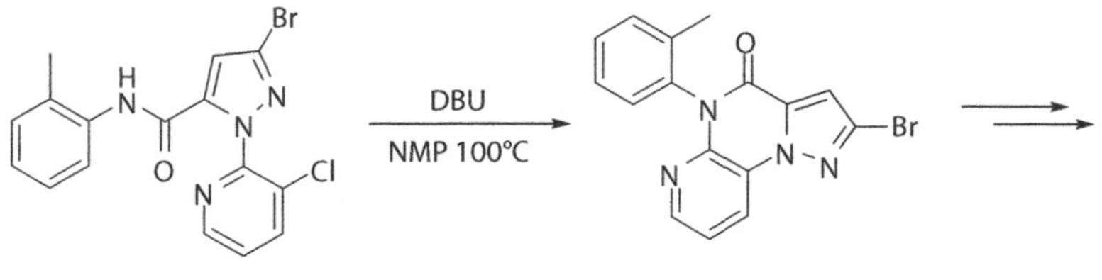

226   **227**

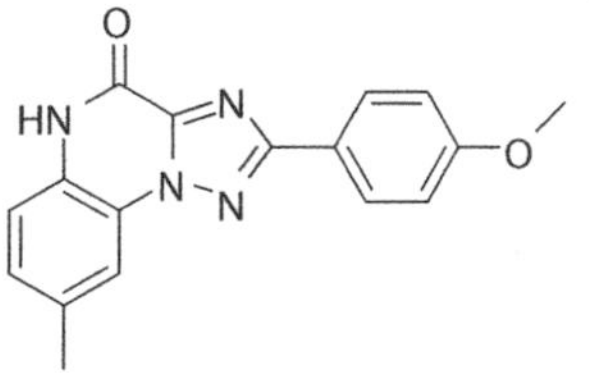

228

6.16.2 STEVENS REARRANGEMENT

Sulfonium and ammonium salts can undergo base-promoted transformation into their corresponding sulfides or tertiary amines via a [1,2]-migration of one of the groups residing on the nitrogen or sulfur atoms.[113]

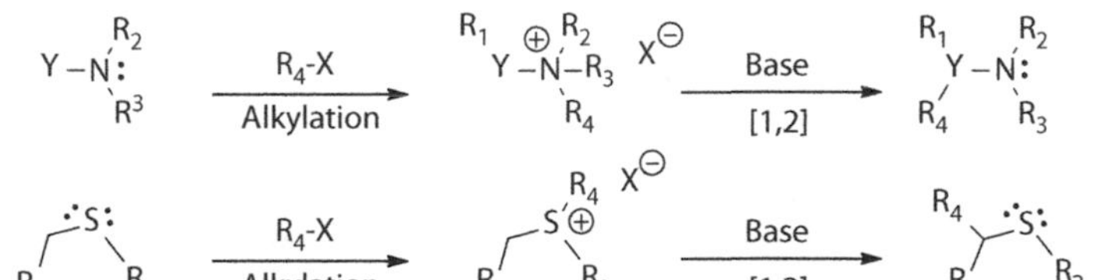

R_1 = EWG = Ar, heteroaryl, COR, COOR, CN; Y=CH_2, CHR, NH
R_{2-3} = alkyl with no β-hydrogen, aryl
R_4 = CH_3, alkyl, allyl, benzyl, CH_2COAr; X=Cl, Br, I. OTs, OMs;
Base = NaH, KH, RLi. RONa. ROK

The synthesis of *Nupharis rhizoma* sesquiterpene thioalkaloids (**232**) is of interest due to their biological activities such as potent antimetastic and apoptosis-inducing activity.[114] All four stereoisomers of thiolane **231** can be synthesized based on Stevens-type ring expansion of readily available thietanes such as **229** in the presence of 2% $Rh_2(OAc)_4$. The synthesis of thiolane **231** provided the central component needed to complete the dimeric natural product.[115]

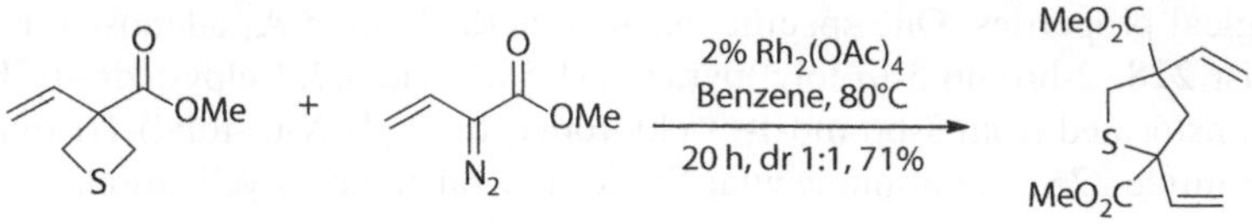

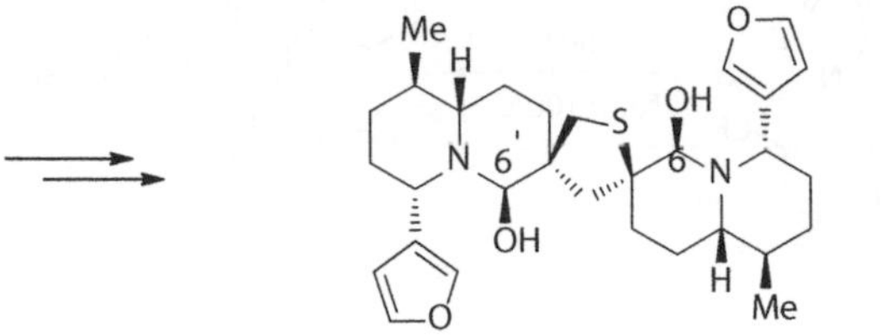

6,6'-dihydroxythiobinupharidine **232**

6.16.3 Pummerer Rearrangement

In 1909, R. Pummerer discovered a general transformation for sulfoxides when they are subjected to an activating agent and acid- or base-co-catalyst.[116] Sulfoxides will transform into their corresponding α-substituted sulfides when the aforementioned conditions are met.[8]

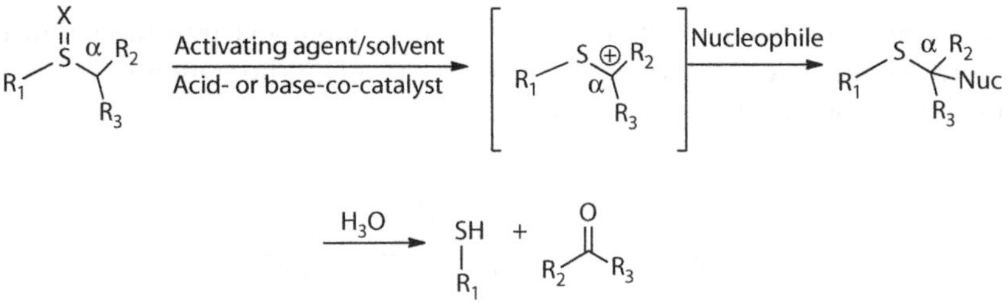

R_1 = alkyl, aryl; R^{2-3}=H, alkyl, aryl; X=O, NR
Activating agent: HCl, H_2SO_4, TsOH, I_2/MeOH, Ac_2O, TFAA, t-BuBr, Me_3SiX, PCl_3, PCl_5
Nucleophile: H_2O, ROH, RCO_2^-nuc: OH, O-alkyl, O_2CR, F, Cl, Br, SR, NR_2
Catalysts: AcOH, TsOH, TFAA, NaOAc

4'-Thionucleosides (**235**) represent a class of thiosugar-containing modified nucleosides, having the furanose ring oxygen replaced with a sulfur atom. These nucleoside analogs are targeted for their potential as antiviral and anticancer therapeutic agents. A series of novel pyrimidine D- and L-apiothionucleosides were synthesized from L- and D- arabinose. A critical point in the synthesis is the utilization of a region- and stereoselective Pummerer rearrangement between a pyrimidine nucleobase and an appropriately functionalized thiosugar moiety (**233**).[117]

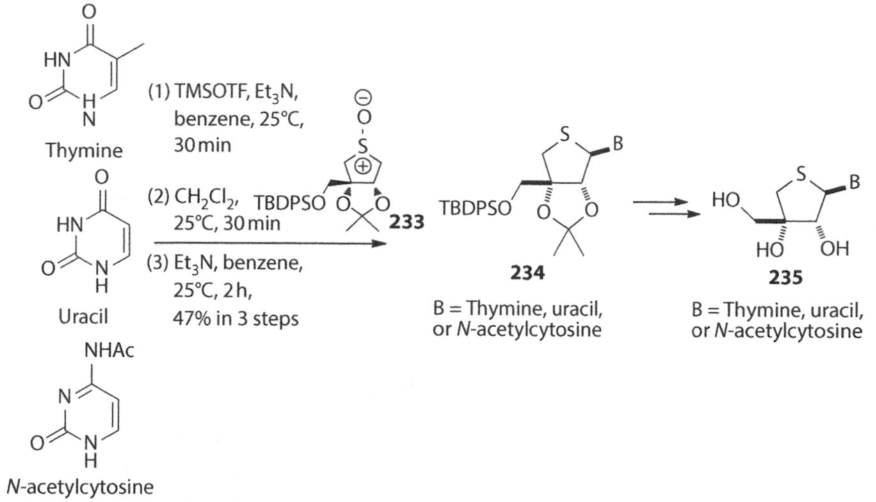

6.16.4 AMADORI REARRANGEMENT

The N-glycosides of aldoses undergo an acid- or base-catalyzed isomerization into 1-amino-1-deoxyketoses.[118] The rearrangement occurs when an aldose is reacted with an amine in the presence of a catalytic amount of acid.

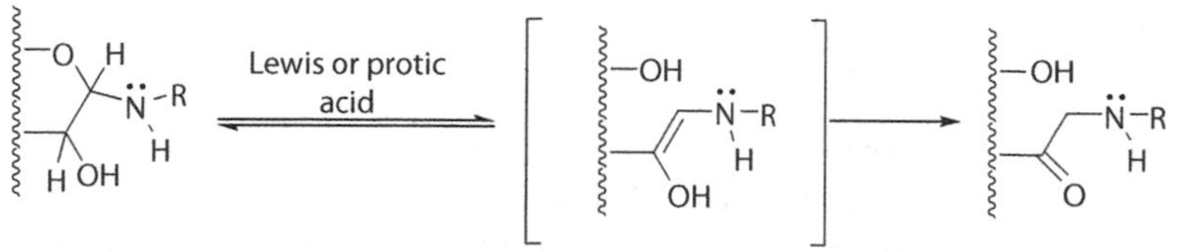

This rearrangement can conveniently be performed in a one-pot reaction motif to produce multibranched australine derivatives. These derivative types are used as selective inhibitors of β-glucosidase and appear to be promising candidates to act as pharmalogical chaperones for Gaucher disease. The Amadori rearrangement proceeds via one-pot reaction following the reaction of a substrate bearing a masked 2-hydroxyaldehyde segment with an amine. The substrate was 3-(N-octylimino)-2-oxacastanospermine derivative **236** and n-butyl,-octyl, and -dodecylamine served as the amine partners. The transformation involved a crucial step with the ring opening of a gem-diamine intermediate to give a transient Schiff base. The presence of the basic isourea functionality facilitates this process by enabling a concerted mechanism involving the intramolecular proton transfer from the amino to the closely located imino group. The Schiff base is then assumed to be in equilibrium with the corresponding enol form that further proceeds to yield the more stable α-aminoketone derivative.[119]

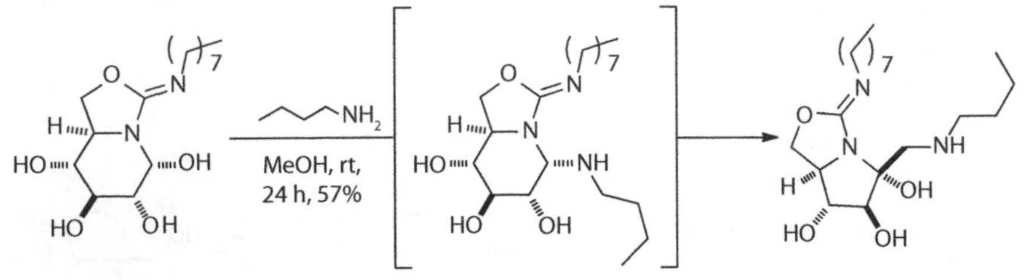

236   237   238

6.17 FRAGMENTATIONS

6.17.1 Eschenmoser–Tanabe Fragmentation

The Eschenmoser fragmentation was devised for the production of muscone (organic compound that is responsible for the smell of musk), as well as other related macrocyclic musks.[120] Published in 1967, this fragmentation describes the chemical reaction of α,β-epoxyketones with aryl sulfonylhydrazines to yield alkynes and carbonyl products.[121]

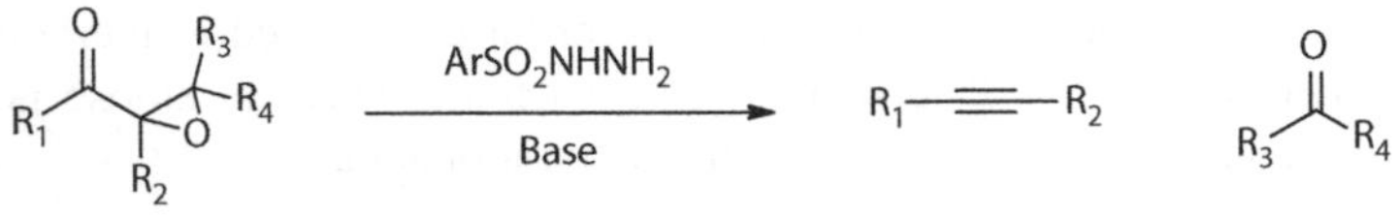

R_{1-4} = alkyl, alkenyl, aryl, H
Base = EtOH, AcOH, etc.

The Eschenmoser–Tanabe fragmentation can also be used for the production of indole alkaloid, mersicarpine (241).[122] The overall structure includes three heterocycles (indole, cyclic imine, and δ-lactam) fused with each other around a tertiary hydroxyl group. This fragmentation method is used in synthesizing the alkyne 240 containing the quaternary carbon center by oxidizing the epoxide 239 with lead tetraacetate. This advanced intermediate then goes through additional reaction scenarios to complete the synthesis of mersicarpine.

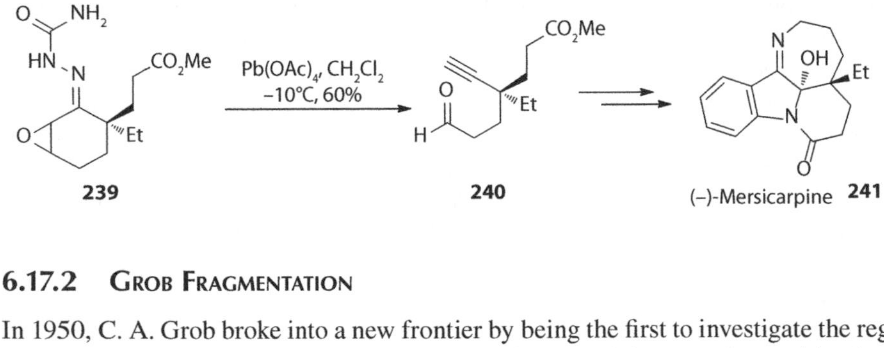

239   240   (–)-Mersicarpine 241

6.17.2 Grob Fragmentation

In 1950, C. A. Grob broke into a new frontier by being the first to investigate the regulated heterolytic cleavage reactions of molecules containing certain combinations of carbon and heteroatoms (B, O, N, S, P, and halogens).[13] The Grob fragmentation is an elimination reaction that breaks a neutral aliphatic chain into three fragments, which consist of the following: a positive ion (electrofuge), an unsaturated neutral fragment, and a negative ion (nucleofuge).[8]

The Grob fragmentation can be applied to the production of the B ring opened 7,8-seco-cyanocobalamins (see structure **244**). Discovery of facile synthetic pathways for 7,8-seco-cyanocobalamins is desirable due to their potential as therapeutic drug carrier molecules. Hydrolysis of the lactone vitamin B12 (**242**) derivative generates a cobalamin with a β-bromo alkoixde subunit that reacts *in situ* via Grob fragmentation to the secocorrin **244**.[123]

REFERENCES

1. Bruckner, R., *Advanced Organic Chemistry: Reaction Mechanisms*, Academic Press: San Diego, 2002, pp. 435–476.
2. Li, J. J., *Name Reactions: A Collection of Detailed Mechanisms and Synthetic Applications*, 4th Ed. Springer: Heidelberg, 2009, pp. 566–567.
3. Olah, G. A., Prakash, G. S., *Carbocation Chemistry*, Wiley-Interscience: Hoboken, 2004, pp. 215–218.
4. Wang, W., Lu, J.-T., Zhang, H.-L., Shi, Z.-F., Wen, J., Cao, X.-P. *J. Org. Chem.* **2013**, *79*, 122–127.
5. Incze, M., Dörnyei, G., Moldvai, I., Temesvári-Major, E., Egyed, O., Szántay, C. *Tetrahedron*, **2008**, *64*, 2924–2929.
6. Petronijevic, F. R., Wipf, P. *J. Am. Chem. Soc.* **2011**, *133*, 7704–7707.
7. Smith, M. B., March, J. *March's Advanced Organic Chemistry: Reactions, Mechanisms, and Structure,* John Wiley & Sons: Danvers, Massachusetts, 2007, pp. 1321–1330.
8. Kurti, L., Czakó, B. *Strategic Applications of Named Reactions in Organic Synthesis,* Elsevier: Burlington, Massachusetts, 2005, pp. 1–500.
9. Dewar, M. J. *Angew. Chem. Int. Ed. Engl.* **1971**, *10*, 761–776.
10. Carey, F. A., Sundberg, R. J., *Advanced Organic Chemistry: Part A: Structure and Mechanisms*, Springer Science & Business Media: New York, 2007, pp. 119–125.
11. Ho, T.-L. *Res. Chem. Intermed.* **1988**, *9*, 117–140.
12. Hoffmann, R. *J. Chem. Phys.* **1963**, *39*, 1397–1412.
13. Grob, C., Schiess, P. *Angew. Chem. Int. Ed.* **1967**, *6*, 1–15.
14. Claisen, L. *Chem. Ber.* **1912**, *45*, 3157–3166.
15. Ziegler, F. E. *Chem. Rev.* **1988**, *88*, 1423–1452.

16. Jamieson, A. G., Sutherland, A. *Org. Lett.* **2007**, *9*, 1609–1611.
17. Wertheim, T. *Justus Liebigs Ann. Chem.* **1856**, *100*, 328–339.
18. Casiraghi, G., Zanardi, F., Rassu, G., Spanu, P. *Chem. Rev.* **1995**, *95*, 1677–1716.
19. Michael, J. P. *Nat. Prod. Rep.* **1997**, *14*, 619–636.
20. Johnson, W. S., Werthemann, L., Bartlett, W. R., Brocksom, T. J., Li, T.-T., Faulkner, D. J., Petersen, M. R. *J. Am. Chem. Soc.* **1970**, *92*, 741–743.
21. Pandey, G., Khamrai, J., Mishra, A. *Org. Lett.* **2015**, *17*, 952–955.
22. Ireland, R. E. Mueller, R. H. *J. Am. Chem. Soc.* **1972**, *94*, 5897–5898.
23. Williams, D. R., Walsh, M. J., Miller, N. A. *J. Am. Chem. Soc.* **2009**, *131*, 9038–9045.
24. Wick, A., Felix, D., Steen, K., Eschenmoser, A. *Helv. Chim. Acta* **1964**, *47*, 2425–2429.
25. Singh, N., Pulukuri, K. K. Chakraborty, T. K. *Tetrahedron* **2015**, *71*, 4608–4615.
26. Cope, A. C., Hardy, E. M. *J. Am. Chem. Soc.* **1940**, *62*, 441–444.
27. Molina-Salinas, G. M., Bórquez, J., Ardiles, A., Said-Fernández, S., Loyola, L. A., Yam-Puc, A., Becerril-Montes, P., Escalante-Erosa, F., San-Martin, A., González-Collado, I. *Phytochem. Rev.* **2010**, *9*, 271–278.
28. Reber, K. P., Xu, J., Guerrero, C. A. *J. Org. Chem.* **2015**, *80*, 2397–2406.
29. Abe, H., Sato, A., Kobayashi, T., Ito, H. *Org. Lett.* **2013**, *15*, 1298–1301.
30. Naengchomnong, W., Thebtaranonth, Y., Wiriyachitra, P., Okamoto, K. T., Clardy, J. *Tetrahedron Lett.* **1986**, *27*, 2439–2442.
31. Yamabe, S., Tsuchida, N., Yamazaki, S. *J. Chem. Theory Comput.* **2006**, *2*, 1379–1387.
32. Liang, D., Wang, Y., Wang, Y., Di, D. *J. Chem. Res.* **2015**, *39*, 105–107.
33. Wang, Y. D., Boschelli, D. H., Johnson, S., Honores, E. *Tetrahedron* **2004**, *60*, 2937–2942.
34. Wang, Z. *Comprehensive Organic Name Reactions and Reagents*, John Wiley & Sons: Hoboken 2009, Vol. 553, pp. 2450–2453.
35. Kadam, V. D., Sudhakar, G. *Tetrahedron* **2015**, *71*, 1058–1067.
36. Baptista, F. R., Pinto, D. C., Silva, A. *Synlett* **2014**, *25*, 1116–1120.
37. Ferrier, R., Prasad, N. *J. Chem. Soc. C: Organic* **1969**, *4*, 570–575.
38. Seth, P. P., Yu, J., Jazayeri, A., Pallan, P. S., Allerson, C. R., Østergaard, M. E., Liu, F., Herdewijn, P., Egli, M., Swayze, E. E. *J. Org. Chem.* **2012**, *77*, 5074–5085.
39. Li, J. J. *Name Reactions: A Collection of Detailed Mechanisms and Synthetic Applications*, Springer: Heidelberg, **2003**, 409–410.
40. Hashimoto, T., Naganawa, Y., Maruoka, K. *J. Am. Chem. Soc.* **2009**, *131*, 6614–6617.
41. Smith, P. A., Baer, D. R. *Org. React.* **1960**, *11*, 157–188.
42. Huang, J.-M., Yokoyama, R., Yang, C.-S., Fukuyama, Y. *Tetrahedron Lett.* **2000**, *41*, 6111–6114.
43. Shi, L., Meyer, K., Greaney, M. F. *Angew. Chem. Int. Ed.* **2010**, *49*, 9250–9253.
44. Martin, R. *Org. Prep. Proced. Int.* **1992**, *24*, 369–435.
45. Gleye, C., Raynaud, S., Hocquemiller, R., Laurens, A., Fourneau, C., Serani, L., Laprévote, O., Roblot, F., Leboeuf, M., Fournet, A. *Phytochemistry* **1998**, *47*, 749–754.
46. Adrian, J., Stark, C. B. *Org. Lett.* **2014**, *16*, 5886–5889.
47. Beretta, R., Giambelli Gallotti, M., Pennè, U., Porta, A., Gil Romero, J. F., Zanoni, G., Vidari, G. *J. Org. Chem.* **2015**, *80*, 1601–1609.
48. Gaillard, S., Slawin, A. M. Z., Nolan, S. P. *Chem. Commun.* **2010**, *46*, 2742–2744.
49. Nakamura, E., Kuwajima, I. *J. Am. Chem. Soc.* **1977**, *99*, 961–963.
50. Toda, F., Shigemasa, T. *J. Chem. Soc. Perkin Trans.* **1989**, *1*, 209–211.
51. Evans, F., Soper, C. *Lloydia* **1978**, *41*, 193.
52. McKerrall, S. J., Jørgensen, L., Kuttruff, C. A., Ungeheuer, F., Baran, P. S. *J. Am. Chem. Soc.* **2014**, *136*, 5799–5810.

53. Snape, T. J. *Chem. Soc. Rev.* **2007**, *36*, 1823–1842.
54. Handore, K. L., Reddy, D. S. *Org. Lett.* **2013**, *15*, 1894–1897.
55. Akhrem, A. A., Ustynyuk, T., Titov, Y. A. *Russ. Chem. Rev.* **1970**, *39*, 732.
56. Chenier, P. J. *J. Chem. Educ.* **1978**, *55*, 286.
57. Keana, J. F., Pou, S., Rosen, G. M. *Magn. Reson. Med.* **1987**, *5*, 525–536.
58. Ullman, E. F., Call, L., Osiecki, J. H. *J. Org. Chem.* **1970**, *35*, 3623–3631.
59. Wu, H., Coble, V., Vasalatiy, O., Swenson, R. E., Krishna, M. C., Mitchell, J. B. *Tetrahedron Lett.* **2014**, *55*, 5570–5571.
60. Wenkert, E., Bakuzis, P., Baumgarten, R. J., Leicht, C. L., Schenk, H. *J. Am. Chem. Soc.* **1971**, *93*, 3208–3216.
61. Zhang, L., Koreeda, M. *Org. Lett.* **2002**, *4*, 3755–3758.
62. Smissma, E. E., Hite, G. *J. Am. Chem. Soc.* **1960**, *82*, 3375–3381.
63. Harmata, M., Wacharasindhu, S. *Org. Lett.* **2005**, *7*, 2563–2565.
64. Selman, S., Eastham, J. F. *Q. Rev. Chem. Soc.* **1960**, *14*, 221–235.
65. Majetich, G., Shimkus, J. M. *J. Nat. Prod.* **2010**, *73*, 284–298.
66. Jana, C. K., Scopelliti, R., Gademann, K. *Chem. Eur. J.* **2010**, *16*, 7692–7695.
67. Tada, N., Shomura, V., Nakayama, H., Miura, T., Itoh, A. *Synlett* **2010**, *13*, 1979–1983.
68. Markad, S. B., Argade, N. P. *Org. Lett.* **2014**, *16*, 5470–5473.
69. Ramsewak, R. S., Nair, M. G., Strasburg, G. M., DeWitt, D. L., Nitiss, J. L. *J. Agric. Food Chem.* **1999**, *47*, 444–447.
70. Evans, D. A., Andrews, G. C. *Acc. Chem. Res.* **1974**, *7*, 147–155.
71. Shinada, T., Fuji, T., Ohtani, Y., Yoshida, Y., Ohfune, Y. *Synlett* **2002**, *8*, 1341–1343.
72. Bach, G., Breiding-Mack, S., Grabley, S., Hammann, P., Hütter, K., Thiericke, R., Uhr, H., Wink, J., Zeeck, A. *Liebigs Ann. Chem.* **1993**, *1993*, 241–250.
73. Bamoharram, F. F., Heravi, M. M., Roshani, M., Gharib, A., Jahangir, M. *J. Chin. Chem. Soc.* **2007**, *54*, 1017–1020.
74. Smith, P. A. *Org. React.* **1946**, *3*, 337–449.
75. Levinson, A. M. *Org. Lett.* **2014**, *16*, 4904–4907.
76. Hoare, D., Olson, A., Koshland Jr., D. *J. Am. Chem. Soc.* **1968**, *90*, 1638–1643.
77. Sulzer-Mosse, S., Cederbaum, F., Lamberth, C., Berthon, G., Umarye, J., Grasso, V., Schlereth, A., Blum, M., Waldmeier, R. *Bioorg. Med. Chem.* **2015**, *23*, 2129–2138.
78. Wallis, E. S., Lane, J. F. *Org. React.* **1946**, *3*, 267–306.
79. Kimishima, A., Umihara, H., Mizoguchi, A., Yokoshima, S., Fukuyama, T. *Org. Lett.* **2014**, *16*, 6244–6247.
80. Caldwell, J., Hale, M., Boyd, R., Hague, J., Iwan, T., Shi, M., Lacouture, P. *J. Rheumatol.* **1999**, *26*, 862–869.
81. Vogel, C. *Synthesis*, **1997**, *5*, 497–505.
82. Parsons, A. F. *Tetrahedron*, **1996**, *52*, 4149–4174.
83. Anderson, J. C., Whiting, M. *J. Org. Chem.* **2003**, *68*, 6160–6163.
84. Aahman, J., Somfai, P. *J. Am. Chem. Soc.* **1994**, *116*, 9781–9782.
85. Loh, C. C., Raabe, G., Enders, D. *Chem. Eur. J.* **2012**, *18*, 13250–13254.
86. Johnstone, R. A. W. In *Mechanism of Molecular Migrations*, Thyagarajan, B. S., (Ed.), Interscience: New York, 1969, Vol. 2, 249pp.
87. Xie, Y., Sun, M., Zhou, H., Cao, Q., Gao, K., Niu, C., Yang, H. *J. Org. Chem.* **2013**, *78*, 10251–10263.
88. Ren, R. X., Zueva, L. D., Ou, W. *Tetrahedron Lett.* **2001**, *42*, 8441–8443.
89. Blatt, A. H., *Chem. Rev.* **1933**, *12*, 215–260.
90. Gaware, R., Khunt, R., Czollner, L., Stanetty, C., Da Cunha, T., Kratschmar, D. V., Odermatt, A., Kosma, P., Jordis, U., Claßen-Houben, D., *Bioorg. Med. Chem.* **2011**, *19*, 1866–1880.
91. Overman, L. E., *J. Am. Chem. Soc.* **1974**, *96*, 597–599.
92. Calder, E. D., Grafton, M. W., Sutherland, A. *Synlett* **2014**, *25*, 1068–1080.

93. Middel, O., Greff, Z., Taylor, N. J., Verboom, W., Reinhoudt, D. N., Snieckus, V. *J. Org. Chem.* **2000**, *65*, 667–675.
94. Sibi, M. P., Snieckus, V. *J. Org. Chem.* **1983**, *48*, 1935–1937.
95. Qin, X. D., Dong, Z. J., Liu, J. K., Yang, L. M., Wang, R. R., Zheng, Y. T., Lu, Y., Wu, Y. S., Zheng, Q. T. *Helv. Chim. Acta* **2006**, *89*, 127–133.
96. Chang, C.-W., Chein, R.-J. *J. Org. Chem.* **2011**, *76*, 4154–4157.
97. Birladeanu, L., *J. Chem. Ed.* **2000**, 77, 858–863.
98. Ding, R., Sun, B.-F., Lin, G.-Q. *Org. Lett.* **2012**, *14*, 4446–4449.
99. OH, H. In *Name Reactions A collection of Detailed Mechanisms and Synthetic Applications*, 4th Ed., Li, J. J. (Ed). Springer: Heidelberg, 2009, 190pp.
100. Murakata, M., Kimura, M. *Tetrahedron Lett.* **2010**, *51*, 4950–4952.
101. Taylor, R. J., Casy, G. *Org. React.* **2003**, *62*, 357–475
102. Yang, G., Franck, R. W., Byun, H.-S., Bittman, R., Samadder, P., Arthur, G. *Org. Lett.* **1999**, *1*, 2149–2151.
103. Petasis, N. A., Lu, S.-P. *J. Am. Chem. Soc.* **1995**, *117*, 6394–6395.
104. Kozioł, A., Grzeszczyk, B., Kozioł, A., Staszewska-Krajewska, O., Furman, B., Chmielewski, M. *J. Org. Chem.* **2010**, *75*, 6990–6993.
105. Cornforth, J. *Chem. Penicillin, Princeton Book Review* **1949**, 688.
106. Dewar, M. J. *J. Am. Chem. Soc.* **1974**, *96*, 6148–6152.
107. Brown, P., Davies, D. T., O'Hanlo, P. J., Wilson, J. M. *J. Med. Chem.* **1996**, *39*, 446–457.
108. Nolt, M. B., Smiley, M. A., Varga, S. L., McClain, R. T., Wolkenberg, S. E., Lindsley, C. W. *Tetrahedron* **2006**, *62*, 4698–4704.
109. El Ashry, E., El Kilany, Y., Rashed, N., Assafir, H. *Adv. Heterocycl. Chem.* **2000**, *75*, 80–166.
110. Zavodskaya, A. V., Bakharev, V. V., Parfenov, V. E., Gidaspov, A. A., Slepukhin, P. A., Isenov, M. L., Eltsov, O. S. *Tetrahedron Lett.* **2015**, *56*, 1103–1106.
111. Plesniak, K., Zarecki, A., Wicha, J. In *Sulfur-Mediated Rearrangements II*, Schaumann, E., (Ed.) Springer: Heidelberg, **2010**, pp. 163–250.
112. Xiao, Y., Zhang, Z., Chen, Y., Shao, X., Li, Z., Xu, X. *Tetrahedron* **2015**, *71*, 1863–1868.
113. Brewster, J. H., Kline, M. W. *J. Am. Chem. Soc.* **1952**, *74*, 5179–5182.
114. Matsuda, H., Shimoda, H., Yoshikawa, M. *Bioorg. Med. Chem.* **2001**, *9*, 1031–1035.
115. Lu, P., Herrmann, A. T., Zakarian, A. *J. Org. Chem.*, **2015**, *80*, 7581–7589.
116. Pummerer, R. *Chem. Ber.* **1909**, *42*, 2282–2291.
117. Kotoulas, S. S., Kojić, V. V., Bogdanović, G. M., Koumbis, A. E. *Tetrahedron* **2015**, *71*, 3396–3403.
118. Hodge, J. E. *Adv. Carbohydr. Chem.* **1954**, *10*, 169–205.
119. Sánchez-Fernández, E. M., Álvarez, E., Ortiz Mellet, C., Garcia Fernandez, J. M. *J. Org. Chem.* **2014**, *79*, 11722–11728.
120. Moser, W. H. *Tetrahedron* **2001**, *57*, 2065–2084.
121. Eschenmoser, A., Felix, D., Ohloff, G. *Helv. Chim. Acta* **1967**, *50*, 708–713.
122. Nakajima, R., Ogino, T., Yokoshima, S., Fukuyama, T. *J. Am. Chem. Soc.* **2010**, *132*, 1236–1237.
123. Oetterli, R. M., Prieto, L., Spingler, B., Zelder, F. *Org. Lett.* **2013**, *15*, 4630–4633.

Index

Printed and bound by CPI Group (UK) Ltd, Croydon, CR0 4YY
01/11/2024
01782614-0004